NANOTECHNOLOGY-ENHANCED SOLID MATERIALS

Design, Synthesis, Properties, Applications, and Perspectives

Innovations in Physical Chemistry: Monograph Series

NANOTECHNOLOGY-ENHANCED SOLID MATERIALS

Design, Synthesis, Properties, Applications, and Perspectives

Edited by

Lionello Pogliani, PhD

Ann Rose Abraham, PhD

A. K. Haghi, PhD

Prabhat Ranjan, PhD

AAP | APPLE ACADEMIC PRESS

First edition published 2024

Apple Academic Press Inc.
1265 Goldenrod Circle, NE,
Palm Bay, FL 32905 USA

760 Laurentian Drive, Unit 19,
Burlington, ON L7N 0A4, CANADA

CRC Press
6000 Broken Sound Parkway NW,
Suite 300, Boca Raton, FL 33487-2742 USA

4 Park Square, Milton Park,
Abingdon, Oxon, OX14 4RN UK

Library and Archives Canada Cataloguing in Publication

Title: Nanotechnology-enhanced solid materials : design, synthesis, properties, applications, and perspectives / edited by Lionello Pogliani, PhD, Ann Rose Abraham, PhD, A.K. Haghi, PhD, Prabhat Ranjan, PhD.
Names: Pogliani, Lionello, editor. | Abraham, Ann Rose, editor. | Haghi, A. K., editor. | Ranjan, Prabhat, editor.
Series: Innovations in physical chemistry.
Description: First edition. | Series statement: Innovations in physical chemistry: monographic series | Includes bibliographical references and index.
Identifiers: Canadiana (print) 20230171206 | Canadiana (ebook) 20230171230 | ISBN 9781774912201 (hardcover) | ISBN 9781774912218 (softcover) | ISBN 9781003333449 (ebook)
Subjects: LCSH: Nanostructured materials. | LCSH: Nanostructured materials—Industrial applications. | LCSH: Nanotechnology.
Classification: LCC TA418.9.N35 N38 2023 | DDC 620.1/15—dc23

Library of Congress Cataloging-in-Publication Data

CIP data on file with US Library of Congress

ISBN: 978-1-77491-220-1 (hbk)
ISBN: 978-1-77491-221-8 (pbk)
ISBN: 978-1-00333-344-9 (ebk)

BOOK SERIES: INNOVATIONS IN PHYSICAL CHEMISTRY: MONOGRAPH SERIES

This book series, Innovations in Physical Chemistry: Monograph Series, offers a comprehensive collection of books on physical principles and mathematical techniques for majors, non-majors, and chemical engineers. Because there many exciting new areas of research involving computational chemistry, nanomaterials, smart materials, high-performance materials, and applications of the recently discovered graphene, there can be no doubt that physical chemistry is a vitally important field. Physical chemistry is considered a daunting branch of chemistry— it is grounded in physics and mathematics and draws on quantum mechanics, thermodynamics, and statistical thermodynamics.

Innovations in Physical Chemistry has been carefully developed to help readers increase their confidence when using physics and mathematics to answer fundamental questions about the structure of molecules, how chemical reactions take place, and why materials behave the way they do.

This series covers: Computational Chemistry; Elementary Chemical Kinetics; Fundamental Concepts of Thermodynamics; Heat, Work, Internal Energy, Enthalpy, and the First Law of Thermodynamics; The Importance of State Functions: Internal Energy and Enthalpy; Thermochemistry; Entropy; and the Second and Third Laws of Thermodynamics; Chemical Equilibrium; The Properties of Real Gases; Phase Diagrams and the Relative Stability of Solids, Liquids, and Gases; Ideal and Real Solutions; Electrolyte Solutions; Electrochemical Cells, Batteries, and Fuel Cells; From Classical to Quantum Mechanics; The Schrödinger Equation; The Quantum Mechanical Postulates; Using Quantum Mechanics on Simple Systems; and more.

Modern applications are drawn from biology, environmental science, and material science. Spectroscopy applications are introduced early in concert with theory; for example, IR and rotational spectroscopy are discussed immediately after the harmonic oscillator.

Modern research is featured throughout, along with new developments in the field, such as scanning tunneling microscopy, bandgap engineering, quantum wells, teleportation, and quantum computing.

The books in the series act as guides that take readers into the subject, in spite of the generally modest background they might have on the subject matter. The physical basis of essential ideas is emphasized, and mathematical developments are simple and closely tied to the physical bases. The overall organization of the titles in the series is such that topics are introduced in stages so students can become familiar with, and grow into each topic, and chapters are written so that there is flexibility in the order in which they may be used.

EDITORS-IN-CHIEF

A. K. Haghi, PhD
Former Editor-in-Chief, *International Journal of Chemoinformatics and Chemical Engineering* and *Polymers Research Journal;* Member, Canadian Research and Development Center of Sciences and Cultures (CRDCSC), Montreal, Quebec, Canada; Email: AKHaghi@Yahoo.com

K. Haghi, PhD, is a retired professor with over 250 academic research-oriented books as well as over 1000 research papers published in various journals and conference proceedings with over 3350 citations and h-index of 30, according to the Google Scholar database. He is currently a senior material engineering consultant in the UK. Professor Haghi has received several grants, consulted for several major corporations, and is a frequent speaker to national and international audiences. He is Founder and former Editor-in-Chief of *the International Journal of Chemoinformatics and Chemical Engineering* and as *Polymers Research Journal.* Professor Haghi has acted as an editorial board member of many international journals. He has served as a member of the Canadian Research and Development Center of Sciences & Cultures and the Research Chemistry Centre, Coimbra, Portugal. He has supervised several PhD and MSc theses at the University of Guilan (UG) and co-supervised international doctoral projects. Professor Haghi holds a BSc in urban and environmental engineering from the University of North Carolina (USA) and holds two MSc, one in mechanical engineering from North Carolina State University (USA) and another one in applied mechanics, acoustics, and materials from the Université de Technologie de Compiègne (France). He was awarded a PhD in engineering sciences at Université de Franche-Comté (France).

Lionello Pogliani, PhD
University of Valencia-Burjassot, Spain
Email: lionello.pogliani@uv.es

Lionello Pogliani, PhD, is a retired Professor of Physical Chemistry. He received his postdoctoral training at the Department of Molecular Biology of the C. E. A. (Centre d'Etudes Atomiques) of Saclay, France, at the Physical Chemistry Institute of the Technical and Free University of Berlin, and at the Pharmaceutical Department of the University of California, San Francisco, USA. He spent his sabbatical years at the Technical University of Lisbon, Portugal, and at the University of Valencia, Spain. He has contributed more than 200 papers in the experimental, theoretical, and didactical fields of physical chemistry, including chapters in specialized books and a book on numbers 0, 1, 2, and 3. A work of his has been awarded with the GM Neural Trauma Research Award. He is a member of the International Academy of Mathematical Chemistry, and he is on the editorial board of many international journals. He is presently a part-time teammate at the Physical Chemistry department of the University of Valencia, Spain.

Ana Cristina Faria Ribeiro, PhD
Researcher, Department of Chemistry, University of Coimbra, Portugal
Email: anacfrib@ci.uc.pt

Ana C. F. Ribeiro, PhD, is a researcher in the Department of Chemistry at the University of Coimbra, Portugal. Her area of scientific activity is physical chemistry and electrochemistry. Her main areas of research interest are transport properties of ionic and non-ionic components in aqueous solutions. She has experience as a scientific adviser and teacher of different practical courses. Dr. Ribeiro has supervised master's degree theses as well as some PhD theses and has been a theses jury member. She has been a referee for various journals as well an expert evaluator of some of the research programs funded by the Romanian government through the National Council for Scientific Research. She has been a member of the organizing committee of scientific conferences, and she is an editorial member of several journals. She has received several grants, consulted for a number of major corporations and is a frequent speaker to national and international audiences. She is a member of the Research Chemistry Centre, Coimbra, Portugal.

BOOKS IN THE SERIES

- **Applied Physical Chemistry with Multidisciplinary Approaches**
  Editors: A. K. Haghi, PhD, Devrim Balköse, PhD, and Sabu Thomas, PhD
- **Biochemistry, Biophysics, and Molecular Chemistry: Applied Research and Interactions**
  Editors: Francisco Torrens, PhD, Debarshi Kar Mahapatra, PhD, and A. K. Haghi, PhD
- **Chemical Technology and Informatics in Chemistry with Applications**
- Editors: Alexander V. Vakhrushev, DSc, Omari V. Mukbaniani, DSc, and Heru Susanto, PhD
- **Chemistry and Chemical Engineering for Sustainable Development: Best Practices and Research Directions**
  Editors: Miguel A. Esteso, PhD, Ana Cristina Faria Ribeiro, A. K. Haghi, PhD
- **Chemistry and Industrial Techniques for Chemical Engineers**
- Editors: Lionello Pogliani, PhD, Suresh C. Ameta, PhD, and A. K. Haghi, PhD
- **Engineering Technologies for Renewable and Recyclable Materials: Physical-Chemical Properties and Functional Aspects**
  Editors: Jithin Joy, Maciej Jaroszewski, PhD, Praveen K. M., Sabu Thomas, PhD, and Reza Haghi, PhD
- **Engineering Technology and Industrial Chemistry with Applications**
  Editors: Reza Haghi, PhD, and Francisco Torrens, PhD
- **High-Performance Materials and Engineered Chemistry**
  Editors: Francisco Torrens, PhD, Devrim Balköse, PhD, and Sabu Thomas, PhD
- **Methodologies and Applications for Analytical and Physical Chemistry**
  Editors: A. K. Haghi, PhD, Sabu Thomas, PhD, Sukanchan Palit, and Priyanka Main
- **Modern Green Chemistry and Heterocyclic Compounds: Molecular Design, Synthesis, and Biological Evaluation**
  Editors: Ravindra S. Shinde, and A. K. Haghi, PhD
- **Modern Physical Chemistry: Engineering Models, Materials, and Methods with Applications**
  Editors: Reza Haghi, PhD, Emili Besalú, PhD, Maciej Jaroszewski, PhD, Sabu Thomas, PhD, and Praveen K. M.

- **Molecular Chemistry and Biomolecular Engineering: Integrating Theory and Research with Practice**
  Editors: Lionello Pogliani, PhD, Francisco Torrens, PhD, and A. K. Haghi, PhD
- **Nanotechnology-Enhanced Solid Materials: Design, Synthesis, Properties, Applications, and Perspectives**
  Editors: Lionello Pogliani, PhD, Ann Rose Abraham, PhD, A. K. Haghi, PhD, and Prabhat Ranjan, PhD
- **Physical Chemistry for Chemists and Chemical Engineers: Multidisciplinary Research Perspectives**
  Editors: Alexander V. Vakhrushev, DSc, Reza Haghi, PhD, and J. V. de Julián-Ortiz, PhD
- **Physical Chemistry for Engineering and Applied Sciences: Theoretical and Methodological Implication**
- Editors: A. K. Haghi, PhD, Cristóbal Noé Aguilar, PhD, Sabu Thomas, PhD, and Praveen K. M.
- **Practical Applications of Physical Chemistry in Food Science and Technology**
- Editors: Cristóbal Noé Aguilar, PhD, Jose Sandoval Cortes, PhD, Juan Alberto Ascacio Valdes, PhD, and A. K. Haghi, PhD
- **Research Methodologies and Practical Applications of Chemistry**
  Editors: Lionello Pogliani, PhD, A. K. Haghi, PhD, and Nazmul Islam, PhD
- **Theoretical and Empirical Analysis in Physical Chemistry: A Framework for Research**
  Editors: Miguel A. Esteso, PhD, Ana Cristina Faria Ribeiro, PhD, and A. K. Haghi, PhD
- **Theoretical Models and Experimental Approaches in Physical Chemistry: Research Methodology and Practical Methods**
  Editors: A. K. Haghi, PhD, Sabu Thomas, PhD, Praveen K. M., and Avinash R. Pai

ABOUT THE EDITORS

Lionello Pogliani, PhD
University of Valencia, Burjassot, Spain

Lionello Pogliani, PhD, is a retired Professor of Physical Chemistry. He received his postdoctoral training at the Department of Molecular Biology of the C. E. A. (Center d'Etudes Atomiques) of Saclay, France, at the Physical Chemistry Institute of the Technical and Free University of Berlin, and at the Pharmaceutical Department of the University of California, San Francisco, USA. He spent his sabbatical years at the Technical University of Lisbon, Portugal, and at the University of Valencia, Spain. He has contributed more than 200 papers in the experimental, theoretical, and didactical fields of physical chemistry, including chapters in specialized books and a book on numbers 0, 1, 2, and 3. A work of his has been awarded with the GM Neural Trauma Research Award. He is a member of the International Academy of Mathematical Chemistry, and he is on the editorial board of many international journals. He is presently a part-time teammate at the Physical Chemistry department of the University of Valencia, Spain.

Ann Rose Abraham, PhD
Assistant Professor, Department of Basic Sciences, Amal Jyothi College of Engineering, Kanjirappally, Kottayam, Kerala, India

Ann Rose Abraham, PhD, is currently an Assistant Professor in the Department of Basic Sciences, Amal Jyothi College of Engineering, Kanjirappally, Kerala, India. She has research experience at various national institutes, including Bose Institute, SAHA Institute of Nuclear Physics, and UGC-DAE CSR Center, Kolkata, and she has collaborated with various international laboratories, such as the University of Johannesburg, South Africa; Institute of Physics, Belgrade; etc. She is a recipient of a Young Researcher Award in the area of physics in the year 2016. Dr. Abraham has delivered invited lectures and sessions at national and international conferences. During her tenure as a doctoral fellow, she taught and mentored students at the postgraduate level at the International and Inter-University Center for Nanoscience and Nanotechnology, India. She has co-authored many book chapters and co-edited books. She has a number of publications to her credit, published in many peer-reviewed journals of international repute. She has expertise in the field of materials science, nanomagnetic materials, multiferroics, and polymeric nanocomposites. Dr. Abraham was involved in teaching and also served as the examiner for the valuation of answer scripts of Engineering Physics at APJ Abdul Kalam Technological University. Dr. Abraham received her MSc, MPhil, and PhD degrees in Physics from the School of Pure and Applied Physics, Mahatma Gandhi University, Kerala, India. Her PhD thesis was on "Development of Hybrid Multiferroic Materials for Tailored Applications."

A. K. Haghi, PhD
Professor Emeritus of Engineering Sciences, Former Editor-in-Chief, International Journal of Chemoinformatics and Chemical Engineering Member, Canadian Research and Development Center of Sciences and Culture

A. K. Haghi, PhD, published over 250 academic research-oriented books as well as over 1,000 research papers published in various journals and conference proceedings. He has received several grants, consulted for a number of major corporations, and is a frequent speaker to national and international audiences. He is the Founder and former Editor-in-Chief of the *International Journal of Chemoinformatics and Chemical Engineering,* published by IGI Global (USA), as well as *Polymers Research Journal,* published by Nova Science Publishers (USA). Professor Haghi has acted as an editorial board member of many international journals. He has served as a member of the Canadian Research and Development Center of Sciences and Cultures (CRDCSC) and the Research Chemistry Center, Coimbra, Portugal. Dr. Haghi holds a BSc in urban and environmental engineering from the University of North Carolina (USA), an MSc in mechanical engineering from North Carolina A&T State University (USA), and an MSc in applied mechanics, acoustics, and materials from the Université de Technologie de Compiègne (France), and a PhD in engineering sciences from Université de Franche-Comté (France).

Prabhat Ranjan, PhD
Associate Professor, Department of Mechatronics Engineering at Manipal University, Jaipur, Rajasthan, India

Prabhat Ranjan, PhD, is an Associate Professor in the Department of Mechatronics Engineering at Manipal University Jaipur. He is working in the domain of energy materials, smart materials, computational materials science, and density functional theory since last 10 years. He has published more than 40 research papers in peer-reviewed journals of high repute and several book chapters in research-edited book. He has also edited many high-end research-based books. He has received the prestigious President Award of Manipal University Jaipur, India (2015); material design scholarship from the Imperial College of London, UK (2014); a DAAD Fellowship (2014); the CFCAM-France Award (2015) and President gold medal for excellence in research (2019), and best research award for various recognized international agencies. Dr. Ranjan has received several grants and also participated in national and international conferences and summer schools. He holds a bachelor of engineering in electronics and communication and a master of technology in instrumentation control system engineering from the Manipal Academy of Higher Education, Manipal, India, as well as a PhD in engineering from Manipal University Jaipur, India.

CONTENTS

CONTRIBUTORS

H. Akhina
International and Inter-University Center for Nanoscience and Nanotechnology, Mahatma Gandhi University, Kottayam, Kerala, India

Ahmed Asanidezhad
Tomas Bata University in Zlín, Zlín, Czech Republic

Savita Bansal
Research Scholar, Punjabi University, Patiala, Punjab, India

Tanmoy Chakraborty
Department of Chemistry and Biochemistry, School of Basic Sciences and Research, Sharda University, Greater Noida, Uttar Pradesh, India

Ivan Chodák
Polymer Institute, Slovak Academy of Sciences, Bratislava, Slovakia

Mihai Ciocoiu
Textiles-Leather and Industrial Management at Gheorghe Asachi Technical University of Iasi, Romania

Shayeri Das
Department of Mechatronics Engineering, Manipal University Jaipur, Dehmi Kalan, Rajasthan, India

Ranjit Debnath
Department of Chemistry, National Institute of Technology, Agartala, Tripura, India

Tariq Ahmad Ganaie
Department of Food Science and Technology, Government Degree College Shopian, Jammu and Kashmir, India

Amir Gull
Department of Food Science and Technology, University of Kashmir, Srinagar, Jammu and Kashmir, India

Anita Gupta
Department of Chemistry, Amity University, Noida, Uttar Pradesh, India

Raees Ul Haq
Department of Food Engineering and Technology, SLIET, Sangrur, Punjab, India

Christine Jeyaseelan
Department of Chemistry, Amity University, Noida, Uttar Pradesh, India

Greeshma Sara John
Department of Physics, Nanotechnology and Advanced Materials Research Center, CMS College (Autonomous), Kottayam, Kerala, India

Athira Maria Johnson
Department of Physics, Nanotechnology and Advanced Materials Research Center, CMS College (Autonomous), Kottayam, Kerala, India

Peter Jurkovič
VIPO, A.S., Gen. Svobodu, Partizánske, Slovakia

Rajpreet Kaur
Department of Chemistry, Amity University, Noida, Uttar Pradesh, India; Department of Chemistry, B.B.K. D.A.V. College for Women, Amritsar, Punjab, India

Poonam Khullar
Department of Chemistry, B.B.K. D.A.V. College for Women, Amritsar, Punjab, India

Shaheen Khursid
Department of Zoology, Government Degree College Tral, Jammu and Kashmir, India

K. V. Arun Kumar
Department of Physics, Nanotechnology and Advanced Materials Research Center, CMS College (Autonomous), Kottayam, Kerala, India

Pancham Kumar
School of Electrical Skills, Bhartiya Skill Development University, Jaipur, Rajasthan, India

Marián Lehocký
Tomas Bata University in Zlín, Zlín, Czech Republic

Mushtaq Ahmad Lone
Directorate of Planning and Monitoring, SKUAST-K, Srinagar, Jammu and Kashmir, India

Aabroo Mahal
Department of Chemistry, Khalsa College for Woman, Amritsar, Punjab, India

Maria Mathew
Department of Chemistry, CMS College, Kottayam, Kerala, India

Ján Matyašovský
VIPO, A.S., Gen. Svobodu, Partizánske, Slovakia

K. A. Naseer
Department of Physics, The Gandhigram Rural Institute–Deemed to be University, Gandhigram, Tamil Nadu, India; Department of Physics, Farook College (Autonomous), Kozhikode, Kerala, India

Igor Novák
Polymer Institute, Slovak Academy of Sciences, Bratislava, Slovakia

Rony Rajan Paul
Department of Chemistry, CMS College, Kottayam, Kerala, India

Tressia Alias Princy Paulose
Post-Graduate and Research Department of Chemistry, Bishop Moore College, Mavelikara, Alappuzha, Kerala, India

Anton Popelka
Center for Advanced Materials, Qatar University, Doha, Qatar

Prabhat Ranjan
Department of Mechatronics Engineering, Manipal University Jaipur, Dehmi Kalan, Rajasthan, India

Mitali Saha
Associate Professor, Department of Chemistry, National Institute of Technology, Agartala, Tripura, India

Ján Sedliačik
Technical University in Zvolen, T.G. Masaryka, Zvolen, Slovakia

Susheela Sharma
Department of Basic Science, Bhartiya Skill Development University, Jaipur, Rajasthan, India

Mohd. Aaqib Sheikh
Department of Food Technology, IUST, Awantipora, Jammu and Kashmir, India

Manisha Sheoran
Department of Basic Science, Bhartiya Skill Development University, Jaipur, Rajasthan, India

Aditi Singh
Department of Chemistry, Amity University, Noida, Uttar Pradesh, India

Satya Bir Singh
Professor, Punjabi University, Patiala, Punjab, India

Vijender Singh
Department of Chemistry, S.K. Government College, Kanwali Rewari, Haryana, India

Ladislav Šoltés
Center of Experimental Medicine SAS, Institute of Experimental Pharmacology and Toxicology, Bratislava, Slovakia

P. Arjun Suresh
Department of Physics, Nanotechnology and Advanced Materials Research Center, CMS College (Autonomous), Kottayam, Kerala, India

Lavanya Tandon
Department of Chemistry, B.B.K. D.A.V. College for Women, Amritsar, Punjab, India

N. V. Unnikrishnan
School of Pure and Applied Physics, Mahatma Gandhi University, Kottayam, Kerala, India

Katarína Valachová
Center of Experimental Medicine SAS, Institute of Experimental Pharmacology and Toxicology, Bratislava, Slovakia

ABBREVIATIONS

AA	acrylic acid
AB	ammonia borane
AC	activated carbon
ACNT	aligned carbon nanotube
AmB	amphotericin B
AOP	advanced oxidation process
ATPSs	aqueous two-phase polymer systems
ATRP	atom-transfer radical polymerization
BC	bacterial cellulose
B-CNT	boron-doped CNT
BHJ	bulk heterojunction
BTEX	benzene, toluene, ethylbenzene, and o-xylene
C	carbon
CB	conduction band
CBL	cathode buffer layers
CE	counter electrode
CF	carbon foam
CHP	cumene hydroperoxide
Cl	chlorine
CMC	critical micelle concentration
CNTFETs	carbon nanotube field-effect transistors
CNTPE	carbon nanotube-based paste electrodes
CNTs	carbon nanotubes
CNTY	CNT yarns
CoAl LDH	cobalt–aluminum layered double hydroxide
CPC	cetylpyridinium chloride
CRT	cathode ray tube
CSCs	charge selective contacts
CTAB	cetyltrimethylammonium bromide
CVD	chemical vapor deposition
CWAO	catalytic wet air oxidation
DA	degree of methylation
DCSBD	diffuse coplanar surface barrier discharge
DA	degree of acetylation

DMMP	dimethyl methylphosphonate
DMSS	dual mesoporous silica sphere
DNA	deoxyribonucleic acid
DOA	dissociated oxygen atoms
DSSCs	dye-sensitized solar cells
DWCNTs	double-walled CNTs
DWNTs	double-walled nanotubes
ECD	electron capture detector
EG	ethylene glycol
EMI	electromagnetic interference
ESD	electrostatic discharge
ETL	electron transport layer
Fcc	face-centered cubic
FCs	fuel cells
FET	field effect transistors
FF	fill factor
FGM	functionally graded material
FID	flame ionization detector
FO	forward osmosis
FTO	fluoride tin oxide
FTS	Fischer Tropsch synthesis
GAG	glycosaminoglycan
GCB	gleaming cresyl blue
GCE	glassy carbon electrodes
GC-MS	gas chromatography/mass spectrometry
GE	general electric
GlcA	D-glucuronic acid
GO	graphene oxide
GO/mSiO_2/PANI	graphene oxide, mesoporous silica, and polyaniline
GO/PANI	graphene oxide and polyaniline
GOBIN	GO-based benzimidazole crosslinked network
Gox	glucose oxidase
GY	graft yield
H_2O_2	hydrogen peroxide
HELs	hole extraction layers
HG	homogalacturonan
HOPG	high-oriented pyrolytic graphite
HPC	hierarchical porous structure
HPLC	high-performance liquid chromatography

HTL	hole transport layer
IO	iron oxide
IOMNPs	iron oxide magnetic nanoparticles
ITO	indium tin oxide
LCC	lignin-carbohydrate complex
LCST	lower critical solution temperatures
LDPE	low-density polyethylene
LIBs	lithium-ion batteries
LW/AB	Lifshitz-van der Waals/acid-base
$MAPI_3$	methylammonium lead iodide
MAPR	microwave-assisted polyol reduction
MB	methylene blue
MD	molecular dynamics
MEHQ	monomethyl ether of hydroquinone
MHPs	metal-halide perovskites
MIS	metal-insulator-semiconductor
MM	mixed matrix
MnO_2	manganese dioxide
MNPs	magnetic nanoparticles
MoS_2	molybdenum disulfide
MS	metal-semiconductor
MSS	mesoporous silica sphere
MW	microwave
MWCNT	multi-walled carbon nanotube
MWNTs	multi-wall carbon nanotubes
N-CNT	nitrogen-doped CNT
NCY	Ni-plated cotton yarn
NiAl-LDHs	nickel aluminum layered double hydroxides
NiCo–CH	NiCo–carbonate hydroxide
NIST	National Institute of Standards and Technology
NMOs	nanosized metal oxides
NMP	nitroxide-mediated radical polymerization
NPs	nanoparticles
NR	neutral red
NTs	nanotubes
NW	nanowire
O	oxygen
OPV	organic photovoltaic
P3HT	poly(3-hexylthiophene)

P_3OT	poly(3-octylthiophene)
PAA	polyacrylic acid
PCBM	phenyl Cn butyric acid methyl ester
PCE	power conversion energy
PEDOT	poly(3,4-ethylene dioxythiophene)
PEDOT/GO	poly(3,4-ethylene dioxythiophene) and graphene oxide
PEI	polyethyleneimine
PEM	polymer electrolyte membrane
PEMFC	proton-exchange membrane fuel cells
PHBV	polyhydroxybutyrate-co-valerate
PL	photoluminescence
PMMA	polymethyl methacrylate
PNIPAAm	poly N-isopropylacrylamide
POAP	poly ortho amino phenol
POM	polyoxometalate
PPF	porous polypropylene films
PPy	polypyrrole
PPy/GO	polypyrrole and graphene oxide
PPy/GO/ZnO	polypyrrole, graphene oxide, and zinc oxide
PS	polystyrene
PSC	perovskite solar cell
PS-PAA	poly(styrene)-block-poly(acrylic acid)
PSS	poly(sodium-4-styrenesulfonate)
PSS	poly(styrenesulfonate)
Pt	platinum
PV	photovoltaic
PVA	polyvinyl alcohol
PVC	poly(vinyl chloride)
QD	quantum dot
RAFT	reversible addition-fragmentation chain transfer
RGO	reduced graphene oxide
RO	reverse osmosis
ROS	reactive oxygen species
SCs	supercapacitors
SEE System	surface energy evaluation system
SEM	scanning electron microscopy
Si	silicon
SMAs	shape memory alloys

SMPs	shape memory polymers
SPIONs	superparamagnetic iron oxide nanoparticles
SSQ-GO	silsesquioxane containing graphene oxide
SWCNTs	single-walled carbon nanotubes
SWNT-FETs	single-walled nanotube-filled effect transistor
SWNTs	single-walled nanotubes
TCO	transparent conducting oxide
TEM	transmission electron micrograph
TEOS	tetraethyl orthosilicate
TFT	thin-film transistors
TMOS	tetramethoxysilane
TOF	turnover-frequency
UYC	unified yield criterion
VA	vertically aligned
VB	valence band
VHSVs	volume hour space velocities
WAO	wet air oxidation
WGS	water–gas shift
WPU	waterborne polyurethane
ZnO NPs	zinc oxide nanoparticles
ZnO	zinc oxide

PREFACE

This advanced research-oriented book was developed to equip postgraduate students, nanotechnologists, materials scientists, chemists, physicists, chemical engineers, and researchers with a deep understanding of nanoscale solid materials by focusing on the principles of nanotechnology and nanoscience. It provides the latest information on the emergence and rapid development of nanotechnology-enhanced solid materials and the ways it has impacted almost every aspect of nano-engineering. In each chapter of this new volume, the authors have given a brief description of the problem along with an introduction and a concise summary of the relevant physical principles. The discussions then continue with emphasizing universal commonalities between seemingly disparate areas and encouraging postgraduate students and scientists to develop an intuitive understanding of this diverse area of study. It also provides the readers with a comprehensive overview of the field, focusing on the design, synthesis, and properties of solid materials, presenting updated, practical, and systematic knowledge on the modification of nanomaterials. Topics covered include a wide range of solid material formulations and self-assemblies.

One of the most important points in this new research-oriented book is the theoretical concepts that are linked to real-world applications, allowing readers to connect theory and practice.

This new volume is an excellent resource for senior postgraduate students with backgrounds in materials science and technology, physics, and chemistry. Because of the integral nature of the topics, it will also be of interest to a broad spectrum of audiences, including industrial scientists, industrial engineers, nanotechnologists, materials scientists, chemists, physicists, biologists, chemical engineers, and all those who are involved and interested in the future frontiers of nanomaterials.

This research-oriented book:

- Features a variety of designs of nanomaterials, helping engineers to make the best use of theories.
- Contains underlying principles of how nanomaterials work and the variety of applications they provide.

- Assesses the viability of various methods for mass production and commercialization.
- Highlights the design of functionalized nanomaterials with respect to recent progress in the industrial arena and their respective applications.

CHAPTER 1

PHOTOVOLTAICS APPLICATIONS OF SOLID CARBONS IN MOLECULAR FORMS

K. V. ARUN KUMAR,[1,2] P. ARJUN SURESH,[1,2] GREESHMA SARA JOHN,[1,2] ATHIRA MARIA JOHNSON,[1,2] K. A. NASEER,[3,4] and N. V. UNNIKRISHNAN[4,5]

[1]*Department of Physics, CMS College (Autonomous), Kottayam, Kerala, India*

[2]*Nanotechnology and Advanced Materials Research Center, CMS College (Autonomous), Kottayam, Kerala, India)*

[3]*Department of Physics, The Gandhigram Rural Institute–Deemed to be University, Gandhigram, Tamil Nadu, India*

[4]*Department of Physics, Farook College (Autonomous), Kozhikode, Kerala, India*

[5]*School of Pure and Applied Physics, Mahatma Gandhi University, Kottayam, Kerala, India*

ABSTRACT

Renewable energy conversion using photovoltaic devices plays a crucial role in converting solar energy into electricity. Improving the efficiency and lifetime of the photovoltaic (PV) cell is one of the intense research activities solving the problem of energy issues and environmental concerns. Carbon-based photovoltaic cells exhibit numerous advantages and efficiency

Nanotechnology-Enhanced Solid Materials: Design, Synthesis, Properties, Applications, and Perspectives. Lionello Pogliani, Ann Rose Abraham, A. K. Haghi, & Prabhat Ranjan (Eds.)

improvements over ordinary photovoltaic cells. Material properties like high surface area, low cost, and high electrical conductivity play a key role in efficiency improvement. Energy generation using carbon-based photovoltaic materials in the future is an alternating, safe, low-cost, and efficient method. Structures like carbon nanotube (CNT), fullerene, graphene, etc., have a wide range of applications like photoanode, cathode in DSSC, electron transport layer (ETL), hole transport layer (HTL), and encapsulant in photovoltaic (PV) cells. The applications of carbon-based nanomaterials devices like photovoltaics in energy generation applications are discussed in this chapter.

1.1 INTRODUCTION

Solar energy is considered one of the largest clean energy sources. Exploration of renewable energies is one of the concerns about climate and environmental issues. With the increased demand for global energy due to the increased population and fast-growing, we need improved technology. Researchers are looking for alternative energy sources with improved efficiency and low cost. However, the excessive use of conventional fossil energy sources has limitations. The most promising route to store and utilize solar energy is converting solar light into electricity, which leads to the considerable development of solar cells. Crystalline silicon solar cell is dominated in the current market, but it has limitations like higher fabrication cost and complicated fabrication technique. It is more expensive than conventional fossil fuels. Another drawback of silicon solar cells is the degradation under illumination, which causes a decrease in lifetime and stability. Therefore, finding a new nature-friendly and cheap energy resource is essential [1–9]. Next-generation solar cells like perovskite and dye sensitize solar cells, etc., have more technological advantages. Their advantages, like elevated power conversion efficiency, easy fabrication, and low cost, are attracted particular attention. Carbon is considered an alternative material, which is highly attractive because of some similar properties like silicon and its highly stable nature. In such a situation, carbon-based materials have a great scope in photovoltaic applications. Carbon is one of the most abundant materials on earth and exists in various stable forms such as semiconducting/insulator diamond, metallic/semi-metallic graphite, conducting/semiconducting fullerenes, carbon nanotubes (CNTs), etc. Different carbon allotrope structures and hybridization states show different chemical, mechanical, thermal, and electrical properties, opening various applications in different fields. The nano-scale cylindrical structure of CNT has a tremendous interest. For high predicted mobility applications in p-type, semiconductor CNTs are used.

Solar energy harvesting combination of CNTs with electron donors is an innovative concept. Introduction C_{60}, CNTs into conjugated polymers have tremendous applications in an organic photovoltaic (OPV) device. In the case of organic solar cells, Fullerene C_{60} has been considered as the most suitable electron-accepting and transporting material. Another vital characteristic of carbon-based materials allows electronic transitions between the conduction and valance bands to proceed without the involvement of phonons. A single-layer structure of two-dimensional graphite, namely graphene, is highly transparent and absorbs less incident white light. Graphene is a material that is extremely strong and conductive. Graphene films obtained from thermally reduced graphite oxide show high conductivity and are more transparent. Here, we discussed the applications of carbon nanomaterials like fullerene, CNTs, and graphene in photovoltaic solar cells. The role of carbon materials due to their properties, unique structure, and potential applications in energy conversion has been a great deal of interest in recent years [1, 10–15].

1.2 CARBON NANOMATERIALS AND SYNTHESIS

1.2.1 GRAPHENE

1.2.1.1 GENERAL STRUCTURE OF GRAPHENE

Graphene is one of the allotropes of carbon and was first discovered by Konstatin Novoseov and Andre Geim in 2004 [16]. Graphene is a sheet of sp^2 hybridized atoms of carbon having one atom thickness, and it exists in a 2D structure [17]. Graphene can also exist in 3D structures by stacking several graphene sheets together [16, 18]. It is considered a fundamental building block of graphite, fullerene, and CNT. Graphite is considered stacked graphene, the CNT is considered rolled graphene, and the fullerenes are wrapped graphene. It is transparent, and it shows good mechanical, optical, and electrical properties. Due to these peculiarities, it is used in solar cells, fuel cells (FCs), supercapacitors (SCs), batteries, etc. Graphene mainly exists in three forms: 3D graphene, Heteroatom-doped graphene, and graphene oxide (GO) (Figure 1.1).

Another critical characteristic of graphene materials is no energy band gap between conduction and valence bands (VBs). The stacking order of graphene is equally crucial in determining its properties. The chirality and orientation of the graphene layer play a significant role in varying its characteristics. Due to the hexagonal structure of graphene, it can have two types of edges – Armchair and Zigzag [19]. The presence of defects while synthesizing graphene also varies its mechanical and conductive properties. The type of defects in graphene can be classified into two – intrinsic and extrinsic defects. Intrinsic defects are formed due to the perturbation of its crystal structure, not due to foreign atoms. Extrinsic defects are due to the presence of foreign atoms (Figure 1.2) [20].

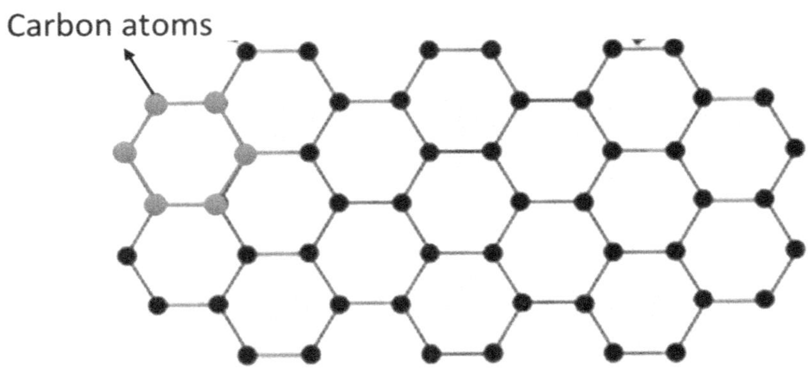

FIGURE 1.1 Structure of graphene.

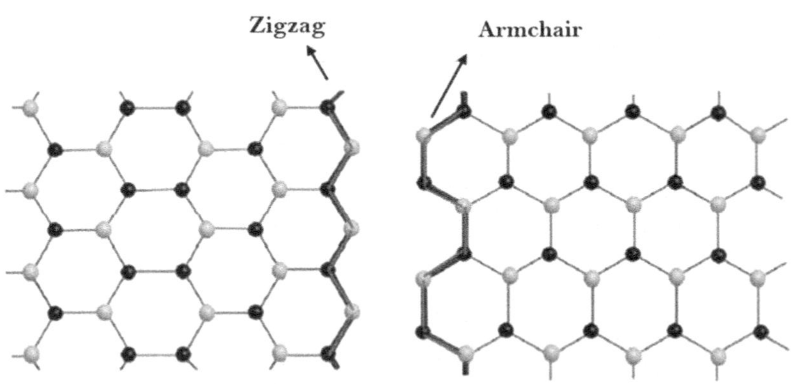

FIGURE 1.2 Types of edges of graphene: (a) zigzag; (b) armchair.

1.2.1.2 SYNTHESIS OF GRAPHENE

Graphene and its derivatives have different applications in various fields. The properties of graphene depend on its preparation methods. Graphene can be synthesized in various methods; research will find a suitable synthesizing technique for mass and cost-effective graphene production.

Graphene can be synthesized by using two approaches, either the top-down approach or by the bottom-up approach. In the top-down approach, graphite is reduced to graphene sheets by dissociating individual layers of graphite using liquid-state oxidation or by solid-state exfoliation of graphite. In the bottom-up approach, carbon molecules create graphene layers by building molecular precursors through different methods [21]. The top-down method is generally used for more economical and mass production of graphene. For the synthesis of graphene, different methods are used, which are mechanical exfoliation, chemical exfoliation, chemical vapor deposition (CVD) method, pulsed laser deposition, etc.

CVD is the most popular method for producing graphene. Graphene that is synthesized through the CVD method shows high purity. In the CVD method, carbon is deposited above metal surfaces. Ni and Cu are the most used metals for the deposition of carbon.

Mechanical exfoliation is considered one of the simplest methods of producing high-quality graphene. The graphene layer can be prepared by simply using scotch tape. The scotch tape is first stuck into graphite, and a few layers of graphene are mechanically peeled from high-oriented pyrolytic graphite (HOPG). Then the graphene layer is transferred to a suitable substrate. The process is repeated at different times [22]. By using this method, we can make a graphene layer of high purity and fewer defects. This method is commonly used for laboratory purposes, and it is not suitable for the large-scale production of graphene [23].

In the liquid exfoliation method, different solvents are used to separate graphene layers with the help of ultrasonication [24]. Different solvents, such as sulfuric acid, acetic acid, etc., are used for this purpose.

This method includes three steps:

- Oxidation of graphite;
- Dispersing the oxidized graphite in the solvent; and
- Reduction of graphite oxide into graphene.

1.2.2 CARBON NANOTUBES (CNTS)

1.2.2.1 STRUCTURE AND SYNTHESIS OF CNTS

Carbon nanotubes (CNTs), being unique nanostructures, have a hexagonal matrix of the carbon atoms in a honeycomb lattice by sp^2 hybridization. CNTs are considered graphene sheets rolled up to make seamless hollow cylinders. They have a high aspect ratio, i.e., hollow cylinders having tens of micrometer length and a small diameter of 0.7 nm. Both the endings of the tubes have a half-fullerene molecule, as caps. They have remarkable electronic and mechanical properties due to their unique structure and morphology. Even though CNTs are closely connected to a 2D graphene sheet, the nanotube curvature, and the quantum confinement of the CNTs in the circumferential direction contribute to having properties different from graphene [25]. CNTs were observed firstly by Iijima in 1991 [15, 25–27] accidentally during the synthesis of fullerenes. Single-Walled CNTs (SWCNTs) or Multi-Walled CNTs (MWCNTs) are the two kinds of CNTs. SWCNTs are fundamental structural units, having a cylindrical shell with only one atom thickness. MWCNTs have multiple coaxial cylinders of SWCNTs about a common axis of increasing diameter (Figure 1.3) [25].

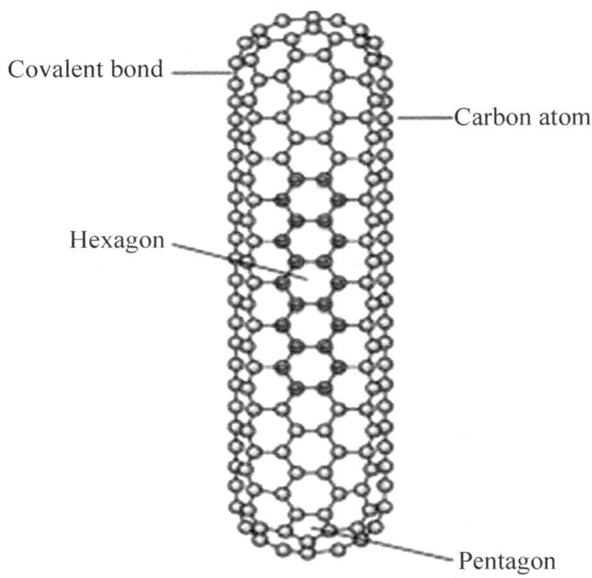

FIGURE 1.3 Structure of carbon nanotube.

1.2.2.1.1 Chemical Vapor Deposition Method (CVD)

In the CVD method, a transition metal catalyst is heated up at a high temperature in a tube furnace while a hydrocarbon gas is flown through the tube reactor for some time. The temperature range is 550–750°C. Iron, nickel, and cobalt are also used as catalysts [25]. A diverse form of CNTs, such as powder, thick or thin films, entangled or aligned, coiled, or straight can be grown by this method. Growing nanotubes of desired architecture at prearranged spots on a patterned substrate is another advantage of CVD [26, 27].

The arc discharge method and laser ablation method are two other synthesis methods of CNTs that can produce high-quality and highly crystalline CNTs [25–27].

1.2.3 FULLERENE

1.2.3.1 FULLERENES: STRUCTURE AND SYNTHESIS

Fullerene, the first synthesized third allotropic form of carbon, was discovered in 1985 by Richard E. Smalley Harold. W. Kroto, and Robert F. Curl. The fullerenes with 60 and 70 carbon atoms, denoted as C_{60} and C_{70}, were the most common out of other higher fullerenes as they were present in 75 and 24%, respectively, in the synthesis conducted Kratschmer et al. in 1990 [28, 30, 33]. The properties of fullerene, such as color, depending on molecular weight and symmetry [28]. C_{60} fullerene, commonly called the Buckminsterfullerene, is arranged in 12 pentagons and 20 hexagons [30]. Fullerenes are essential due to their exciting electronic and structural properties, such as high electron mobility, increased electron affinity, and low reorganization energy [39]. This allotrope's essential properties that enhance its use in energy conversion applications are high electrochemical stability, small size, well-ordered structure, and specific morphology [31]. The most common fullerene derivatives employed in different applications are PCBM (phenyl Cn butyric acid methyl ester), a dimer of fullerene that exists in dumbbell-shape, which is a connection of two fullerene cages [32], etc. (Figure 1.4).

The synthesis method employed to produce fullerenes in large quantity is the arc-discharge method discovered by Kratschmer et al. in 1990 as a peak in the mass spectra of quenched carbon atoms [33]. This method, commonly called the Hufmann-Kra¨tschmer method, utilizes carbon soot produced by evaporated graphite rods in contact as the precursor for producing bulk fullerenes [28, 30, 33]. The advantage of this method is that graphite is the only by-product, which can be reformed into rods and recycled [28]. The process involves arc discharge employed between graphite rods kept in an atmosphere of helium, at a very high temperature. The smoke particles collected as a result of the evaporation and condensation of the precursor were analyzed by ultra-violet, infrared, and visible spectral studies at 100 Torr quenching gas pressure, which confirmed the existence of strong bands [34]. The confirmation of the production of C_{60} fullerenes was done as the experimental result of the infrared spectroscopy showed a good agreement with the theoretical symmetry results [34]. Other methods employed to synthesize fullerenes are benzene combustion in an oxygen-deficient atmosphere, the condensation of polycyclic aromatic hydrocarbons [28], and the Laser-furnace method [35].

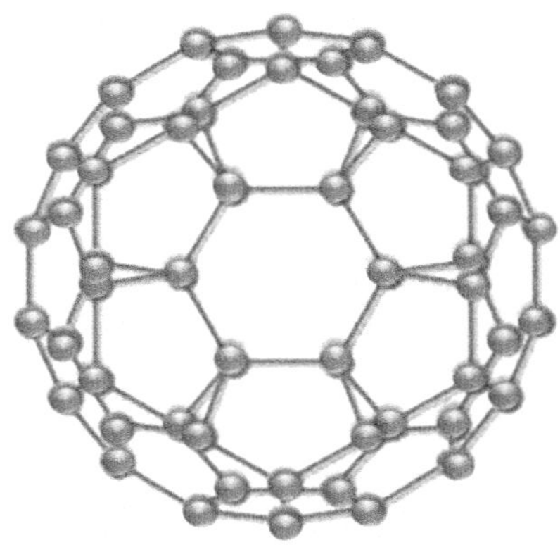

FIGURE 1.4 Structure of fullerene.

1.3 PHOTOVOLTAIC SOLAR CELL

1.3.1 SILICON SOLAR CELL

Traditional silicon solar cells are p-n junctions illuminated under sunlight. The p-n junctions are synthesized by merging a p-doped area of silicon into an n-doped area within the same lattice, where the p and n regions are unsymmetrically doped. At the junction of both layers, there exists the space charge region [36]. The n-doped region is called the emitter through which light enters the solar cell. Even now, crystalline silicon-based solar cells dominate the photovoltaics market due to the high module efficiency and enduring stability, and highly established processing technology. Still, the cost of production is enormous and has a complicated fabrication [37].

1.3.2 DSSC

Dye-sensitized solar cells (DSSCs) belonging to third-generation solar cells were reported back in 1990 [38]. Through these years, DSSC has attained more efficiency and prominence as they could be used in our daily lives like decoration, iPads, etc., as announced by industries like Sony and Apple [38].

DSSC comprises a photoelectrode and a counter electrode (CE) arranged in a sandwich architecture [38]. The former is a semiconductor layer, while the latter consist of a light-harvesting dye adsorbed on an electrode [38]. Semiconductors employed are NiO (p-type) and ZnO or TiO_2 (n-type) [38]. The platinum (Pt) layers cover the substrate electrode, which uses an electrolyte to complete the circuit [38]. The challenges possessed by DSSC are classified as: (i) lack of highly efficient iodine-free electrolytes; (ii) efficient and Pt-free counter-electrodes; (iii) solid-state electrolytes with appreciable ion diffusion and conductivity factors; (iv) panchromatic dyes; and (v) increased charge collection yields at the photoelectrodes [38]. These challenges are being overcome by incorporating nanocarbon such as fullerenes, CNT, or graphene as they possess multifunctional behavior [38]. The first challenge disappears by using nanocarbons as interlayers or dopants on the photoelectrodes [38]. The second challenge could be rectified by employing multifunctional nanocarbons into solid-state electrolytes depending on ionic liquids. The third challenge vanishes as the multifunctional nanocarbons could work as efficient electrolytes without iodine [38]. The fourth and fifth challenges could be overcome by employing porphyrin, ITO, and PEDOT [38].

1.3.3 PEROVSKITE SOLAR CELL (PSC)

Being a new addition to the third-generation solar cells, perovskite solar cells (PSCs) have attained great interest as they could contribute to the large-scale production of solar energy due to their high-power conversion energy (PCE). The perovskite solar cells (PSCs) evolved from the DSSC when the dyes were replaced with $CH_3NH_3PbI_3$ and $CH_3NH_3PbBr_3$ NCs. Perovskite was employed as it could perform as light-harvesting material. Perovskite with a cubic structure possesses the general formula of ABX_3, where A and B are monovalent and divalent ions, respectively [8]. X is either N, O, C, or a halogen. For light-absorbing perovskites A, B, and X are $CH3NH_3^+$, Cs^+, etc., Pb^{2+}, Sn^{2+}, etc.; I^-, Br^-, respectively [39]. PSCs with the advantage of having great optoelectronic properties, such as long carrier diffusion length, high carrier mobility, and high absorption coefficient [39], act as an ideal candidates for third-generation solar cells. Studies have reported that the PCE has enhanced from 3.8% to 22.1% during the past years [39].

Two types of PSCs exist planar heterojunction and mesoscopic structures. The latter could be further divided into p-i-n and n-i-p structures based on the electron/hole transport layer (HTL) [39].

1.3.4 ORGANIC SOLAR CELL

An ample amount of research was already done in the field of organic solar cells. It is also one of the leading devices in photovoltaics. Silicon solar cells are the most popular devices in solar cell technology and the consumer market [40]. It offers a lifetime of nearly 25 years, but the drawback of silicon solar cells is their high cost of production, and it is difficult to manufacture in large sizes industrially. It is imperative to find an alternative material; organic solar cells are such a device that can be manufactured in simple and cost-effective methods. It can be prepared by using spray deposition spin coating, printing, etc. [41].

The OPV can be divided into polymer solar cells and small-molecule solar cells [42]. The basic structure of organic solar cells, including a substrate, HTL, an electron transport layer (ETL), a photoactive layer, ITO, TCO, etc., are generally used as transparent substrates. It will act as a transparent window for passing light to the photoactive layer and also act as an anode. ETL is used to conduct electrons and block holes emitted from the photoactive layer, while HTL conduct holes and block electrons. Generally, PEDOT: PSS is used as an electron-blocking layer.

1.4 APPLICATIONS OF CARBON NANOMATERIALS IN PHOTOVOLTAIC SOLAR CELLS

1.4.1 GRAPHENE AS PHOTOANODE IN DSSC

Due to the peculiar characteristics of graphene, it is used in various fields of research. In the DSSC, things are not different. Due to its high surface area, outstanding transparency, and conductivity, graphene is used for various purposes [43]. Graphene offers a transparency rate of about 97.7%, and its conductivity makes it a worthy candidate for DSSC solar cells (Figure 1.5).

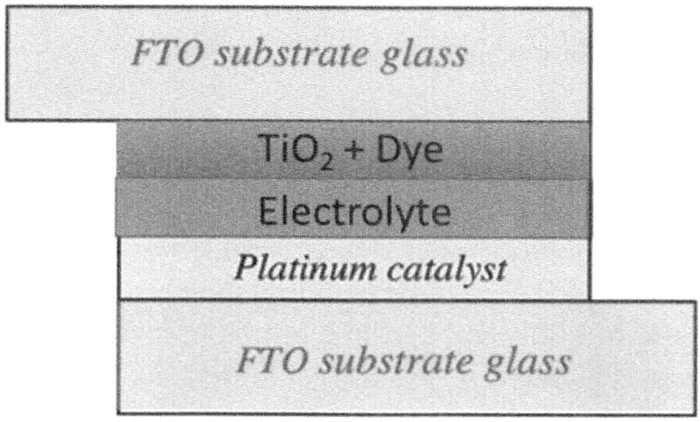

FIGURE 1.5 Structure of DSSC.

Photoanode is considered an electron transport material that transfers electrons from excited dye to the outer circuit. If graphene is placed between FTO and TiO_2, it will increase the charge transfer to the photoanode. The band position of graphene [–4.4 eV] lies in between the FTO [–4.7 eV] and TiO_2 [–4.2 eV], and hence it will act as an easy pathway for electron transfer [44]. Graphene also helps make a crack-free TiO_2 layer by acting as an auxiliary binder without decreasing its efficiency. But it is reported that excess graphene will decrease the transmittance of light into the dye and decrease its photoconversion efficiency [45].

1.4.2 GRAPHENE AS THE CATHODE IN DSSC

The CE is another important part of the DSSC solar cell. These electrodes have the role of transferring electrons from the outer circuit to the electrolyte. The process that happens is called triiodide reduction; in this, I_3^- will convert into I^-. The reaction that happens here is $I_3^- + 2e^- = 3I^-$ [44].

Generally, Pt is used as a CE in DSSC due to its low resistivity and high catalytic activity. The main issues of using Pt as CE are its high cost and its low chemical inertness. Researches are going on to replace Pt with other electrodes. Graphene is considered suitable for replacing Pt due to its high surface area, transparency, and conductive nature. It is reported that using functionalized graphene sheet as CE in DSSC, a power conversion efficiency (PCE) of 3.83% and that of HNO_3 doped graphene was about 3.21% [44].

Graphene shows transparency in all wavelengths of the solar spectrum, including the IR region, so it can also make tandem solar cells. Tandem solar cells can absorb the wavelength of the entire spectrum, increasing the overall efficiencies of the solar cell [44, 46].

1.4.3 EFFECT OF GRAPHENE IN PEROVSKITE SOLAR CELL

In earlier studies, PSC has shown a considerable development in efficiency from 3.8% to 22.1% [47]. The most common materials used for the development of PSCs are $MAPbI_3$ and $FAPbI_3$. In this material, Spiro-OMeTAD is generally used as HTL and TiO_2 as ETL (Figure 1.6).

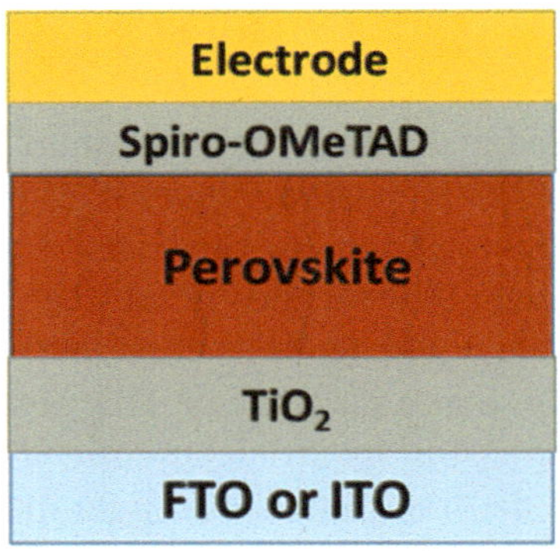

FIGURE 1.6 Structure of PSC.

1.4.4 GRAPHENE IN THE ELECTRON TRANSPORT LAYER (ETL)

TiO_2 is commonly used as ETL in PSCs. The instability of the TiO_2 layer with UV radiation limits its applications. Thus, researchers are going on to find an alternative material. Graphene is a material that exhibits good conductivity and mobility. It is reported that adding a small amount of graphene after the device fabrication will enhance the electron collection and give an efficiency of about 15.6% compared with that of 14.1% of ordinary samples [48]. The optimized amount of graphene that can be used is about 0.6–0.8 wt.%, and excess graphene does not favor increasing efficiency. Another benefit of adding graphene is that the annealing temperature can be reduced from 500°C to 150°C. This reduces its fabrication cost to a great extent.

1.4.5 GRAPHENE AS HOLE TRANSPORT LAYER (HTL)

The critical role of the HTL is to conduct holes and block electrons emitted from the active layer. In PSCs, Spiro-OMeTAD is commonly used as HTL. Due to the high surface area, transparency, and conductive property of graphene, it gains more attention in this also. Researches are going on regarding the change in graphene and GO efficiency by spin coating on FTO and ITO substrates. The perovskite film spin-coated above the GO layer shows more film coverage compared to ITO [49].

Compared to commonly used HTL-Spiro-OMeTAD, GO offers less efficiency [50]. But the advantage of GO is its high stability; graphene-based PSC shows a 14% higher resistance against moisture than Spiro-OMeTAD. The efficiency of Spiro-OMeTAD significantly reduced by about 41% in 1,987 minutes, while the case of PSCs with GO shows an increase in efficiency from 4.87% to 6.62% in the meantime [51].

PSC, which uses rGO and Spiro-OMeTAD together as HTL, shows efficiency of about 13.01%. The composite structure of rGO and Spiro-OMeTAD shows higher efficiency compared to individual ones.

1.4.6 GRAPHENE AS ENCAPSULANT

PSCs are emerging photovoltaic devices that show excellent efficiency comparable to silicon solar cells. The manufacturing cost of PSC is comparatively lesser compared to other photovoltaic devices. Even though it shows higher efficiency, the critical problem they are facing is its stability. The PSC is a device that is sensitive to moisture and oxygen in the outer atmosphere. This decreases its efficiency to a great extent [52].

The instability of PSCs may be due to a lack of intrinsic and extrinsic stability. The intrinsic stability is due to internal factors, including changes between the PSC layers. Extrinsic stability indicates the changes due to external factors like moisture, oxygen, etc., in the outer atmosphere [53]. It is found that the rate of degradation dramatically depends on the humidity content. The PSC materials show faster degradation in humid conditions than expected, and the degradation process will be faster in a brighter environment than in the dark [52]. Various research has been carried out to decrease the instability of perovskite materials. Graphene encapsulation is considered one of the most suitable methods to decrease its degradation. Due to the hydrophobic property of graphene, the graphene layer acts as an encapsulant against the oxygen and moisture content. The graphene layer can be added to PSC using the spin coating method or the CVD method [52, 54].

1.4.7 CNTS IN SILICON SOLAR CELLS

A hybrid heterojunction device of silicon and CNT is an alternative due to the advantageous qualities of simple structure, ease of fabrication, promising performance, and low cost [55]. The superior optoelectronic properties of CNTs, together with conventional silicon photovoltaic technologies, is a highly promising photovoltaic devices [56].

In CNT/Si hybrid solar cells, the transparent CNT films implanted on the Si wafer act as conductive electrodes for charge carrier accumulation. It sets up a built-in voltage for the separation of photo carriers [56]. The silicon's maximum absorbance of the incident light occurs due to the CNT film's noteworthy optical transparency. Thus, CNT films work as window electrodes for accumulating holes in the CNT/Si solar cells [55]. SWCNTs are more suitable for being used as the electrodes than the MWCNTs or double-walled CNTs (DWCNTs) because of their tunable/direct bandgap energies that match a broad range of the solar spectrum and superior charge carrier transporting properties [56].

The p-n junction heterojunction, metal-insulator-semiconductor (MIS) Schottky junction, which acts as a p-n homojunction, and metal-semiconductor (MS) Schottky junction, are three operation models. In the p-n junction design, the p-type emitter is SWCNT thin film; photocarriers got separated at the SWCNT (p-type)/Si (n-type) junction by the activity of the barrier potential. A thin insulator is embedded between SWCNT (M) and Si (S) in the MIS Schottky model. The Schottky barrier is generated at the junction of SWCNT (M)/Si (S) in the MS Schottky junction, which is accountable for the majority of carrier-driven modification [56].

Jung et al. demonstrated SWCNT/Si solar cells with an innate power conversion efficiency (PCE) of 11.2% and a fill factor (FF) of 74.1%. The superacid sliding coating method was followed for the synthesis of high-quality SWCNT thin films [56].

Sun et al. fabricated flexible CNT/Si solar cells by the usage of ultra-thin monocrystalline Si films to overcome the problem of the high rate of recombination of the electron-hole pairs inside the Si side and the high cost of the materials in the case of CNT/Si cells fabricated from amorphous silicon. An improved CVD method was followed to prepare CNT films, including a mixture of SWCNTs and DWCNTs. The efficiency obtained was approximately 4.2%. An improvement in efficiency is obtained by dropping HNO_3 at the interface between CNTs and Si thin films. A PCE of 6% was obtained after acid dropping, showing the heterojunction interface's significance on the solar cell's performance. The reflectance declined evidently, and thereby absorption of light at the visible region was enhanced by the presence of CNTs. The CNT/Si thin-film solar cells also have appealing stability through the bending-recovery cycles [37].

Hu et al. reported CNT/Si solar cells using small-bundle SWCNT film of high quality. These CNTs were produced by the floating catalyst CVD. They have a minimal sheet resistance of 180 Ω sq^{-1} when the transmittance is 90%. Devices with active areas of 9.0 mm^2 and 2.3 mm^2 exhibit exceptional PCEs of 11.8% and 14.2%, respectively. Also, the cell showcased impressive stability in the air without any encapsulation [57].

Xu et al. explored the possibility of expanding the CNT-Si solar cell area above 2 cm^2, keeping the efficiencies high by about 10%. They introduced narrow CNT strips of superior conductivity to act like self-similar top electrodes corresponding to the metal grids employed in traditional silicon solar cells. Adding an optimum number of CNT strips boosted the cell efficiency by enhancing the charge transportation and collection at the CNT side. Also, CNT/Si junctions have improved. Adding six strips increased the efficiency from 3.97% to 6.52%. The efficiency was further enhanced to 10.11% after TiO_2 coating and acid doping. Even though there is efficiency degradation after storing in the air for two days, the cell performance could be recovered by the HNO_3 treatment [58].

1.4.8 CNTS IN DSSC

The large surface area of the nanoporous TiO_2 photoanode increases the adsorption of dye molecules, but it decreases electron transportation. The resistance of the host material is reduced by introducing CNTs into the semiconductor material. CNTs also enhance the mechanical strength, electrical conductivity, thermal conductivity, and flexibility of the device. The main intention of using CNTs is to transmit electrons from the semiconducting component to the transparent conducting oxide (TCO) through CNT by experiencing no resistance in the grain boundaries [60].

Mehmood et al. fabricated DSSCs utilizing TiO_2/MWCNT composite with CNTs of varying concentration and using N_3 dye as a photosensitizer. An optimum concentration of 0.03% MWCNTs/TiO_2 gave the highest PCE of 0.85%, approximately 30% more than pure TiO_2 photoanodes (0.65%).

The increment in the surface area of the photoanode enhances the efficacy of the cell, with an outcome of improved dye loading. The PCE enhancement is the result of improved injection efficiency of electrons. The TiO_2/MWCNT photoanode exhibits a greater redshift than the pure TiO_2 electrode pertaining to the photosensitizing properties of CNTs, as a fact of which the photovoltaic properties of the cell are extended to the visible region of the spectrum. Increasing the concentration of MWCNT above the optimum value negatively affects the working of the cell [61].

Kabir et al. reported TiO_2/SWCNT-based DSSC using a mixture of two natural dyes, betalain, and curcumin. An optimum mixture of betalain and curcumin dye in the volume ratio of 1:2 yields an efficiency of 0.649% for pure TiO_2 DSSC. Incorporating SWCNT into the TiO_2 photoanode gives an improved efficiency of 0.903%. $TiCl_4$-treated SWCNT/TiO_2 further enhances the efficiency to 1.108%. SWCNT/TiO_2 hybrid material allocates the co-occurrence of dual pores, satisfying the high surface area required for dye adsorption. Also, it supports a high-quality junction between semiconductor TiO_2 material and the dyes. The degradation study of the cell under normal atmospheric conditions for 360 hours results in significant degradation (16.3–27.5%) of the cell performance [60].

Kilic successfully fabricated DSSC using CNT/ZnO nanowire (NW) hybrid photoelectrode. The efficiency obtained was 5.55%. The efficiency of the DSSC without CNT was 5.05%. The N719 ruthenium dye absorption on the surface of hybrid composites enhances, and the short circuit current density increases significantly due to better electron lifetime. All these happen due to the presence of CNT. The power conversion efficiency of the DSSC is improved with an increased device voltage due to the carrier density increment by the CNTs, and decremented recombination of charge carriers [59].

Pt CEs used in DSSC have the drawback of high cost. CNTs having high electrochemical activity, and low cost is a better alternative [62] because a CE must have excellent electronic conductivity, large active surface area, high catalytic activity, high transparency, and excellent compliance with the transparent TCO support [63], for achieving a high PCE of DSSCs [62].

Nam et al. reported the functioning of DSSC using CNTs as CEs. Two methods were used for the CNT preparation. Screen printing of randomly dispersed MWCNT was the first method. For the second CNT preparation, the catalytic CVD method was followed, and reasonably well-aligned CNTs were grown directly from this method. DSSC with reference Pt CE yielded an efficiency of 8.80%. The efficiency corresponding to the CE from the randomly dispersed screen-printed CNT was 8.03%. It is due to the decreased short circuit current. But the directly grown CNT as CE yielded an efficiency of 10.04% for DSSC from the remarkably improving FF, which is due to the fast redox reaction on the CEs [62].

Ma et al. developed a novel bilayer CEs of reduced graphene oxide (RGO) and aligned carbon nanotube (ACNT). The ACNT underlayer acts as a transition layer for RGO. The overlayer of RGO plays the role of catalytic layer. The directional arrangement of ACNT makes it easier for the electrons to pass through the transport channel. Also, the ACNT CE yields higher PCE than the DSSC based on the MWCNT CE. The surfactant functionalization of RGO by a cationic surfactant cetylpyridinium chloride (CPC) enhances cell performance. The CPC-RGO/ACNT-based DSSC gives a PCE of 3.9%, a 1.24-fold increase than the ACNT-based DSSC [64].

Fiber-shaped DSSC has attracted significant attention due to its portable and flexible nature. Modified titanium wire-based photoanode and Pt-based CEs are the essential components for a fiber-shaped DSSC. Kim et al. used CNT yarns (CNTY) for the CE for the fiber-shaped DSSC. The CNTY was fabricated by dry spinning from vertically aligned (VA) MWCNTs. CNTY-based CE showed increased electrocatalytic activity in the solid-state fiber DSSCs. The PCE enhanced up to 4.00% was obtained, competing with 2.64% PCE of the Pt wire electrodes. Also, the device exhibits a decreased series resistance which indicates a decrease in the leakage current and an increase in the extraction efficiency. The device exhibits a high level of flexibility and washability and better long-term stability over 35 days. These can be accredited to the large specific area of CNTY, that derives from high graphitization, the mesoporous structure of the spun CNTY, and the long length of the constituent CNT arrays [63].

1.4.9 CNTS IN PEROVSKITE SOLAR CELLS (PSCS)

Habisreutinger et al. studied the successful incorporation of CNTs based upon metal-halide perovskites (MHPs). CNTs can work as contact materials which have an essential part in the functioning of MHP solar cells because the contact materials introduce electronic symmetry in the apparatus. Wu et al. used a thin semi-transparent electronically connected matrix of SWCNTs as alternative contact material that could maintain a light balance between high bandwidth light transmittance and low electrical resistance. The contacts should have good charge selectivity and improve the long-term stability of the cells [65].

CNTs have been found out to have the usage as charge selective contacts (CSCs) both for the p-type and n-type sides of perovskite devices. Habisreutinger et al. studied the usage of SWCNT-based HTL combination with various configurations like that with polymers which could meet the stability requirements of the p-type contact. But further studies proved that SWCNTs used as the sole HTL could not build up a solid built-in field to lessen the recombination losses at the interfaces, resulting in a substantial photovoltage deficit and thus lower steady-state performance. The CNTs can work as charge extraction routes from the absorber to the HTL when CNTs are combined with secondary hole transporters. The CNT minimizes the recombination of charge carriers at the interfaces by the accelerated charge extraction into the secondary material. The poor charge carrier transportation properties of various hole-transporting materials are compensated by combining SWCNT with them. Studies have been done on hybrid systems showing the advancement of carrier transport inside the absorber layer by directly constituting SWCNT into the absorber layer. SWCNTs have the possibility of usage as bridging elements to boost up the inter-grain charge carrier transport mechanism [65].

Wang et al. used hydrophobic poly(3-hexylthiophene) (P3HT) combining MWCNTs as the HTL and developed a carbon electrode-based $CsPbI_2Br$ inorganic PSC, which showcased excellent cell performance as well as excellent stability. A PCE up to 10.01% with a stabilized efficiency of 8.85% was attained. The P3HT-MWCNT protects the $CsPbI_2Br$ from atmospheric moisture and prevents moisture inrush that improves the stability of the cell. MWCNTs significantly boost up the charge-transporting mechanism of the P3HT matrix, whilst P3HT forbids the transfer of electrons from the $CsPbI_2Br$ layer to MWCNTs. The device exhibits exceptional long-term stability as it retains ~85% of the beginning value of PCE over 240 hours under ambient conditions with no need for encapsulation [66].

Mohammed et al. reported CNT-incorporated methylammonium lead iodide ($MAPI_3$) perovskite film made in atmospheric air. It was used as an effective absorber layer in $MAPI_3$-based solar cells giving a PCE up to 15% [67].

1.4.10 CNTS IN ORGANIC SOLAR CELLS

The intention of incorporating CNTs into BHJ devices is mainly to replace fullerenes that are expensive, less conducting, and low air-stable with CNTs to combine their unique charge transport characteristics with the physicochemical properties of conductive polymers. An improved percolation network with exceptionally high exciton separation in polymeric BHJs is obtained [68].

CNTs can be used as the photoactive components of BHJ devices [68], enhancing device efficiency by several orders of magnitude. CNTs act as exciton dissociation centers; the CNTs and the conducting polymers transport electrons and holes, respectively to the electrodes. Even though there is an enhancement of PCE, it remains lower than the PCE of inorganic PV cells. CNTs should be uniformly dispersed in the polymer for efficient charge dissociation. The uniform dispersion of CNTs in polymer has another advantage of forming a percolation path through the composites, making the transportation of electrons easier and reducing the recombination of charge carriers in the active layer [69].

CNTs can also have the application as transparent conductive or hole extraction layers (HELs) to replace indium tin oxide (ITO). This application is inferred from the substantial inherent charge mobility of individual CNTs, amenability to solution processing, and large optical transparency over a broad range of frequencies [68].

Lee et al. presented BHJ solar cells from nitrogen-doped CNT (N-CNT) or boron-doped CNT (B-CNT) as an exceedingly selective hole or electron-transport enhancing materials. The semiconducting polymer used was P3HT/PCBM, and the CNTs were the multi-walled ones. Both the CNTs showed enhanced cell performances. B-doped CNT-based solar cells, in particular, showed more improvement in short circuit current and PCE. PCE obtained was 4.1% for 1.0% wt. of B-CNT. It results from the observed enhancement in carrier mobility [70].

1.4.11 APPLICATIONS OF FULLERENE IN PHOTOVOLTAICS

The environmental hazards and the high cost of fossil fuels paved an opening for a renewable energy substitute for them, which resulted in the invention of many energy conversion devices such as FCs, photovoltaic cells, SCs, etc. Our area of interest is photovoltaics, which converts the abundant energy resource, solar energy, into electrical energy [72, 73]. The basic phenomenon associated with photovoltaic devices is the production of electron-hole pairs due to photon absorption [74]. The branch of photovoltaics could be broadly divided into organic solar cells and inorganic solar cells [3]. Organic solar cells are metal-free solar cells that possess advantages such as (i) tunable electrochemical and optical properties; (ii) their technology has high flexibility and low production cost; and (iii) increased light-harvesting ranges [74]. Organic solar cells work based on the formation of excitons under irradiation of white light and are thus called excitonic solar cells [72]. Although inorganic solar cells are characterized by their higher efficiency, they possess the disadvantage of high cost [73], rigidity, and weight [74]. The lower efficiency faced by organic solar cells could be increased by incorporating the appropriate donor-acceptor pair and the device architecture [72]. The significant affinity for electrons and increased ability to transport charge [72] make fullerenes the electron-accepting component in organic solar cells.

Thus, here we are discussing in detail the role of fullerenes in the field of photovoltaics, specifically, organic-inorganic perovskites solar cells, organic solar cells, and DSSCs.

1.4.12 FULLERENES IN ORGANIC SOLAR CELLS

The organic solar cells employed in the present scenario are the bulk heterojunction (BHJ) solar cells, which possess polymer donors and fullerene acceptors. Several factors, such as donor-acceptor interaction, morphology, etc., are being studied to understand the effect of fullerene derivatives as electron acceptors [72]. Fullerene-based polymers are examples of previously synthesized BHJ solar cells [72, 73].

Since the fundamental phenomenon to enhance efficiency is incorporating fullerenes as the electron acceptor, studies on the donor-acceptor interactions should be considered. Thus, the electronic properties of polymer-fullerene interactions are tuned to absorb a more significant amount of light, produce free electrons, and transport charges to respective electrodes [72]. Fullerenes are capable of acting as electron acceptors as they possess: (a) energetically deep-lying LUMO with high electron affinity [72]; (b) they are capable of stabilizing negative charges; (c) they have very high photoinduced charge transfer; (d) possess high electron mobility; (e) it possesses a constant electronic structure (does not depend on the functionalization).

Morphology is another factor responsible for increasing the efficiency of BHJ solar as the physical interaction of donor and acceptor has a significant role in the BHJ solar cells [72]. The factors affecting the morphology of the polymer-fullerene interaction are the crystallinity and miscibility of the materials used, the solvent we chose, and its annealing [72]. At present, the most suitable combination among MDMO-PPV/PCBM and P3HT/PCBM combinations is P3HT/PCBM [72]. Chlorobenzene and toluene are the two critical solvents employed to improve the efficiency of organic solar cells. Moreover, thermal healing helps to attain morphology to a greater extent [72].

C_{70} derivative of PCBM and Porphyrin-fullerene combination of donor-acceptor are composites used in organic solar cells to increase efficiency [73].

1.4.13 EFFECTS OF FULLERENES IN DSSC

The role of fullerenes is introduced in the solid-state electrolyte to produce iodine-free electrolytes [38]. The fullerene-based electrolyte was first reported in 2012 [38]. Fullerene-based electrodes possess the advantages of insignificant absorption features in the visible region compared to iodine-based electrolytes and the efficient regeneration of ruthenium dyes. The addition of fullerene-based electrolytes improved the efficiency by up to 2% [38]. It also enhances the device's photocurrent [38]. Fullerene-based dyes were regarded as potential dye-based on multifunctional nanocarbons, which could enhance the performance of the DSSC [38]. The addition of fullerene-based dye date back to 2007, where fullerene attached via diaminohydrocarbon linkers were employed [38]. Reports suggest that metallotetraphenylporphyrin – fullerene could also be used as a fullerene-based dye [29]. Thus, fullerene derivatives could improve the efficiency of DSSC by causing a change in the electrolytes and dyes.

1.4.14 EFFECTS OF FULLERENES IN PEROVSKITE SOLAR CELLS (PSCS)

The PSC faces many disadvantages of which, the major one being the usage of toxic lead. Apart from this, they encounter the problem of instability and hysteresis [39]. Thus, studies report that fullerenes could be employed in PSCs due to their highly-matched energy levels with that of perovskites and also since they possess other merits of low reorganization energy, increased electron affinity, and high electron mobility. Fullerenes and fullerene derivatives could be used as trap-state passivation, interfacial modification materials, and electron transport materials [39].

The fullerenes could enhance the efficiency of the PSC by reducing the hysteresis as they could successfully passivate the trap states at the grain boundaries and surfaces of the perovskite layer concerning the formation of fullerene–halide radicals that inhibit the formation of trap states [39]. The chemical structures of fullerenes could also affect the efficiency of the PSC.

The role of fullerenes in PSCs could be discussed as the effect of fullerenes in the ETL, the effect of fullerenes in the interlayers, and the studies on fullerene–perovskite heterojunctions [39].

Fullerenes incorporated in the ETL of p-i-n structure: Studies reported affirm that the efficiency of PSC is improved by employing fullerene derivatives in the ETL. Many research groups have worked on the same field, of which one group in 2014 employed double fullerene layers as layers capable of extracting electrons, thus increasing the FF to 80% [39]. These double fullerene layers were capable of lowering the dark current leakage and also could passivate charge traps. Studies could also elucidate the effect of electron mobility of fullerene in enhancing photovoltaic performance [39]. A fulleropyrrolidine with a hydrophilic triethylene glycol monoethyl ether side chain (PTEG-1) was used as the ETL to overcome the light-soaking phenomenon. The Voc of PSCs is improved by using fullerene derivatives with increased LUMO energy levels as ETLs [39] and high bandgap perovskites as absorbers [39].

Fullerenes employed in the cathode buffer layer: fullerenes owing to the advantage of chemical tunability and solution processability, could be used as cathode buffer layers (CBL) [39]. Amine-functionalized fullerene derivatives are majorly used as CBL since the amine group has the potential to behave as an interfacial dipole layer that can decrease the work function of metal electrodes and enable ohmic contact at ETL/metal electrode interfaces [39].

Fullerenes employed in the interfacial layer were used in the interfacial layer of the n-i-p structure, which used TiO_2 as the ETL, which had the disadvantage of lower stability and anomalous hysteresis. A self-assembled fullerene monolayer (C_{60}-SAM) was used to overcome the difficulties possessed by TiO_2 as it could inhibit electron transfer from the perovskite to TiO_2 and decrease Voc Loss [39]. PC61BM is the fullerenes derivative commonly used as the interfacial layer. Fullerene derivative has the added advantage of acting as surface modifiers on other metal oxide materials [39].

Effect of fullerene in perovskite-fullerene heterojunction: Due to the small diffusion length of electrons (shorter than that of holes,) the latter could be extracted more efficiently than electrons in perovskites. Therefore, studies are done to enhance the electron extraction efficiency to increase the device's performance. Fullerene and fullerene-based polymers could improve the electron extraction efficiency in perovskites [39]. PC61BM is widely employed for this purpose. Pristine fullerenes are also being used to form perovskite-fullerene heterojunctions.

Thus, fullerenes could improve the efficiency of PSCs by acting as ETL, CBL, and interfacial layer, and also it could be employed in perovskite heterojunctions.

KEYWORDS

- **carbon nanotube**
- **fullerene**
- **graphene**
- **multi-walled CNTs**
- **nanomaterials**
- **photovoltaic**
- **solar cell**

REFERENCES

1. Moore, K., & Wei, W. (2021). Applications of carbon nanomaterials in perovskite solar cells for solar energy conversion. *Nano Materials Science, 3*(3), 276–290.
2. Zhu, H., Wei, J., Wang, K., & Wu, D., (2009). *Solar Energy Materials and Solar Cells, 93*(9), 1461–1470.
3. Shah, A., Torres, P., Tscharner, R., Wyrsch, N., & Keppner, H., (1999). *Science, 285* (5428), 692–698.
4. Parida, B., Iniyan, S., & Goic, R., (2011). *Renewable and Sustainable Energy Reviews, 15*(3), 1625–1636.
5. El Chaar, L., & El Zein, N., (2011). Review of photovoltaic technologies. *Renewable and Sustainable Energy Reviews, 15*(5), 2165–2175.
6. Guldi, D. M., (2007). *Physical Chemistry Chemical Physics, 9*(12), 1400–1420.
7. Notarianni, M., Liu, J., Vernon, K., & Motta, N., (2016). *Beilstein Journal of Nanotechnology, 7*(1), 149–196.
8. Choi, H. J., Jung, S. M., Seo, J. M., Chang, D. W., Dai, L., & Baek, J. B., (2012). *Nano Energy, 1*(4), 534–551.
9. Pumera, M., (2011). *Energy & Environmental Science, 4*(3), 668–674.
10. Robertson, J., (2002). *Materials Science and Engineering: R: Reports, 37*(4–6), 129–281.
11. Novoselov, K. S., Geim, A. K., Morozov, S. V., Jiang, D. E., Zhang, Y., Dubonos, S. V., & Firsov, A. A., (2004). *Science, 306*(5696), 666–669.
12. Geim, A. K., & Novoselov, K. S., (2010). *A Collection of Reviews from Nature Journals*, 11–19.
13. Kroto, H. W., Heath, J. R., O'Brien, S. C., Curl, R. F., & Smalley, R. E., (1985). C 60: Buckminsterfullerene. *Nature, 318*(6042), 162, 163.

14. Hirsch, A., & Brettreich, M. (2005). *Fullerenes: Chemistry and Reactions*. John Wiley & Sons, 1–412; ISBN 3-527-30820-2.
15. Iijima, S., (1991). *Nature, 354*(6348), 56–58.
16. Das, S., Pandey, D., Thomas, J., & Roy, T., (2019). *Advanced Materials, 31*(1), 1802722.
17. Verdejo, R., Bernal, M. M., Romasanta, L. J., & Lopez-Manchado, M. A., (2011). *Journal of Materials Chemistry, 21*(10), 3301–3310.
18. Agrawal, N., Parihar, A. S., Singh, J. P., Goswami, T. H., & Tripathi, D. N., (2015). *Procedia Mater. Sci., 10*.
19. Pan, B., Xu, G., Zhang, B., Ma, X., Li, H., & Zhang, Y. (2012). Preparation and tribological properties of polyamide 11/graphene coatings. *Polymer-Plastics Technology and Engineering, 51*(11), 1163–1166.
20. Yang, M., Zhang, Z., Zhu, X., Men, X., & Ren, G., (2015). *Friction, 3*, 72.
21. Mahmoudi, T., Wang, Y., & Hahn, Y. B., (2018). *Nano Energy, 47*, 51–65.
22. Lee, H. C., Liu, W. W., Chai, S. P., Mohamed, A. R., Aziz, A., Khe, C. S., & Hashim, U., (2017). *RSC Advances, 7*(26), 15644–15693.
23. Rosli, N. N., Ibrahim, M. A., Ludin, N. A., Teridi, M. A. M., & Sopian, K., (2019). *Renewable and Sustainable Energy Reviews, 99*, 83–99.
24. Hernandez, Y., Nicolosi, V., Lotya, M., Blighe, F. M., Sun, Z., De, S., & Coleman, J. N., (2008). *Nature Nanotechnology, 3*(9), 563–568.
25. Mildred S. Dresselhaus, Phaedon Avouris (auth.), Mildred S. Dresselhaus, Gene Dresselhaus, Phaedon Avouris (eds.). Smalley, R. E. (Foreword) (2001). *Carbon Nanotubes: Synthesis, Structure, Properties, and Applications.* 1–421. ISBN 3-540-41086-4.
26. Gupta, N., Gupta, S. M., & Sharma, S. K., (2019). *Carbon Letters, 29*(5), 419–447.
27. Ando, Y., Zhao, X., Sugai, T., & Kumar, M., (2004). *Materials Today, 7*(10), 22–29.
28. De La Puente, F. L., & Nierengarten, J. F. (Eds.). (2012). *Fullerenes: Principles and Applications.* Royal Society of Chemistry, 3–594. ISBN: 978-1-84973-136-2.
29. El Mahdy, A. M., Halim, S. A., & Taha, H. O., (2018). *Journal of Molecular Structure, 1160*, 415–427.
30. Yadav, B. C., & Kumar, R., (2008). *International Journal of Nanotechnology and Applications, 2*(1), 15–24.
31. Coro, J., Suarez, M., Silva, L. S., Eguiluz, K. I., & Salazar-Banda, G. R., (2016). *International Journal of Hydrogen Energy, 41*(40), 17944–17959.
32. Zhang, R., Murata, M., Wakamiya, A., & Murata, Y., (2013). *Chemistry Letters, 42*(8), 879–881.
33. Tanigaki, K., Kuroshima, S., & Ebbesen, T. W., (1995). *Thin Solid Films, 257*(2), 154–165.
34. Krätschmer, W., Fostiropoulos, K., & Huffman, D. R., (1990). *Chemical Physics Letters, 170*(2, 3), 167–170.
35. Smalley, R. E., (1992). *Accounts of Chemical Research, 25*(3), 98–105.
36. Goetzberger, A., Knobloch, J., & Voß, B., (2014). *Crystalline Silicon Solar Cells*, 9–48.
37. Sun, H., Wei, J., Jia, Y., Cui, X., Wang, K., & Wu, D., (2014). *Nano-Scale Research Letters, 9*(1), 1–6.
38. Costa, R. D., Lodermeyer, F., Casillas, R., & Guldi, D. M., (2014). *Energy & Environmental Science, 7*(4), 1281–1296.
39. Deng, L. L., Xie, S. Y., & Gao, F., (2018). *Advanced Electronic Materials, 4*(10), 1700435.
40. Tao, M., (2008). *The Electrochemical Society Interface, 17*(4), 30.
41. Tang, C. W., (1986). Two-layer organic photovoltaic cell. *Applied Physics Letters, 48*(2), 183–185.

42. Shin, R. Y. C., Kietzke, T., Sudhakar, S., Dodabalapur, A., Chen, Z. K., & Sellinger, A., (2007). *Chemistry of Materials, 19*(8), 1892–1894.
43. Das, S., Sudhagar, P., Verma, V., Song, D., Ito, E., Lee, S. Y., & Choi, W., (2011). *Advanced Functional Materials, 21*(19), 3729–3736.
44. Das, S., Sudhagar, P., Kang, Y. S., & Choi, W., (2014). *Journal of Materials Research, 29*(3), 299–319.
45. Kim, H. N., Yoo, H., & Moon, J. H., (2013). *Nanoscale, 5*(10), 4200–4204.
46. Reina, A., Jia, X., Ho, J., Nezich, D., Son, H., Bulovic, V., & Kong, J., (2009). *Nano Letters, 9*(1), 30–35.
47. Saliba, M., Matsui, T., Seo, J. Y., Domanski, K., Correa-Baena, J. P., Nazeeruddin, M. K., & Grätzel, M., (2016). *Energy & Environmental Science, 9*(6), 1989–1997.
48. Han, G. S., Song, Y. H., Jin, Y. U., Lee, J. W., Park, N. G., Kang, B. K., & Jung, H. S., (2015). *ACS Applied Materials & Interfaces, 7*(42), 23521–23526.
49. Liu, Z., Robinson, J. T., Sun, X., & Dai, H., (2008). *Journal of the American Chemical Society, 130*(33), 10876–10877.
50. Yun, J. M., Yeo, J. S., Kim, J., Jeong, H. G., Kim, D. Y., Noh, Y. J., & Na, S. I., (2011). *Advanced Materials, 23*(42), 4923–4928.
51. Shi, Z., & Jayatissa, A. H., (2018). *Materials, 11*(1), 36.
52. Berhe, T. A., Su, W. N., Chen, C. H., Pan, C. J., Cheng, J. H., Chen, H. M., & Hwang, B. J., (2016). *Energy & Environmental Science, 9*(2), 323–356.
53. Uddin, A., Upama, M. B., Yi, H., & Duan, L., (2019). *Coatings, 9*(2), 65.
54. Cao, J., Liu, Y. M., Jing, X., Yin, J., Li, J., Xu, B., & Zheng, N., (2015). *Journal of the American Chemical Society, 137*(34), 10914–10917.
55. Hu, X., Hou, P., Liu, C., & Cheng, H., (2019). *Nano Materials Science, 1*(3), 156–172.
56. Jung, Y., Li, X., Rajan, N. K., Taylor, A. D., & Reed, M. A., (2013). *Nano Letters, 13*(1), 95–99.
57. Hu, X. G., Hou, P. X., Liu, C., Zhang, F., Liu, G., & Cheng, H. M., (2018). *Nano Energy, 50*, 521–527.
58. Xu, W., Wu, S., Li, X., Zou, M., Yang, L., Zhang, Z., & Cao, A., (2016). *Advanced Energy Materials, 6*(12), 1600095.
59. Kilic, B., (2019). *Journal of Materials Science: Materials in Electronics, 30*(4), 3482–3487.
60. Kabir, F., Sakib, S. N., Uddin, S. S., Efaz, E. T., & Himel, M. T. F., (2019). *Sustainable Energy Technologies and Assessments, 35*, 298–307.
61. Mehmood, U., Hussein, I. A., Harrabi, K., Mekki, M. B., Ahmed, S., & Tabet, N., (2015). *Solar Energy Materials and Solar Cells, 140*, 174–179.
62. Nam, J. G., Park, Y. J., Kim, B. S., & Lee, J. S., (2010). *Scripta Materialia, 62*(3), 148–150.
63. Kim, J. H., Hong, S. K., Yoo, S. J., Woo, C. Y., Choi, J. W., Lee, D., & Song, M., (2021). *Dyes and Pigments, 185*, 108855.
64. Ma, J., Yang, H. L., & Ren, W. H., (2020). *Journal of Nanoscience and Nanotechnology, 20*(3), 1749–1755.
65. Habisreutinger, S. N., & Blackburn, J. L., (2021). *Journal of Applied Physics, 129*(1), 010903.
66. Wang, G., Liu, J., Chen, K., Pathak, R., Gurung, A., & Qiao, Q., (2019). *Journal of Colloid and Interface Science, 555*, 180–186.
67. Mohammed, M. K., Sarusi, G., Sakthivel, P., Ravi, G., & Younis, U., (2021). *Materials Research Bulletin, 137*, 111182.

68. Cataldo, S., Salice, P., Menna, E., & Pignataro, B., (2012). *Energy & Environmental Science, 5*(3), 5919–5940.
69. Abdulrazzaq, O. A., Saini, V., Bourdo, S., Dervishi, E., & Biris, A. S., (2013). *Particulate Science and Technology, 31*(5), 427–442.
70. Lee, J. M., Park, J. S., Lee, S. H., Kim, H., Yoo, S., & Kim, S. O., (2011). *Advanced Materials, 23*(5), 629–633.
71. Yan, J., & Saunders, B. R., (2014). *RSC Advances, 4*(82), 43286–43314.
72. Thompson, B. C., & Fréchet, J. M., (2008). *Angewandte Chemie International Edition, 47*(1), 58–77.
73. Imahori, H., & Fukuzumi, S., (2004). *Advanced Functional Materials, 14*(6), 525–536.
74. Echeverry, C. A., & Castro, E., (2018). *Int. J. Chem. Res., 1*(1), 1–8.

CHAPTER 2

ROLE OF CARBON NANOTUBES AS SUPPORTED METAL-NANOPARTICLES CATALYSIS

H. AKHINA[1] and TRESSIA ALIAS PRINCY PAULOSE[2]

[1]*International and Inter-University Center for Nanoscience and Nanotechnology, Mahatma Gandhi University, Kottayam, Kerala, India*

[2]*Post-Graduate and Research Department of Chemistry, Bishop Moore College, Mavelikara, Alappuzha, Kerala, India*

ABSTRACT

The unique tubular morphology of carbon nanotubes (CNTs) has fascinated scientists and attracted widespread research interest in the last few decades. Ever since its discovery in 1991, CNTs have found diverse commercial applications in almost all areas of science and technology, ranging from rechargeable batteries, energy storage, environmental remediation, and biotechnology to automotive parts and sporting goods. Their unique features, including a large specific surface area, chemical inertness, and high oxidative stability, have contributed to their widespread use as catalysis supports. The contributions of CNTs as supports in heterogeneous catalysis are discussed in this chapter. Recent investigations on the preparation and morphological characterization of various metal/metal oxide particles on CNTs are discussed. The potential applications of CNT as catalyst supports in heterogeneous catalysis are also explored in this chapter. According to reports in the literature, CNTs are a promising material for use as supports in various catalytic processes, contributing to a greener environment by providing ligandless, recyclable,

Nanotechnology-Enhanced Solid Materials: Design, Synthesis, Properties, Applications, and Perspectives. Lionello Pogliani, Ann Rose Abraham, A. K. Haghi, & Prabhat Ranjan (Eds.)

cheaper catalysts with good activity in catalytic reactions using solventless/ environmentally benign solvents at lower/ambient pressure/temperature conditions.

2.1 INTRODUCTION

The year 1991 witnessed the beginning of a new era in the field of nanoscience and nanotechnology with the entry of the fourth allotrope of carbon which exhibited unique physical and chemical properties. The material consisted of graphitic sheets, i.e., an aromatic mono-atomic layer made up of a hexagonal arrangement of sp^2 hybridized carbon atoms, rolled up into a tubular structure. Iijima, who discovered these extremely thin needle-like structures while examining carbon materials under an electron microscope, named them carbon nanotubes (CNTs) [1, 2]. Although tubular structures were known for some silicate minerals, tubular carbon structures had never been synthesized or found in nature. With his dual publication in the journal *Nature,* Iijima announced the preparation of these magnificent structures.

CNTs can be produced in two forms: Single-wall carbon nanotubes (SWCNTs) and multi-wall carbon nanotubes (MWCNTs). SWCNTs are formed from a graphene sheet that has been coiled into a hollow tube with a hemisphere of fullerene at one or both ends. CNTs range in length from a few microns to several millimeters, with an interior diameter of 0.4 to 2.5 nanometers. MWCNTs can be regarded as concentric SWCNTs with an increasing diameter, which is coaxially disposed of like 'Russian dolls.' The number of concentric tubes present can vary from two (double-wall carbon nanotubes) to 100, such that the external diameter can reach 100 nm. Depending on the number of layers and the helicity created by rolling up each graphitic sheet, CNTs have conducting or semi-conducting qualities. SWCNTs display superior properties, such as higher mechanical strength, higher thermal/electrical conductivity, higher specific surface area, higher thermal resistance, and unique optical properties when compared to MWCNTS. SWCNTs also have lesser defects on their surfaces and therefore are more stable than MWCNTs. SWCNTs have a tendency to form bundles or ropes due to van der Waals attraction between the tubes, with each rope containing tens to hundreds of tightly packed tubes of similar diameter that are entangled together in the solid state, giving rise to a highly complex network; and thus, are difficult to disperse. Despite its excellent features, SWCNT applications have been limited due to difficulties in synthesizing synthetically pure SWCNTs and their dispersibility.

Driven by their extraordinary electronic, mechanical, and thermal properties, CNTs have found diverse applications as composite materials, in energy storage, in the field of photovoltaics, integrated circuits, biosensors, nanomedicine, etc. However, one of the major applications of CNTs has been in the field of catalysis, especially as support.

2.2 PREPARATION AND MORPHOLOGICAL STUDIES OF CNT-SUPPORTED CATALYST

Supported catalysts can be prepared by depositing catalytic material onto the inert support and then optionally treating them before feeding them into the reaction chamber. Catalytic materials supported or bonded to CNTs are referred to as CNT-supported catalysts. The literature review reveals that both the exterior as well as the interior wall of the CNT have been utilized for the bounding of catalytic materials. The catalytic materials generally used are precious metals like Pt, Pd, Ru, Rh, Ir, Au, etc., transition metals like Fe, Co, Ni, Mo, etc., and metal oxides, hybrids [3], bifunctional catalyst materials, etc. [4–7]. The catalytic property of a supported metal catalyst has been reported to depend on the degree of dispersion of the metal on the support. Various methods have been utilized for the preparation of CNT-supported catalysts, including wet impregnation [3, 5, 8–10], deposition/precipitation techniques [11], microemulsion-templated synthesis [12], liquid phase reduction [13–15], microwave (MW) assisted synthesis [16, 17], strong electrostatic adsorption method [18], etc. As CNT is inert support, pre-treatment is needed in most of the preparation methods to attain optimal interaction between the support and the catalyst material. Some of the recent studies dealing with the preparation of CNT-supported catalysts using MW assistance are discussed here.

The MW-assisted approach has been regarded as one of the most effective methods for chemical reactions and the manufacture of CNT-supported catalysts [19, 20]. In terms of energy use, MW heating is more efficient. The literature review reveals that one can tailor the particle size of the catalyst materials by changing the duration of MW heating. It is an advantage of this method over conventional preparation methods. Also, in most cases, only a very short reaction time is required for MW-assisted synthesis. Moreover, MW-assisted synthesis is seen as an effective, fast, simpler, homogeneous method, and it can increase the reaction rate greatly [16, 21].

Using a MW heating technique, multi-walled carbon nanotube (MWCNT)-supported catalysts were recently prepared by Lee et al. [22]. With a MW heating for 10 minutes, they were able to generate a uniform dispersion of nanoparticles (NPs) with an average size of roughly 5 nm. They have selected the precursors $H_2PtCl_6 \cdot 6H_2O$, $RuCl_3$, and $NiCl_2$ and were mixed at the atomic ratio of 6:3:1 to obtain 0.1 g of Pt–Ru–Ni nanoparticles supported on MWCNTs. The functionalized MWCNTs were disseminated in ultra-pure water and isopropyl alcohol at first. The pH of the mixture was adjusted to 10 by adding 0.3 M KOH. The solution was transferred into a container and was then MW-treated at 800 W for 10, 20, and 30 min. The resulting catalysts were using a vacuum oven at 80°C for 24 h. Morphology of as prepared catalysts was analyzed using TEM, and they come in to the conclusion that an increase in the MW heating time accelerated the aggregation of the PtRuNi nanoparticles.

Devrim and Arica [23] used MW-assisted synthesis for the preparation CNT supported Pt catalyst. They have varied the time of heating, and MW power obtained a particle size in the range of 2.6–2.9 nanometer was obtained. For the synthesis of Pt/MWCNT catalyst with the Pt loading of 50 wt.%, they have ultrasonicated the mixture containing 50 mg MWCNT, 24 ml EG and 6 mL IPA with ultrasonic bath for 1 h. After 1 h, H_2PtCl_6 (100 mg) was added into the solution and ultrasonic treatment continued for 10 min. 1M NaOH in EG was used to adjust the pH of the mixture above 12. The resultant mixture was then kept in MW oven for different times and MW power. Finally, the resulting mixture was filtered, washed with acetone, deionized water, and dried in the oven at 100°C overnight. The catalysts thus prepared are denoted as PT/MWCNT-1, PtMWCNT-3, and Pt/MWCNT-4 based on the time of heating, MW power, and Pt loading.

TEM images (Figure 2.1) expose the successful attachment of Pt on the MWCNT and state of dispersion. The Pt catalyst nanoparticles are almost spherically formed. The average size of the dispersed Pt particles was estimated at approximately 2.9, 2.6, and 2.8 nm for Pt/MWCNT-1, Pt/MWCNT-3, and Pt/MWCNT-4 catalysts, respectively. All catalyst has a narrow Pt particle size distribution.

Xing et al. [24] prepared nitrogen doped-mildly oxidized MWCNT-supported Pt catalyst using MW-assisted polyol process. The modified Hummers method was used for the oxidation of CNT, and then this mildly oxidized MWCNT was treated with $NH_3.H_2O$, followed by hydrothermal reduction at 180° for 12 h. Thus, nitrogen-doped-mildly oxidized MWCNT (catalyst support) was obtained. Then 20 mg support was dispersed in 40 mL of ethylene glycol (EG) under ultrasonic treatment for 1 h. Subsequently, the proper amount of H_2PtCl_6 solution was added to the uniform ink under vigorous stirring. At room temperature, the mixture was stirred for 3 hours.

The pH value was then tuned to 10 and the suspension was subjected to consecutive MW heating for 60 s. After microwaving, the solution was stirred again for another 8 h, after that the product was filtrated, washed with alcohol, and distilled water, and dried at 80°C in a vacuum oven for 8 h. They have obtained a particle size of about 1.84 nm. Following is TEM image (Figure 2.2) and size distribution of Pt on nitrogen doped-mildly oxidized MWCNT.

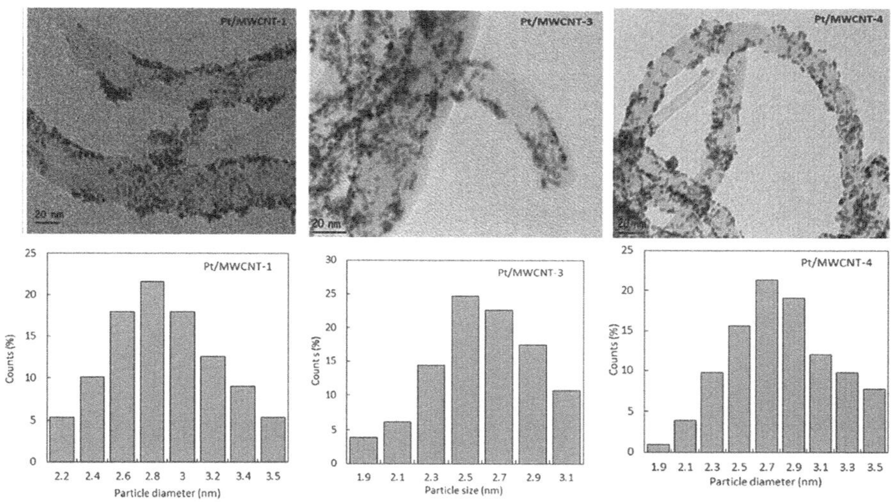

FIGURE 2.1 TEM images of CNT supported Pt catalyst and particle size distribution.

Source: Reprinted with permission from Ref. [23]. © 2019 Elsevier.

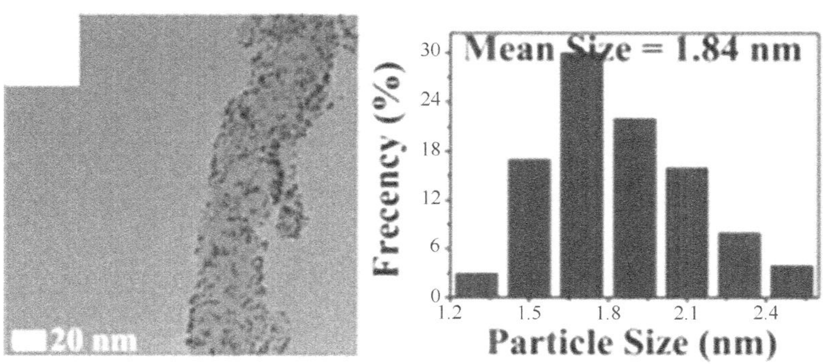

FIGURE 2.2 TEM image and size distribution of Pt on nitrogen doped-mildly oxidized MWCNT.

Source: Reprinted with permission from Ref. [24]. © 2019 Elsevier.

MWCNT -supported Ni catalysts were prepared by Kim et al. [20] using MW-assisted synthesis. Typically, CNT-supported Ni catalysts were prepared by a combined impregnation-chemical reduction method. Green homogeneous $NiCl_2.H_2O$ solution having good stability was prepared by dissolving 0.42 g in 70 ml EG. Then, after 1.5 hours of ultrasonic dispersion, sufficient amounts of oxidized CNTs were suspended in the as-prepared solution, followed by 2 hours of vigorous stirring. Then, with constant stirring, 3.5 ml hydrazine hydrate was added dropwise to the solution, followed by suitable volumes of NaOH. The solution was heated in a MW oven for 1 minute at 300 W, resulting in decreased Ni particles being formed from the solution, which was then washed and dried.

An interesting work done by Rudd et al. [21] that demonstrated a one-pot synthesis of copper oxide decorated CNTs using MW treatment. In this method, firstly, 10 mg of MWCNTs were taken in a glass vial. To establish a range of CuAc loadings on MWCNTs, a stock solution of copper acetate in water (concentration 1 mg/mL) was prepared and the appropriate volume of solution was applied. After that, the CuAc and MWCNTs vial was sonicated for 15 minutes. Following that, the vial was placed in an oven for 4 hours at 80°C to evaporate the water and dry the sample. After allowing the vial to cool to ambient temperature, it was placed in a MW for the appropriate number of 1-minute MW treatments at 1,000 W. They have placed two Pyrex beakers containing water in the MW during the treatments, to prevent over-heating. TEM images of the CuO-decorated CNTs are shown in Figure 2.3.

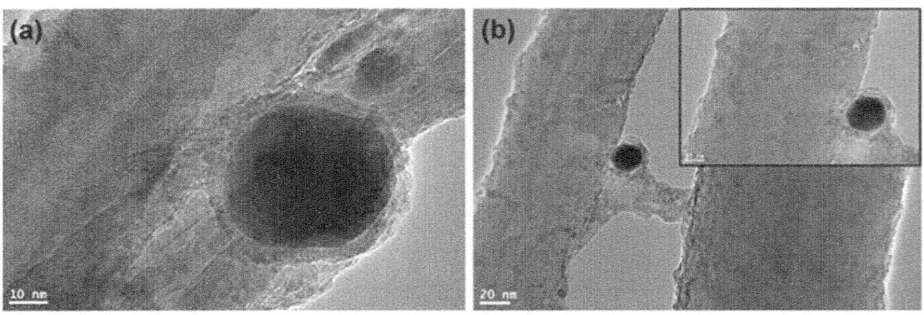

FIGURE 2.3 TEM images of tenorite nanoparticles decorating MWCNTs from: (a) 9.1 wt.%; and (b) 47.4 wt.% CuAc initial loading. Inset, higher magnification image of the CuO nanoparticle.

Source: Reprinted with permission from Ref. [21]. © 2019 Elsevier.

Wallace et al. [17] prepared a SWCNT-supported Pt catalyst using MW-polyol synthesis. In the typical process, 40 mg of acid-treated SWCNTs were dispersed in 25 mL EG solvent. Then 1.0 mL 0.05 M H_2PtCl_6 in EG was added and the pH of the solution was tuned to 12 by adding 0.5 mL of 0.05 M NaOH aqueous solution. The solution was heated for two 75-second periods at 700 W in a domestic MW oven (National 700 W, Japan). The solution was then agitated overnight to ensure that the particles were completely deposited on the CNT supports. The solid Pt/FCNTs catalyst was centrifuged and washed four times with acetone before being dried in a vacuum oven at 85°C overnight. The S-TEM image indicated the uniform distribution of Pt nanoparticles with an average size of 2.5–4 nm on SWCNT support.

2.3 APPLICATION OF CARBON NANOTUBES (CNTS) AS SUPPORTS IN CATALYSIS

The activity and selectivity of a supported heterogeneous catalyst are widely influenced by choice of its support. Catalyst supports serve to improve the mechanical strength, distribution, stability, reactivity, and selectivity of catalysts. Granular, powdered, colloidal, co-precipitated, extruded, pelleted, spherical, wires, honeycombs, and skeletal supports are examples of support types. They can be either inert or active in the reactions; most supports also act as promoters. The goal of the support is to create a solid granular material covered with the catalytic component, which will provide a hard and sturdy structure that will withstand disintegration under reaction conditions. Supports reduce the amount of catalyst needed by dispersing the catalyst onto a larger area, thus making it environment-friendly and economical. Supports can also act as stabilizers to prevent lower-melting-point materials from agglomeration. There are many different types of supports available, ranging from the traditional alumina and silica to the new carbon supports. CNTs have been increasingly used as supports in literature due to their exceptional physical [25] and chemical properties such as high mechanical strength, excellent chemical and thermal stability, thermal/electrical conductivity, large specific surface area because of its hollow tubular structure, decreased mass-transfer limitations, good chemical inertness, relatively high oxidation stability, resistance to acid/basic media, controllable porosity and surface chemistry to certain limits, and easy recovery of the precious metals supported on CNTs by burning off the support in a low environmental impact. Depending on their helicity and diameter, CNTs may have either metallic nature [26, 27] or semiconductive nature [28, 29], and this property can greatly affect charge transfer processes. Their excellent thermal conductivity [30] helps to prevent the agglomeration and growth of small nanoparticles during post-annealing treatments, and stabilize newly formed phases.

CNTs have been used as catalytic supports for a variety of nanoparticles in many catalytic reactions such as hydrogenation, oxidation, hydrodehalogenation, etc. CNTs usually contain residual metal impurities, carried over from the catalysts used in their production process, which are usually removed by a purification step which involves a concentrated nitric acid treatment. Pristine CNTs are highly hydrophobic as they do not have oxygen containing functional groups and hence are unsuitable for supporting metal nanoparticles. Oxidation of CNTs by various oxidants is therefore commonly employed as an effective way to introduce various functional groups, such as –COOH, –OH and –C=O on the surface of the CNTs, which acts as nucleation sites for the metal, resulting in a high dispersion of metal particles. The surface oxygen-containing groups, although initially essential to enhance the loading and dispersion of metal precursors, are later removed after the metal loading by a high temperature treatment to avoid unnecessary side reactions during the catalytic process.

2.3.1 OXIDATION REACTIONS

Carbon monoxide to carbon dioxide oxidation in the presence of a catalyst is one of the most studied reactions in heterogeneous catalysis, and it has sparked a lot of interest in recent years because of its potential applications in fields like pollution control in the auto industry, gaseous waste reduction in petrochemical industries, ethanol or other fuel production, and pure hydrogen production for proton-exchange membrane fuel cells (FCs) [31]. Platinum (Pt), palladium, and rhodium are considered the most efficient transition metal catalysts due to their ability to separate molecular oxygen at a low temperature and to combine strongly with both atomic oxygen and CO. Due to the high cost of noble metals, researchers are now turning towards employing relatively cheaper transition metals such as Mn, Fe, Co, Ni, Cu, or their combinations as catalysts. Co/CNT have demonstrated high catalytic activity for CO oxidation at low temperatures and a thermal stability that is superior to Co/AC which is attributed to the highly dispersed Co particles on the CNTs (20–30 nm) with nanosized cobalt particles (10–15 nm) [32]. The optimum reaction conditions for CO conversion are O_2 concentration 3%, operation temperature 250°C, CO concentration 5,000 ppm, and space velocity 1,56,000 h^{-1}. Moreover, as gaseous NO and CO coexist in incineration flue gas, the simultaneous removal efficiency of CO and NO was also assessed, and the results demonstrated that regardless of whether NO coexisted with CO, the efficiency of CO removal was not impacted by the presence of oxygen. At 250°C, CO may act as a reductant for NO reduction over Co/CNT in the presence of oxygen, with a maximum NO reduction of 30% at CO/NO = 2.5. As the electrodes in proton-exchange membrane FCs are extremely sensitive to even low levels of CO, the efficiency of CO oxidation under H_2 rich conditions was also assessed. Under H_2 rich conditions, the optimum reaction temperature for CO conversion was under 300C, and CO_2 selectivity was better at 200°C than at 250°C as the oxygen concentration increased from 0.5% to 0.7%.

The clean-burning nature of hydrogen made it an important energy carrier for sustainable energy consumption. It can be used as a fuel either directly in internal combustion engines or indirectly in polymer electrolyte membrane (PEM) FCs to generate power. However, storage and safety concerns have restricted the use of H_2 for mobile applications such as fuel cell vehicles. However, these problems can be overcome by the production of hydrogen *in situ* from a suitable H_2 rich liquid fuel, such as methanol. Selective synthesis of hydrogen from methanol by partial oxidation has significant advantages over other methods since it is an exothermic reaction with a greater reaction rate, which reduces the reaction time to reach the working temperature from the cold start-up conditions. Conventional Cu/ZnO-based catalysts supported on alumina and silica have issues with long-term stability, limited resistance to impurities, and the creation of toxic CO as a by-product, despite their better activity and selectivity for H_2 synthesis from methanol. The production of hydrogen from the partial oxidation of methanol has been studied using Cu-Zn catalysts supported on high-quality CNTs (POM) [33]. Using these CNTs, synthesized using acetylene as carbon source over anodic aluminum oxide template by chemical vapor deposition (CVD) method, as support, Cu-Zn catalysts with varying metal loading were prepared by co-precipitation method. The methanol conversion rate and selectivity of H_2 improved from 0.066 to 0.11 mol/h/g cat and 57 to 70.6%, respectively, when increasing Cu loading from 5 to 12 wt.% at 260°C and any further increase shows a fall in activity. The enhanced activity is due to the improved metal dispersion, narrow particle size distribution, and almost complete reduction of Cu particles. The XRD study revealed that the active Cu^0 species is slowly transformed into CuO during the POM reaction, resulting in a decrease in activity. Cu^0 species has a high activity for hydrogen synthesis with inhibited CO creation, but unreduced species Cu^+ and Cu^{2+} hinder hydrogen production from methanol, according to the findings.

Hydrogen is usually produced through the steam reforming of hydrocarbons or alcohols, followed by the water–gas shift (WGS) reaction, with the subsequent formation of trace amount of carbon monoxide. Due to the poisonous nature of CO for the Pt-based anodes used in fuel cell technology at the operation temperature (~353 K), it has to be selectively removed without modifying the hydrogen stream obtained after the WGS reaction. Over Pt catalysts supported on MWCNTs, preferential oxidation of CO in excess hydrogen (PROX) was investigated [34]. Pt/CNT catalyst exhibited a superior performance compared to a Pt/AC catalyst prepared under similar conditions. More active catalysts having increased particle size were yielded by increasing the thermal treatment temperature. The addition of oxygen functions to the carbon support, regardless of the carbon support employed, was detrimental to the CO oxidation reaction. Both catalytic activity and selectivity of Pt/CNT catalyst can be further improved by the addition of CeO_2 at low temperatures.

Toxic, dangerous, and highly concentrated organic pollutants are typically found in wastewater from chemical, petrochemical, pharmaceutical, agricultural, and textile manufacturing companies, posing a serious threat to water bodies and human health. Catalytic wet air oxidation (CWAO) is one of the popular wastewater treatment methods, which aims at effectively oxidizing hazardous, toxic, and highly concentrated organic pollutants in wastewater to innocuous end products without any toxic emissions using oxygen as the oxidant. Nitric acid oxidized MWCNT-supported Pt [35–37], Cu [37], and Ru [37, 38] catalysts have displayed good activities for the CWAO of aniline aqueous solutions in the order of Pt/CNT > Ru/CNT, Cu>CNT. Pt catalysts have been made from bare oxidized CNTs and carboxylated CNTs treated with sodium carbonate to create carboxylate groups on the outer surface. These catalysts have been tested in the CWAO of phenol aqueous solutions in a continuous trickle-bed reactor at 2.0 MPa of total pressure and temperatures of 160 and 200°C [39]. Good conversion rate for phenol and total organic carbon was obtained with the activity per catalyst weight of 30.0 $mmol_{phenol}\ g_{cat}^{-1}\ h^{-1}$ in steady-state conditions. At temperatures ranging from 40°C to 150°C, Pt/MWCNT catalysts manufactured using a molecular-level mixing approach have been used in the catalytic oxidation of volatile organic molecules such as benzene, toluene, ethylbenzene, and o-xylene (BTEX) [40]. Catalytic oxidation using a BTEX concentration ranging from 100 to 500 ppmv in air at volume hour space velocities (VHSVs) of approximately $7.5 \times 10^4\ h^{-1} - 3.4 \times 10^5\ h^{-1}$ has been conducted. With 30 wt.% Pt/CNT, full oxidation was achieved at temperatures as low as 115°C, significantly below the temperatures required in previously researched Pt-based catalysts. These VOCs were completely oxidized into CO_2 and H_2O over the Pt/CNT catalysts at low temperatures without the formation of by-products. The oxidation activity was likely boosted by the greater surface BTEX concentration provided by MWCNTs' adsorption capacity.

Surface-functionalization of CNTs can highly promote the catalytic performance of CNT-supported catalysts for oxidation reactions. In the solvent-free selective aerobic oxidation of benzyl alcohol to benzaldehyde, the catalytic performance of palladium nanoparticles supported on CNTs functionalized with various organosilane modifiers, prepared through a post-synthesis grafting method followed by a metal adsorption-reduction approach, has been tested [41]. The variation in type and amount of surface-functional groups played a significant role in controlling catalytic behavior. An appropriate amount of 3-aminopropyl triethoxysilane surface-modified CNT-supported Pd catalyst turned out to be the optimal catalyst as a result of the finely-tuned surface basicity, small Pd particle size with a narrow size distribution, high electron density, and enhanced metal-support electronic interaction, with a remarkably high quasi-turnover frequency of 2,88,755 h^{-1} for benzyl alcohol oxidation. The catalytic activity of Pd/MnO_x/CNT has also been tested in solvent-free benzyl alcohol oxidation. Palladium nanoparticles were homogeneously deposited onto MnO_x/MWCNT composite, synthesized via a surface deposition method, followed by a reduction under H_2 atmosphere [42]. The catalytic activity was correlated with MnO_x loading and dispersion. Due to the disturbance of the electronic structure of the active sites, the reducible MnOx on CNT enables electron transport at the interfacial region of Pd and MnOx, changing the physical and chemical properties of the catalyst. It also enhances the oxygen activation by transporting lattice oxygen to Pd catalytic active site. During the catalytic reaction, the oxygen vacancies within MnOx clusters can be generated and reloaded by molecular oxygen due to the highly reducible nature of transition metal oxide. The Pd/MnOx/CNT catalysts were extremely durable against deactivation due to the strong metal support contact.

Ruthenium (IV) oxide (RuO_2) nanoparticles supported by herringbone-shaped CNTs have demonstrated high activity and selectivity in the aerobic oxidation of alcohols to aldehydes and ketones in the liquid phase under mild conditions [43]. A homogeneous oxidation-precipitation (HOP) approach using hydrogen peroxide (H_2O_2) as both oxidant and precipitant at room temperature was used to make herringbone CNT-supported amorphous hydrous RuO_2 nanoparticles. When compared to commonly used supports such as γ-Al_2O_3 and activated carbon (AC), CNTs had a better ability to induce the formation of highly dispersed nanoparticles, resulting in high activity and excellent selectivity as demonstrated by the oxidation of various aromatic, saturated, and cyclic alcohols.

2.3.2 HYDROGENATION REACTIONS

Because of its over-reduction to cyclohexanol, direct selective hydrogenation of phenol to cyclohexanone under green circumstances is difficult. The direct hydrogenation of phenol to cyclohexanone was shown to be very active using palladium nanoparticles supported on polyaniline-functionalized carbon nanotubes (Pd–PANI/CNT) [44]. Selective hydrogenation of other hydroxylated aromatic compounds with similar performance, again under green and moderate circumstances, was used to prove the catalyst's applicability for this process. Pd–N interactions and polymeric stabilization are required for the creation of stable and widely dispersed palladium nanoparticles on the conducting composite material PANI/CNT. The results also revealed a link between phenol conversion rate and PANI/conductive CNT's characteristic, while cyclohexanone selectivity is ascribed to PANI/nitrogen-containing CNT's nature. Furthermore, the aggressive and selective phenol hydrogenation is caused by the high adsorption of phenol in a "nonplanar" form on nitrogen-containing basic sites. Thus, it is obvious from the above-mentioned literature that the catalyst support can alter hydrogenation via steric or electronic contact, which is usually mediated by the support material's surface features.

Pd–ZnO catalysts supported on MWCNTs exhibited excellent performance for CO_2 hydrogenation to methanol at 3.0 MPa and 523 K, with an observed turnover-frequency (TOF) of 1.15×10^{-2} s^{-1} over the 16% $Pd_{0.1}Zn_1$/CNTs (h-type) higher than that obtained for the 35% $Pd_{0.1}Zn_1$/AC and 20% $Pd_{0.1}Zn_1$/Υ-Al_2O_3 catalysts with the respective optimal $Pd_{0.1}Zn_1$-loading [45]. Using MWCNTs as the catalyst support instead of AC or Υ-Al_2O_3 had no effect on the apparent activation energy for CO_2 hydrogenation, but it did result in an increase in the relative content of the catalytically active Pd^0-species closely associated with methanol generation at the surface of the functioning catalyst. At temperatures ranging from ambient temperature to 623 K, the MWCNT-supported Pd–ZnO catalyst could reversibly adsorb a larger amount of hydrogen. This particular property would aid in the creation of a micro-environment at the surface of the functional catalyst with a larger concentration of H-adsorbed species, hence enhancing the rate of surface hydrogenation processes. Silica-coated Ru/CNT catalysts have displayed high activity for the hydrogenation of CO and 1-hexene [46]. In order to improve the stability, they covered the Ru/CNT catalysts with silica layers by the successive hydrolysis of 3-aminopropyltriethoxysilane and tetraethoxysilane followed by reduction with hydrogen at 623 K. Coverage of the Ru/CNT with these silica layers improved its durability towards Ru sintering at high temperatures and prevented the detachment of Ru metal particles during the repeated hydrogenation of 1-hexene with the recycled catalyst. The catalytic activity of functionalized MWCNT-supported Pd, Rh, and Rh/Pd nanoparticle catalysts prepared using water-in-oil microemulsions containing metal ions as starting materials were studied in hydrogenation reactions [12]. MWCNT-supported Rh nanoparticles was found to be an active catalyst for the hydrogenation of arenes, and the CNT-supported bimetallic Pd/Rh nanoparticles exhibited unusually high catalytic activity for the hydrogenation of anthracene. Pt/CNT exhibited higher catalytic activity than the commercial Pt/AC catalyst in the hydrogenation of *trans*-β-methyl styrene and *trans*-stilbene in toluene at room temperature under the atmospheric pressure of hydrogen [47]. Pt was anchored onto SWCNTs using a two-step process of wet oxidation followed by an ion exchange reaction with a Pt precursor salt. Nitrobenzene hydrogenation under atmospheric pressure and ambient temperature has been carried out using Pt/CNT catalysts prepared by impregnation and reduction–precipitation method with chloroplatinic acid as metal precursor [48]. Both lower and higher Pt loading exhibited high activity for direct hydrogenation of nitrobenzene to aniline under mild conditions. Extraordinary activity of catalysts can be attributed to the good dispersion and mesoporosity.

Selective hydrogenation of cinnamaldehyde to cinnamyl alcohol is a very important reaction in the pharmaceutical and fragrance industry. Due to the thermodynamically favorable hydrogenation of C=C double bonds in – unsaturated aldehydes, hydrogenation of cinnamaldehyde produces additional by-products such as hydrocinnamaldehyde and hydrocinnamyl alcohol. The most effective catalysts for the selective hydrogenation of cinnamaldehyde appear to be supported Pt nanoparticles, according to considerable research. Pt-based bimetallic catalysts supported on MWCNTs prepared by a facile microwave-assisted polyol reduction (MAPR) method was found to be active in the selective hydrogenation of cinnamaldehyde to cinnamyl alcohol and superior than that prepared by the conventional impregnation method [49]. The surface features of CNTs, as well as the solvent used and the transition metal promoter used, were shown to be crucial in fine-tuning the catalytic performance. The highest catalytic results were obtained on

Fe-and Co-modified Pt catalysts in the presence of ethyl acetate as the solvent after removing oxygen-containing groups from CNT surfaces. By removing oxygen-containing species from CNTs, acid-catalyzed side reactions are suppressed, and electron transport between CNTs and Pt nanoparticles is accelerated, enhancing cinnamaldehyde hydrogenation activity and selectivity. Whereas, further functionalization with iron and cobalt nanoparticles on MWCNT surfaces after removing oxygen-containing groups leads to a metal–metal synergic effect and therefore elevates the activity and selectivity in cinnamaldehyde hydrogenation. SWCNTs grown by arc method, containing Pt and cobalt, have also been as catalysts in the selective hydrogenation of cinnamaldehyde to cinnamyl alcohol [50]. This catalytic system containing 7% (w/w) cobalt and 5.4% (w/w) Pt and presenting a mean particle size(s) of 8 nm exhibited 80–85% molar selectivity towards cinnamyl alcohol. Furthermore, the initial cinnamyl alcohol selectivity of close to 100%, was not reported for the PtCo/charcoal catalytic system. This might indicate the presence of some peculiar SWNT–PtCo(s) interactions. Unpurified nanotubes coming from direct current arc discharge evaporation of graphite were used to support ruthenium to conduct the liquid-phase cinnamaldehyde hydrogenation reaction [51]. This unpurified MWCNT-supported Ru catalytic system was compared to Ru/alumina and Ru/AC catalysts presenting similar metallic particle sizes (3.5 nm): the selectivity for cinnamyl alcohol was 92, 20–30, and 30–40%, respectively.

Rhodium catalysts supported on MWCNTs have been used for the liquid-phase hydrogenation of cinnamaldehyde to hydrocinnamaldehyde [52]. MWCNTs were first oxidized via the nitric acid treatment to introduce carboxylic acid groups on the surface and their corresponding sodium carboxylate salt was then made to undergo an organometallic transformation with $[Rh_2Cl_2(CO)_4]$, which led to the production of highly dispersed Rh nanoparticles on the MWCNTs (with a specific surface area of 180 m^2/g) after a decomposition/reduction step. In the liquid phase hydrogenation of cinnamaldehyde, this 1% (w/w) Rh/MWCNTs showed 100% selectivity for hydrocinnamaldehyde, and their catalytic activity was three times higher than that of a 1% (w/w) Rh supported on AC with a surface area of 700 m^2/g. Rhodium catalysts prepared without the HNO_3 treatment on MWCNTs (particle size up to 100 nm) showed no activity, nitric acid-treated MWCNTs (particle size 2.5–5 nm) had an activity of 27 $g_{substrate}\ g_{Rh}^{-1}\ h^{-1}$ and MWCNTs functionalized with –COONa groups (particle size 1.5–2.5 nm) which had the most highly dispersed Rh nanoparticles was the most active with a value of 78 $g_{substrate}\ g_{Rh}^{-1}\ h^{-1}$. In the liquid-phase hydrogenation of cinnamaldehyde, nitrogen-doped CNTs (N-CNTs) produced using a CVD method at temperatures ranging from 600°C to 850°C have been employed as supports for palladium catalysts [53]. When nitrogen atoms were introduced into the carbon matrix, the number of surface nucleation sites increased, resulting in a considerable increase in metal dispersion and consequently catalytic activity in the hydrogenation reaction when compared to undoped ones. Palladium particle size on the nitrogen-doped CNTs (N/C: 3 at.%) was centered at around 3 nm versus 10 nm on the undoped CNTs. The researchers also discovered that the type of nitrogen species inserted in the structure of CNTs might affect the catalytic activity. The catalyst's exceptional stability was confirmed by catalytic recycling testing, which revealed no palladium leaching or deactivation.

2.3.3 FISCHER TROPSCH SYNTHESIS (FTS)

Fischer Tropsch synthesis (FTS) is a well-established eco-friendly, industrial process, in which synthesis gas ($CO + H_2$) is converted into liquid hydrocarbons. This is a potentially attractive technology for production of high-quality liquid fuels and chemicals from coal, natural gas or biogas and has received attention as a source of low-sulfur diesel fuel [54]. The most common catalysts used for FTS are iron, cobalt, manganese, and ruthenium. Usually, these metals are supported on silica, alumina or titania, and these supported catalysts are used for FTS. It has been revealed from literature that the catalyst activity and selectivity in FTS is influenced by nature and structure of the support, nature of the active metal, catalyst dispersion, metal loading, and catalyst preparation method [55]. Recently, CNTs have been identified as a promising support for catalysts used in FTS, because CNTs have a relatively large surface area and are able to improve the dispersion of the catalytically active nanoparticles.

In 2002, van Steen and Prinsloo [11], first used CNTs as a support for Fe-based FT catalysts to investigate the effect of catalyst preparation method and promoters on the catalyst performance during FTS. They have synthesized CNT supported Fe-based catalyst system using three methods viz-incipient wetness, deposition/precipitation using K_2CO_3, and deposition/precipitation using urea. The catalyst prepared by initial dampness was the most active. The selectivity of the three catalysts in the FT synthesis appears to be independent of the method of production. CNT-supported cobalt FT catalyst is a preferred commercial catalyst over iron due to its low water-gas shifting activity, high long-chain hydrocarbon selectivity and high stability. Another advantage of using Co-based catalysts is the production of less amount of unwanted CO_2 in FTS compared with Fe-based catalysts. Recently, Haghtalab et al. [56] prepared CNT-supported Co and Co-Ru bifunctional catalyst systems and investigated the effect on FT synthesis. The results showed that the Co-Ru/CNTs catalysts exhibited high catalytic activity with the maximum long-chain hydrocarbons ($C5^+$) selectivity of 74.76%, which was greater than the Co-Ru/γ-Al_2O_3 (61.20%), Co/CNTs (43.68%), and Co/γ-Al_2O_3 (61.20%) catalysts (37.69%). However, as the CNTs are inherently inert and hydrophobic in nature, surface modification is necessary to attain high dispersion of catalytically active metal particles. Oxidation/activation of CNT using HNO_3/H_2SO_4 is the most accepted modification [57, 58]. Nitrogen-doped CNTs can be obtained by treating oxidized CNTs in flowing ammonia [59, 60]. Introduced functional groups can act as anchoring sites for metal nanoparticles, and hence, metals supported on oxidized CNT typically display increased resistance towards sintering under reaction conditions [58].

2.3.4 DEHYDROGENATION

CNT-supported catalysts have also been investigated for catalytic dehydrogenation reactions and hydrogen fuel production. CNT-supported Pt exhibited the highest yield toward H_2 in the microwave (MW)-assisted decalin dehydrogenation, compared to carbon black, AC, graphite, and carbon nanofibers, due to the relatively small Pt nanoparticles formed on CNT surfaces with relatively large surface area as well as the highest temperature attained by the system during the microwave-assisted dehydrogenation, shown in Figure 2.4 [61]. CNTs' high aspect ratio and low bulk density result in strong electrical conductivity and, as a result, a considerable dielectric loss in CNT-containing suspension, resulting in the suspension's remarkable thermal performance under the microwave (MW). It is revealed that MW can heat the Pt/CNT-catalyzed reaction system to a temperature above the boiling point of decalin.

Defect-rich CNTs obtained after an acid oxidation followed by a high temperature treatment in an Argon atmosphere at 800°C, were used as supports for Pt nanoparticles and was found to have high catalytic activity in the ammonia borane (AB) hydrolytic dehydrogenation reaction [62]. Acid-treated CNTs possess a number of defects and oxygen-containing groups. The defects have been proved to be responsible for the improvement of the Pt catalytic activity in the AB hydrolytic dehydrogenation instead of oxygen-containing groups. An exceptional H_2 generation turnover frequency as high as 567 mol_{H_2} mol_{Pt}^{-1} min^{-1} at 30°C was obtained, significantly higher than those of Pt/AC. Higher cyclohexanone selectivity in cyclohexanol dehydrogenation has been reported for Co/CNT than Co/AC catalysts because of the electronic properties of the CNTs as well as its lesser acid sites [63]. The effect of K addition, attributed to electronic promotion, was also stronger on Co–K/CNT than on Co–K/AC. The catalytic performance of Co supported on nitric acid oxidized MWCNTs and unoxidized MWCNTs have been studied in the selective dehydrogenation of cyclohexanol to cyclohexanone which is an important reaction in the industrial production of caprolactam [64]. While the selectivity of both the catalytic systems were almost similar (95%), the conversions of Co catalysts supported on the nitric acid oxidized MWCNTs were higher than the unoxidized MWCNTs catalyst system due to the smaller particle size obtained on the nitric acid-treated MWNT, 5 nm instead of 100 nm on untreated support. The initial conversions of cyclohexanol over two catalysts are 75.2 and 56.8%, respectively.

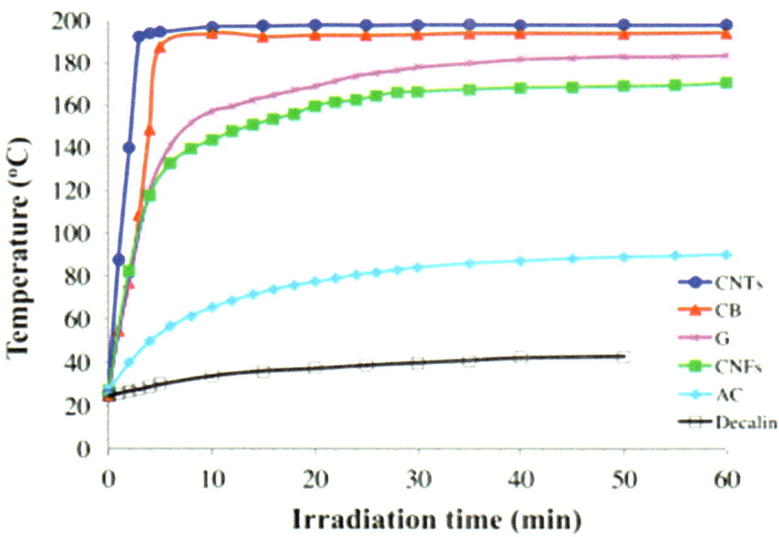

FIGURE 2.4 Temperature profiles of different carbon materials suspended in decalin under MW (input power 250 ± 15 W).

Source: Reprinted with permission from Ref. [61]. © 2014 Elsevier.

2.3.5 AMMONIA/METHANE DECOMPOSITION

The ammonia decomposition to produce H_2 without CO_x for FCs has attracted great attention, due to its feasibility, cleanness, superior hydrogen storage capacity of 17.6 wt.% and is also more economical than methanol decomposition. Supported transition metals have been used as catalysts for the decomposition of ammonia to release hydrogen. CNTs have been reported to be an attractive support for transition metal nanoparticles. The effects of active components (Ru, Rh, Pt, Pd, Ni, Fe) and support (CNTs, AC, Al_2O_3, MgO, ZrO_2, TiO_2) on the catalysis of ammonia decomposition have been investigated for the generation of CO_x-free hydrogen [65, 66]. The Ru/CNT catalyst exhibited the highest conversion and TOF. In the range of 2–5 nm, the particle size of Ru on CNTs is the smallest with the highest dispersion among the supported Ru catalysts. The catalytic activity was further improved by modifying CNTs with KOH which indicated that strong basicity of support is essential for high catalytic performance. The influences of the porous and graphitic structures of carbon supports on the catalytic ammonia decomposition have also been reported [67]. The catalytic activity over supported Ru catalysts is ranked as Ru/graphitic carbon > Ru/CNTs > Ru/carbon black > Ru/AC. Ru particles are highly dispersed on carbon supports with an optimum particle size of around 3–4 nm. In comparison to the pore architectures of the carbon supports, the graphitic structure of the carbons is crucial to the catalytic activity of the supported Ru catalyst.

The high dispersion of Ru, as well as the excellent graphitization and purity of CNTs, when Ru/CNTs are modified with potassium ions, result in a significant increase in catalytic activity [66, 68]. The influence of different rare earth, alkali, and alkaline earth compounds as promoters with Ru/CNTs have also been studied. These promoters decrease the pore volume and surface area but improves the thermal stability of Ru/CNTs [68]. These modifications have no effect on the size and morphology of Ru particles. The activities of metal nitrate-modified catalysts were observed in the following order: K–Ru > Na–Ru > Li–Ru > Ce–Ru > Ba–Ru > La–Ru > Ca–Ru > Ru, indicating that within the same groups (K, Na, and Li; Ba and Ca), the greater the electronegativity of the promoter, the lower the NH_3 conversion. The electron-withdrawing groups, F^-, Cl^-, Br^-, SO_4^{2-}, and PO_4^{3-} were found to be inhibitors of Ru catalysts. A modifier's promotional effects are the consequence of a combination of increased combinative desorption of nitrogen atoms and a lower apparent activation energy of the decomposition reaction. Despite having similar Ru particle sizes, nitrogen doping of MWCNT supports using NH_4OH solution increases the activity of Ru nanoparticles for low temperature ammonia decomposition, with TOF numbers of 6,200 L_{H_2} mol_{Ru}^{-1} h^{-1} at 400°C, which is higher than the values of unmodified CNT support under the same conditions (4,400 L_{H_2} mol_{Ru}^{-1} h^{-1}) [69]. Nitrogen doping causes an electronic modification of Ru due to improved conductivity and basicity of the support, promoting a strong metal-support interaction, which results in higher TOF in the ammonia decomposition reaction than the Ru catalysts supported on unmodified commercial MWCNT. CNTs doped with nitrogen using N_2 MW plasma at the powers of 200 and 400 W have also been used as supports for Ru catalyst in ammonia decomposition [70]. However, due to the low nitrogen content, doping using this method does not dramatically alter the bulk or the surface texture of CNTs or improve the TOF.

Low-temperature hydrogen production via ammonia decomposition has been reported by graphitized MWCNT-supported and cesium-promoted Ru nanoparticles at as low as 450 K [71]. Commercial MWCNTs are graphitized to remove any residual metal catalysts from the CNT synthesis, as well as surface functional groups and defects on the CNTs that are known to act as anchoring points for metal active nanoparticles, and to lower the surface area of the material slightly. The high conductivity of graphitized CNTs allows for more electronic alteration of the Ru nanoparticles by cesium which is in proximity but without direct contact, avoiding active site blockage and thereby boosting the metal-support interaction. This advancement in low-temperature catalytic activity is a major step forward in the use of ammonia as a chemical storage medium for *in-situ* hydrogen generation in FCs. Addition of sodium as a promoter in the Ru/CNT catalyzed low-temperature hydrogen production via ammonia decomposition was found to enhance the catalytic activity (316.7 mol_{H_2} mol_{Ru}^{-1} h^{-1} at 350°C) due to the electronic modification of the Ru nanoparticles by the electron-donating sodium promoter [72]. The activity of Ru/CNTs in ammonia decomposition is dependent on the dispersion of Ru particles and the amount of nitrogen on the surface of CNTs. Commercial CNT-supported Co-Mo nanoparticles have also exhibited catalytic activity for hydrogen production from ammonia. The hollow channels of CNTs were found to efficiently limit the particles at the appropriate size to keep the balance between dispersion and intrinsic activity. The metal-carbon intercalation on the interface area is expected to be responsible for the stabilization of the CNTs. The nature of Co-Mo synergic effect improves the long-term stability as well as keeps the activity [73].

2.3.6 HYDRODEHALOGENATION REACTIONS

CNT-supported Pd catalysts have been widely used to catalyze the hydrodehalogenation of aryl halides, a hazardous pollutant, because Pd promotes the cleavage of carbon–halogen bonds and enables hydrogenation. Pd/MWCNT prepared by traditional impregnation method was active in the liquid-phase hydrodehalogenation of bromobenzene in milder conditions. Compared to traditional supports such as AC, Al_2O_3, SiO_2, and MgO, MWCNTs offer improved performance with lesser amounts of Pd as shown in Table 2.1. The significant activity of Pd/CNTs in liquid-phase bromobenzene hydrodehalogenation was attributed to their high exterior surface area and absence of micropores in MWCNTs, which eliminated significant mass transfer limitations [74]. In contrast, despite the high specific surface area, the presence of a large fraction of micropores inside AC considerably hampered reactant and product diffusion, resulting in decreased activity. In addition, the electron transfer from the nanotube support to metal particles increases the electronic density around Pd nanoparticles which promotes the adsorption and activation of bromobenzene, thus improving catalyst activity. The use of CNTs as a catalytic support could result in unusual activity and selectivity in bromobenzene hydrodehalogenation.

TABLE 2.1 Comparison of the Catalytic Conversion and TOF Using Different Catalyst Supports

Support	Pd Loading (wt.%)[a]	Conversion (%)[b]	TOF (h^{-1})[c]
CNTs	2.04	69	691
AC	1.85	21	232
SiO_2	1.72	12	143
Al_2O_3	1.74	18	211
MgO	1.37	6	90

[a] *Determined by inductively coupled plasma atomic emission spectrometry.*
[b] *Reaction time*: 30 min.
[c] *Mol bromobenzene converted per mol Pd per hour.*

Source: Reprinted with permission from Ref. [19]. © 2008 Elsevier.

Pd/CNT nanocomposites made by depositing tris(dibenzylideneacetone) dipalladium(0) [$Pd_2(dba)_3$] on thiolated nanotube surfaces where the thiol groups act as linkers to prevent the Pd nanoparticles from agglomeration, were active in the hydrodehalogenation of aryl halides at a low Pd content (2.3%) and without any additional ligand [75]. CNTs significantly influenced the catalytic activities of CNT-supported metal catalysts for hydrodehalogenation than the reference systems. MWCNTs were oxidatively functionalized using a 3:1 (v/v) mixture of concentrated H_2SO_4/HNO_3, sonicated at 60°C, resulting in a large number of surface oxide groups which favored the dispersion of Pd nanoparticles, thus improving its catalytic performance in the hydrodehalogenation of bromobenzene [76].

2.3.7 CARBON-CARBON COUPLING REACTIONS

Carbon-carbon coupling reactions on supported Pd nanoparticles can decrease product contamination and promote the development of greener routes in organic synthesis. However, the catalytic reaction mechanisms on supported Pd catalysts remain a controversial topic with regards to the leaching of Pd species into solution and the true heterogeneous nature of the reaction. MWCNTs supported Pd^0 have efficiently catalyzed Suzuki-Miyaura cross-coupling of aryl bromides with arylboronic acids in aqueous solvents or neat water without the assistance of any ligands or additives, and the catalyst could be recovered by simple filtration, thus meeting the criteria of green chemistry [77]. The low catalyst loading as well as simple reaction setup is also economically attractive. Pd/MWCNT prepared by a simple approach based on supercritical fluid deposition has high catalytic activity in Suzuki coupling reactions. Furthermore, the catalysts can be reused at least six times without losing their effectiveness [78]. Suzuki coupling of phenyl-boronic acid and 1-iodo-4-nitrobenzene to nitro-biphenyl was attained with 94% conversion after 30 min and with a TOF of 709 h^{-1}. For the Heck reaction of styrene and iodobenzene and the Suzuki coupling of phenylboronic and iodobenzene, palladium nanoparticles (2–10 nm) produced on SWCNTs by spontaneous reduction from $Pd(OAc)_2$ or from oxime carbapalladacycle displayed stronger catalytic activity than Pd/AC [79]. The activity of palladium-supported catalysts depends dramatically on the metal dispersion and the particle size. Leaching of palladium is a problem which questioned the true heterogeneous nature and reusability of the catalyst. However, in spite of the occurrence of the palladium leaching during the reaction, no substantial drop in the activity of the recycled catalyst was observed which was explained by the re-deposition of most of the palladium nanoparticles on the SWNT as the reaction ceased and the system cooled down.

Pd/SWCNT and Pd/MWCNT were made using a fast, solvent-free method that did not require any reducing agents or electricity and involved a direct dry mixing of a precursor Pd salt (e.g., palladium acetate) with CNT at room temperature by ball-milling or followed with annealing at 300°C in an inert atmosphere [80]. The nanoparticles made by ball-milling at room temperature had higher catalytic activity towards the Suzuki cross-coupling reactions than those prepared under thermal conditions, mainly due to the smaller particle size, uniform dispersion of Pd nanoparticles on the surface of SWCNTs and MWCNTs, and lack of agglomeration of these nanoparticles during the mechanochemical preparation at room temperature. While both the catalysts exhibited excellent catalytic activity in Suzuki cross-coupling reaction under ligand free ambient conditions in an environmentally benign solvent system, Pd/MWCNT prepared by ball-milling showed a slightly better reactivity and recyclability with the TON of 7,250 and TOF of 2,17,500 h^{-1}. The superior catalytic activity and increased recyclability was attributed to the larger diameter of the MWCNTs (~20–150 nm), which provides stronger surface contacts and anchor for the Pd particles, thus facilitating the deposition of the greater number of Pd nanoparticles on the surface of MWCNTs. In addition, the higher Pd content of 8.5 wt.% in this sample, smaller particle size (1–3 nm), and the more uniform distribution of the Pd nanoparticles on the surface are also considered to be contributing factors to the overall catalyst performance.

After an annealing treatment at 700°C and 1,500°C to improve the graphitization and the concentrated nitric acid treatment at 120°C to improve the surface functionalization, CNTS were used as supports for Pd nanoparticles [81]. CNTs annealed at a lower temperature 700°C, had more defects than CNTs annealed at 1,500°C and the acid treatment introduced a high functionalization on the defective CNTs when compared to the graphitized CNTs. The functionalities act as sites for anchoring metal cations during impregnation. The Pd supported on highly functionalized CNTs was found to have higher activity in the Suzuki-Miyaura reaction of iodobenzene with phenylboronic acid with a TOF of 190 h^{-1} at completion of the reaction when compared to the Pd supported on low functionalized CNTs with a TOF of 124 h^{-1}. Nanopalladium clusters supported on MWCNTs prepared using the direct reaction between the MWCNTs and $Pd(dba)_2$ (dba: dibenzylideneacetone) and removal of the ligand dba demonstrated high efficiency in the Suzuki Miyaura, Heck, and Sonogashira cross-coupling reactions as well as in the hydrogenation of unsaturated hydrocarbons [82]. The catalyst is recyclable and can work in aqueous media.

Pt supported on thiolated MWCNTs demonstrated high activity and recyclability for the ligand-free Suzuki coupling of 4-iodoanisole and 4-methylbenzene boronic acid [83]. The catalyst achieved >99% conversion within 24 h, while traditional Pd or Pt complex catalysts showed <5% conversion. Moreover, even after 12 recycles it gave 82% yield. This high activity was attributed to a combination of the electronic state of Pt in almost zero-valent state, the large surface area due to the extreme degree of particle dispersion, and the fast charge compensation of Pt by the sulfur ligand. Pd-CNT prepared from thiolated MWNT and Na_2PdCl_4 showed good activities without any additional ligand in the decarboxylative coupling reactions of aryl iodides and aryl alkynyl carboxylic acids to form the diaryl alkynes in good yields [84]. The Pd nanoparticles had an average diameter of 2.9 nm and a limited size dispersion. Without the need of an expensive palladium catalyst or toxic ligands, Co nanoparticles supported on imidazolium ionic liquid-functionalized MWCNTs have been used as an environmentally safe and cost-effective catalyst for the Heck coupling reaction in a significantly shorter reaction time [85]. The catalyst could be recycled at least six times without losing its effectiveness. A Pd(II) salen complex supported on MWCNTs showed excellent catalytic activity and stability for the Mizoroki-Heck and Hiyama cross-couplings of aryl halides with olefins and phenylsiloxanes [86]. Furthermore, the catalytic activity of this air/moisture and thermally stable heterogeneous catalyst could be recovered readily following simple modifications, allowing it to be reused up to four times.

Furthermore, the catalytic activity of this air/moisture and thermally stable heterogeneous catalyst may be easily recovered following simple modifications, allowing it to be reused up to four times [87]. The catalytic activity of Pd(0) deposited on MWCNT is higher than that of commercially available Pd(0) on charcoal for Sonogashira-type cross-coupling reactions. When Pd (10%)/MWCNTs are utilized, the creation of new enzyme cross-coupled products is regio- and stereoselective. Pd nanoparticles supported on MWCNTs functionalized with mercapto-melamine groups by covalent grafting of CNTs with ethylenediamine, cyanuric chloride, and 2-mercapto-ethanol were successfully used in the Buchwald-Hartwig C–N cross-coupling reaction of aryl halides and amines with a wide variety of functionalized substrates to give high yields of aryl amines [88]. The catalyst was recycled six times without much change in its catalytic activity.

2.4 CONCLUSIONS

The morphology of CNTs is very crucial for its application as support in catalysis. CNTs can be modified with a variety of surface functionalities which play a significant role in binding and dispersing the active catalytic phases on the CNTs. In many studies, CNTs have been found to be much superior as supports in enhancing the activity of the catalyst, when compared to other conventional and carbonaceous supports like alumina, silica, ACs, carbon black, charcoal, etc. Although many applications of CNTs as support have been reviewed in this chapter, it is by no means an exhaustive list. Also, the applications of CNTs as catalyst supports in many fields, such as photocatalysis, electrocatalysis, FCs, etc., including catalysts confined inside the CNTs have not been discussed. Even though the major use of CNTs in catalysis is as supports, they have also been employed as catalysts in various chemical transformations, either as such or after modifications by introducing surface functional groups via different oxidation treatments. CNTs as catalysts have shown remarkable activity in wet air oxidation (WAO), wet peroxide oxidation, ozonation, oxidative dehydrogenation, hydroxylation, etc. In the last three decades, CNTs have been widely explored as catalysts and as supports in various fields such as wastewater treatment, pollution control, hydrogen storage, FCs, etc.; however, the full potential of CNTs in catalysis yet remains to be unraveled.

KEYWORDS

- **carbon nanotube**
- **catalyst support**
- **functionalization**
- **green environment**
- **heterogeneous catalysis**
- **morphology**

REFERENCES

1. Iijima, S., & Ichihashi, T., (1993). *Nature, 363*, 603–605.
2. Iijima, S., (1991). *Nature, 354*, 56–58.
3. Mierczynski, Pawel, Krasimir Vasilev, Agnieszka Mierczynska, Waldemar Maniukiewicz, Malgorzata I. Szynkowska, & Tomasz P. Maniecki. (2016). *Applied Catalysis B: Environmental 185*, 281–294.
4. Oosthuizen, R. S., & Nyamori, V. O., (2011). *Platin. Met. Rev, 55*, 154–169.
5. Rather, R. A., Siddiqui, S., Khan, W. A., & Siddiqui, Z. N., (2020). *Mol. Catal., 490*, 110975.
6. Liu, Z., Li, Z., Wang, F., Liu, J., Ji, J., Wang, J., & Wang, W., (2011). *Mater. Lett., 65*, 3396–3398.
7. Ying, Q., Naidoo, S., & Vaivars, G., (2015). *Phys. Scr., 90*, 94021.
8. Aliana-Nasharuddin, N., Asikin-Mijan, N., Abdulkareem-Alsultan, G., Mohd Izham Saiman, Fahad A. Alharthi, Abdulaziz Ali Alghamdi, & Taufiq-Yap, Y. H. (2020). *RSC Adv., 10*, 626–642.
9. Hjorth, I., Wang, Y., Li, Y., Melandsø, B. M. E., Nord, M., Rønning, M., Yang, J., & Chen, D., (2020). *Catal. Today,* 1–10.
10. Li, M., Sun, Y., Tang, Y., Sun, J., Xu, Z., & Zheng, S., (2020). *J. Hazard. Mater., 388*, 121745.

11. Van, S. E., & Prinsloo, F. F., (2002). *Catal. Today, 71*, 327–334.
12. Yoon, B., & Wai, C. M., (2005). *J. Am. Chem. Soc., 127*, 17174–17175.
13. Ulas, B., Caglar, A., Kivrak, A., & Kivrak, H., (2019). *Chem. Pap., 73*, 425–434.
14. Bhuvanendran, Narayanamoorthy, Sabarinathan Ravichandran, Weiqi Zhang, Qiang Ma, Qian Xu, Lindiwe Khotseng, & Huaneng Su, (2020). *Int. J. Hydrogen Energy, 45*, 6447–6460.
15. Er, O. F., Caglar, A., Ulas, B., Kivrak, H., & Kivrak, A., (2020). *Mater. Chem. Phys., 241*, 122422.
16. Xu, Q. C., Lin, J. D., Li, J., Fu, X. Z., Liang, Y., & Liao, D. W., (2007). *Catal. Commun., 8*, 1881–1885.
17. Zhang, W., Chen, J., Swiegers, G. F., Ma, Z. F., & Wallace, G. G., (2010). *Nanoscale, 2*, 282–286.
18. Tropsch, F., Akbarzadeh, O., Asmawati, N., Zabidi, M., Wang, G., Kordijazi, A., Sadabadi, H., et al., (2020). *Symmetry (Basel), 12*, 1–14.
19. Chen, H., Tang, Q., Chen, Y., Yan, Y., Zhou, C., Guo, Z., Jia, X., & Yang, Y., (2013). Carbon nanotube supported Pd catalyst for liquid-phase hydrodehalogenation of bromobenzene. *Catal. Sci. Technol., 3*, 328–338.
20. Li, F., Li, Q., & Kim, H., (2013). *Adv. Mater. Res., 622*, 816–820.
21. Rudd, J. A., Gowenlock, C. E., Gomez, V., Kazimierska, E., Al-Enizi, A. M., Andreoli, E., & Barron, A. R., (2019). Solvent-free microwave-assisted synthesis of tenorite nanoparticle-decorated multi-walled carbon nanotubes. *J. Mater. Sci. Technol., 35*, 1121–1127.
22. Woo, S., Cho, H., Kim, J., Lee, Y., & Lee, S., (2020). *J. Korean Ceram. Soc., 57*, 192–199.
23. Devrim, Y., & Arıca, E. D., (2019). Multi-walled carbon nanotubes decorated by platinum catalyst for high temperature PEM fuel cell. *Int. J. Hydrogen Energy, 44*, 18951–18966.
24. Liang, L., Xiao, M., Zhu, J., Ge, J., Liu, C., & Xing, W., (2019). Low-temperature synthesis of nitrogen doped carbon nanotubes as promising catalyst support for methanol oxidation. *J. Energy Chem., 28*, 118–122.
25. Dresselhaus, M. S., Dresselhaus, G., Charlier, J. C., & Hernández, E., (2004). *Philos. Trans. R. Soc. London. Ser. A Math. Phys. Eng. Sci., 362*, 2065–2098.
26. Mintmire, J. W., Dunlab, B. I., & White, C. T., (1992). *Phys. Rev. Lett., 68*, 631–634.
27. Tans, S. J., Devoret, M. H., Dai, H., Thess, A., Smalley, R. E., Geerligs, L. J., & Dekker, C., (1997). *Nature, 386*, 474–477.
28. Hamada, N., Sawada, S., & Oshiyama, A., (1992). *Phys. Rev. Lett., 68*, 1579–1581.
29. Hassanien, A., Tokumoto, M., Ohshima, S., Kuriki, Y., Ikazaki, F., Uchida, K., & Yumura, M., (1999). *Appl. Phys. Lett., 75*, 2755–2757.
30. Berber, S., Kwon, Y., & Tománek, D., (2000). *Phys. Rev. Lett., 84*, 4613–4616.
31. Al Soubaihi, R. M., Saoud, K. M., & Dutta, J., (2018). *Catalysts, 8*, 660.
32. Lu, C. Y., & Wey, M. Y., (2007). *Fuel, 86*, 1153–1161.
33. Eswaramoorthi, I., Sundaramurthy, V., & Dalai, A. K., (2006). *Appl. Catal. A Gen., 313*, 22–34.
34. Jardim, E. O., Gonçalves, M., Rico-Francés, S., Sepúlveda-Escribano, A., & Silvestre-Albero, J., (2012). *Appl. Catal. B Environ., 113, 114*, 72–78.
35. Gomes, H. T., Samant, P. V., Serp, P., Kalck, P., Figueiredo, J. L., & Faria, J. L., (2004). *Appl. Catal. B Environ., 54*, 175–182.

36. Garcia, J., Gomes, H. T., Serp, P., Kalck, P., Figueiredo, J. L., & Faria, J. L., (2005). *Catal. Today, 102, 103*, 101–109.
37. Ovejero, G., Sotelo, J. L., Romero, M. D., Rodríguez, A., Ocaña, M. A., Rodríguez, G., & García, J., (2006). *Ind. Eng. Chem. Res., 45*, 2206–2212.
38. Garcia, J., Gomes, H. T., Serp, P., Kalck, P., Figueiredo, J. L., & Faria, J. L., (2006). *Carbon N. Y., 44*, 2384–2391.
39. Ovejero, G., Sotelo, J. L., Rodrı, A., Diaz, C., Sanz, R., & Garcia, J., (2007). *Ind. Eng. Chem. Res., 46*, 6449–6455.
40. Joung, H. J., Kim, J. H., Oh, J. S., You, D. W., Park, H. O., & Jung, K. W., (2014). *Appl. Surf. Sci., 290*, 267–273.
41. Yan, Y., Chen, Y., Jia, X., & Yang, Y., (2014). *Applied Catal. B, Environ., 156, 157*, 385–397.
42. Tan, H. T., Chen, Y., Zhou, C., Jia, X., Zhu, J., Chen, J., Rui, X., Yan, Q., & Yang, Y., (2012). *Appl. Catal. B Environ., 119, 120*, 166–174.
43. Fu, X., Yu, H., Peng, F., Wang, H., & Qian, Y., (2007). *Appl. Catal. A Gen., 321*, 190–197.
44. Chen, J., Zhang, W., Chen, L., Ma, L., Gao, H., & Wang, T., (2013). *Chempluschem, 78*, 142–148.
45. Liang, X. L., Dong, X., Lin, G. D., & Zhang, H. B., (2009). *Appl. Catal. B Environ., 88*, 315–322.
46. Arike, M., T., Takenaka, S., Matsune, H., & Kishida, (2010). *Bull. Chem. Soc. Jpn., 959*, 953–959.
47. Onoe, T., Iwamoto, S., & Inoue, M., (2007). *Catal. Commun., 8*, 701–706.
48. Li, C. H., Yu, Z. X., Yao, K. F., Ji, S. F., & Liang, J., (2005). *J. Mol. Catal. A Chem., 226*, 101–105.
49. Guo, Z., Chen, Y., Li, L., Wang, X., Haller, G. L., & Yang, Y., (2010). *J. Catal., 276*, 314–326.
50. Brotons, V., Coq, B., & Planeix, J. M., (1997). *J. Mol. Catal. A Chem., 116*, 397–403.
51. Planeix, J. M., Coustel, N., Coq, B., Brotons, V., Kumbhar, P. S., Dutartre, R., Geneste, P., Bernier, P., & P. M. Ajayan. (1994). *Journal of the American Chemical Society, 116*(17), 7935–7936.
52. Giordano, R., Serp, P., Kalck, P., Kihn, Y., Schreiber, J., Marhic, C., & Duvail, J., (2003). *Eur. J. Inorg. Chem., 4*, 610–617.
53. Chizari, K., Janowska, I., Houllé, M., Florea, I., Ersen, O., Romero, T., Bernhardt, P., Ledoux, M. J., & Pham-Huu, C., (2010). *Appl. Catal. A Gen., 380*, 72–80.
54. Dry, M. E., (1999). *Appl. Catal. A Gen., 189*, 185–190.
55. Bukur, D. B., Lang, X., Mukesh, D., Zimmerman, W. H., Rosynek, M. P., & Li, C., (1990). *Ind. Eng. Chem. Res., 29*, 1588–1599.
56. Shariati, J., Haghtalab, A., & Mosayebi, A., (2019). *J. Energy Chem.*, 9–22.
57. Eschemann, T. O., Lamme, W. S., Manchester, R. L., Parmentier, T. E., Cognigni, A., Rønning, M., & De Jong, K. P., (2015). *J. Catal., 328*, 130–138.
58. Van, D. T. W., Yoshida, H., Oord, R., Zečević, J., Weckhuysen, B. M., & De Jong, K. P., (2020). *Appl. Catal. A Gen., 593*, 117441.
59. Feng, L., Yan, Y., Chen, Y., & Wang, L., (2011). *Energy Environ. Sci., 4*, 1892–1899.
60. Mabena, L. F., Sinha, R. S., Mhlanga, S. D., & Coville, N. J., (2011). *Appl. Nanosci., 1*, 67–77.
61. Li, X., Tuo, Y., Li, P., Duan, X., Jiang, H., & Zhou, X., (2014). Effects of carbon support on microwave-assisted catalytic dehydrogenation of decalin. *Carbon N. Y., 67*, 775–783.

62. Chen, W., Ji, J., Duan, X., Qian, G., Li, P., Zhou, X., Chen, D., & Yuan, W., (2014). *Chem. Commun., 50*, 2142–2144.
63. Liu, Z. J., Xu, Z., Yuan, Z. Y., Lu, D., Chen, W., & Zhou, W., (2001). *Catal. Letters, 72*, 203–206.
64. Liu, Z. J., Yuan, Z. Y., Zhou, W., Peng, L. M., & Xu, Z., (2001). *Phys. Chem. Chem. Phys., 3*, 2518–2521.
65. Yin, S. F., Zhang, Q. H., Xu, B. Q., Zhu, W. X., Ng, C. F., & Au, C. T., (2004). *J. Catal., 224*, 384–396.
66. Yin, S. F., Xu, B. Q., Zhu, W. X., Ng, C. F., Zhou, X. P., & Au, C. T., (2004). *Catal. Today, 93, 95*, 27–38.
67. Li, L., Zhu, Z. H., Yan, Z. F., Lu, G. Q., & Rintoul, L., (2007). *Appl. Catal. A Gen., 320*, 166–172.
68. Wang, S. J., Yin, S. F., Li, L., Xu, B. Q., Ng, C. F., & Au, C. T., (2004). *Appl. Catal. B Environ., 52*, 287–299.
69. Bell, T. E., Zhan, G., Wu, K., Zeng, H. C., & Torrente-Murciano, L., (2017). *Top. Catal., 60*, 1251–1259.
70. Chen, J., Zhu, Z. H., Wang, S., Ma, Q., Rudolph, V., & Lu, G. Q., (2010). *Chem. Eng. J., 156*, 404–410.
71. Hill, A. K., & Torrente-Murciano, L., *Appl. Catal. B Environ.*, 2015, *172–173*, 129–135.
72. Hu, Z., Mahin, J., Datta, S., Bell, T. E., & Torrente-Murciano, L., (2019). *Top. Catal., 62*, 1169–1177.
73. Zhang, J., Su, D. S., & Schlögl, R., (2009). *Phys. Status Solidi Basic Res., 246*, 2502–2506.
74. Chen, L., Yang, K., Liu, H., & Wang, X., (2008). *Carbon N. Y., 46*, 2137–2143.
75. Kim, J. Y., Jo, Y., Lee, S., & Choi, H. C., (2009). *Tetrahedron Lett., 50*, 6290–6292.
76. Xu, Y., Chen, L., & Wang, X., (2013). *Can. J. Chem., 91*, 307–314.
77. Zhang, P. P., Zhang, X. X., Sun, H. X., Liu, R. H., Wang, B., & Lin, Y. H., (2009). *Tetrahedron Lett., 50*, 4455–4458.
78. Bin, P. H., Yen, C. H., Yoon, B., Sato, M., & Wai, C. M., (2006). *Synth. Commun., 36*, 3473–3478.
79. Corma, A., Garcia, H., & Leyva, A., (2005). *J. Mol. Catal. A Chem., 230*, 97–105.
80. Siamaki, A. R., Lin, Y., Woodberry, K., Connell, J. W., & Gupton, B. F., (2013). *J. Mater. Chem. A, 1*, 12909–12918.
81. Shao, L., Zhang, B., Zhang, W., Hong, S. Y., Schlçgl, R., & Su, D. S., (2013). *Angew. Chemie - Int. Ed., 52*, 1–5.
82. Sokolov, V. I., Rakov, E. G., Bumagin, N. A., & Vinogradov, M. G., (2010). *Fullerenes, Nanotubes. Carbon Nanostructures, 18*, 558–563.
83. Lee, E., Park, S. A., Woo, H., Hyun, K., Wook, D., Lim, H., & Kim, Y. T., (2017). *J. Catal., 352*, 388–393.
84. Pyo, A., Kim, J. D., Choi, H. C., & Lee, S., (2013). *J. Organomet. Chem., 724*, 271–274.
85. Hajipour, A. R., Khorsandi, Z., & Karimi, H., (2015). *Appl. Organomet. Chem., 29*, 805–808.
86. Movassagh, B., Parvis, F. S., & Navidi, M., (2015). *Appl. Organomet. Chem., 29*, 40–44.
87. Olivier, J. H., Camerel, F., Ziessel, R., Retailleau, P., Amadou, J., & Pham-Huu, C., (2008). *New J. Chem., 32*, 920–924.
88. Veisi, H., Safarimehr, P., & Hemmati, S., (2019). *Mater. Sci. Eng. C, 96*, 310–318.

CHAPTER 3

MESOPOROUS SILICA NANOMATERIALS AND POROUS SOLIDS

LAVANYA TANDON[1], VIJENDER SINGH[2], AABROO MAHAL[3], and POONAM KHULLAR[1]

[1]*Department of Chemistry, B.B.K. D.A.V. College for Women, Amritsar, Punjab, India*

[2]*Department of Chemistry, S.K. Government College, Kanwali Rewari, Haryana, India*

[3]*Department of Chemistry, Khalsa College for Woman, Amritsar, Punjab, India*

ABSTRACT

The morphology of the surface depends upon the use of solvent during Stöber synthesis. After treating hydrothermally, obtained silica spheres possess smooth surfaces on the use of the surfactant cetyltrimethylammonium bromide (CTAB), whereas corrugated shells are obtained on using isopropanol or butanol. Using static acidic conditions mesoporous spheres of silica whose size lies within the range of 2 to 6 μm have been synthesized. By using cetyl trimethyl ammonium bromide (CTAB) and amphiphilic block copolymer (polystyrene-*b*-poly(acrylic acid), PS-*b*-PAA) as the co-templates, dual-mesoporous silica spheres (DMSS) with pore size 2.0 nm can be synthesized. By varying the TEOS amount and length of the hydrophobic block, the size of the pore and the thickness of the shell can

Nanotechnology-Enhanced Solid Materials: Design, Synthesis, Properties, Applications, and Perspectives.
Lionello Pogliani, Ann Rose Abraham, A. K. Haghi, & Prabhat Ranjan (Eds.)

be tuned. In the case of a heterogeneous system, small particles kept on emerging during the synthesis. By increasing the amount of silica precursor particle size was found to be enlarged. On adding different silica precursor mesoporous sphere of silica having hydrophobic shell and hydrophilic core was obtained. In mesoporous silica spheres, immobilization of enzymes is obtained by layer-by-layer assembly of the thin shells. The use of enzyme provides the enhanced stability of the reaction by increasing the number of layers and reducing the stability of the reaction.

3.1 INTRODUCTION

Currently, modern nanotechnology is a part of science. It possesses major applications in the field of biotechnology and plays a major role in disease diagnoses. The medicine evolution has been a great boon for society. Non-toxic, biodegradable, and biocompatible nanocarriers are being developed. For infectious and cancerous diseases polymeric and lipid base nanocarriers are being developed. Inorganic nanoparticles also have a vital role in the biomedical field. The commercially available particles are the nanoparticles of gold and carbon dots. Among these, the particles of silica have attained great attention in phase I trials on humans, which is necessary during the investigation of the new drug. In recent years, the mesoporous particles of silica have acquired great attraction as it has uniform and tunable pore size. The functionalization on its surface takes place independently. It serves as the promising carrier for the drug. This carrier has been used for loading drugs to various macromolecules like RNA, DNA, and proteins. The present given review focuses on the synthesis of the mesoporous silica sphere under different conditions.

Mesoporous spheres of silica having tunable pore size, large surface area is used in different applications like catalysis [1, 2], delivery of drug [3–5], and various other applications [6–9].

3.2 SYNTHESIS OF MESOPOROUS SILICA SPHERES

Many methods have been developed for synthesizing silica mesoporous sphere in which surfactants are used as templates for the formation of spheres [6, 10–15]. Stöber synthesis is used to produce ordered mesoporous spheres possessing high monodispersity, smooth surface and tunable pore size [16]. Mesoporosity can also be introduced into the Stöber spheres, via pseudomorphic transformations. These transformations are the reaction which involves the change in chemical composition. Reprecipitation followed by dissolution results in perseverance of the solid materials shape. Hydrothermal treatment is involved for the pseudomorphic transformation of amorphous and dense silica and this is an alternative method for synthesizing mesoporous silica spheres.

During transformation, size and monodispersity of original spheres of silica is retained [17, 18]. There is a great influence of solvent in synthesizing silica spheres. The use of solvent affects the condensation occurring in silica spheres and this determines the silica dissolution and the architecture of the spheres obtained. When surfactant having long chain like alkylammonium is obtained then pseudomorphic products of silica spheres are synthesized. When EtOH is used then silica spheres having smooth surface is obtained. When BuOH or iPrOH is used then silica spheres with corrugated surface is obtained. For nanocasting, silica spheres having corrugated surfaces are used as template and mesoporous spheres of carbon with hollow cores resembling to mesoporous silica are obtained as the products and these provides insights into the surfactant molecules diffusion depth in the parent silica sphere. It is revealed that the cores of mesoporous spheres remain solid. The parent silica spheres as shown in Figure 3.1, having smooth surfaces were obtained by Stöber syntheses using various solvents like iO_2-iPrOH, SiO_2-BuOH and SiO_2-EtOH [19].

The smooth surface can be seen in Figure 3.1(A), (D), and (G). After undergoing reactions of pseudomorphic transformation, the mesoporous product average diameter of all the spheres were found to be increased HT-SiO_2-BuOH (733, 40 nm), SiO_2-EtOH (436, 18 nm), SiO_2-iPrOH (623, 37 nm), SiO_2-BuOH (494, 23 nm), HT-SiO_2-EtOH (527, 20 nm); HT-SiO_2-iPrOH (730, 45 nm). The product obtained have been found to be more aggregated as compared to parents' sphere. The morphologies of the surface have been found to be different than those obtained after treating hydrothermally (can be seen in Figure 3.2). In Figure 3.1(B) and (C) smooth surfaces are obtained in which EtOH has been used as a surfactant. In these, mesoporous channels are directed on the outside surface. These results have been sufficiently observed through TEM. It can be seen in given Figure 3.2(A).

The products which are obtained from parent spheres of silica using BuOH or iPrOH solvent are found to possess deep cavities along with the corrugation at the surface of the product, it can be seen in Figure 3.1(E) and (F), in which iPrOH is used as solvent and in Figure 3.1(H) and (I) in which BuOH is used as solvent. The cavities possess elongated openings. Figure 3.2(B)–(D) represents the spheres outer parts. It has been found that the products obtained after the hydrothermal treatment have been directed outwards from core to outside. As seen in the TEM images of Figure 3.2(B) and (C), the cores of HT-SiO_2-BuOH and HT-SiO_2-iPrOH are found to be denser.

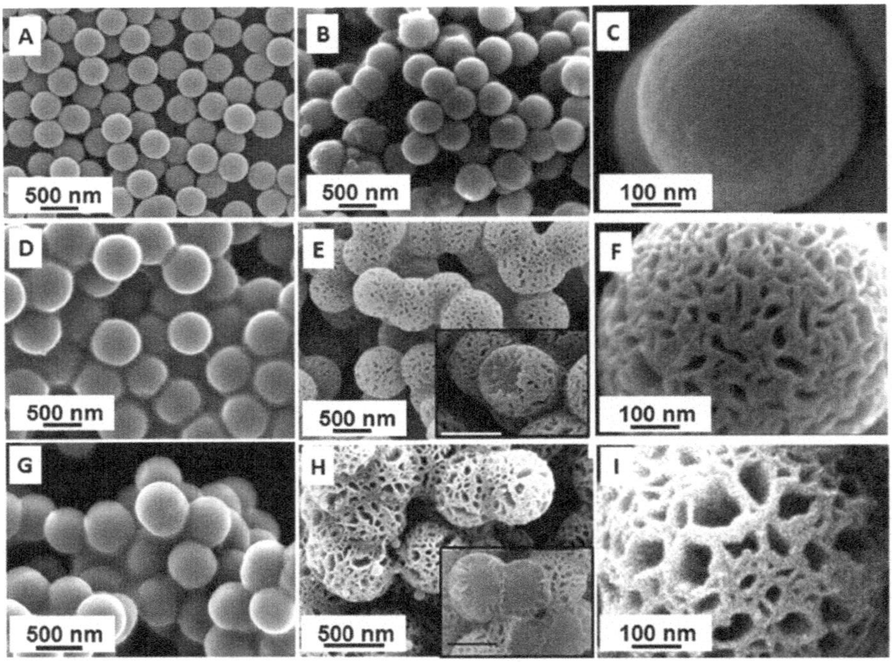

FIGURE 3.1 SEM images of parent silica and hydrothermal silica spheres: (A) parent silica spheres produced in EtOH (SiO_2–EtOH); (B) hydrothermal product HT–SiO_2–EtOH; (C) magnified image of HT–SiO_2–EtOH; (D) parent silica spheres produced in iPrOH (SiO_2–iPrOH); (E) hydrothermal product HT–SiO_2–iPrOH and broken structures of HT–SiO_2–iPrOH in the inset with a scale bar of 500 nm; (F) magnified image of HT–SiO_2–iPrOH; (G) parent silica spheres produced in BuOH (SiO_2–BuOH); (H) hydrothermal product HT–SiO_2–BuOH and broken structures of HT–SiO_2–iPrOH in the inset with a scale bar of 500 nm; (I) magnified image of HT–SiO_2–BuOH.

Source: Reproduced with permission from Ref. [19]. © 2011 American Chemical Society.

The SEM image, as shown in Figure 3.1(E) and (H) shows that spheres are broken and the shell structure is corrugated. In Figure 3.2(B), small aggregate particles along with larger spheres have been observed. The side products have been obtained during the hydrothermal treatment via surface templating of silica and through multiple centrifugation it can be removed.

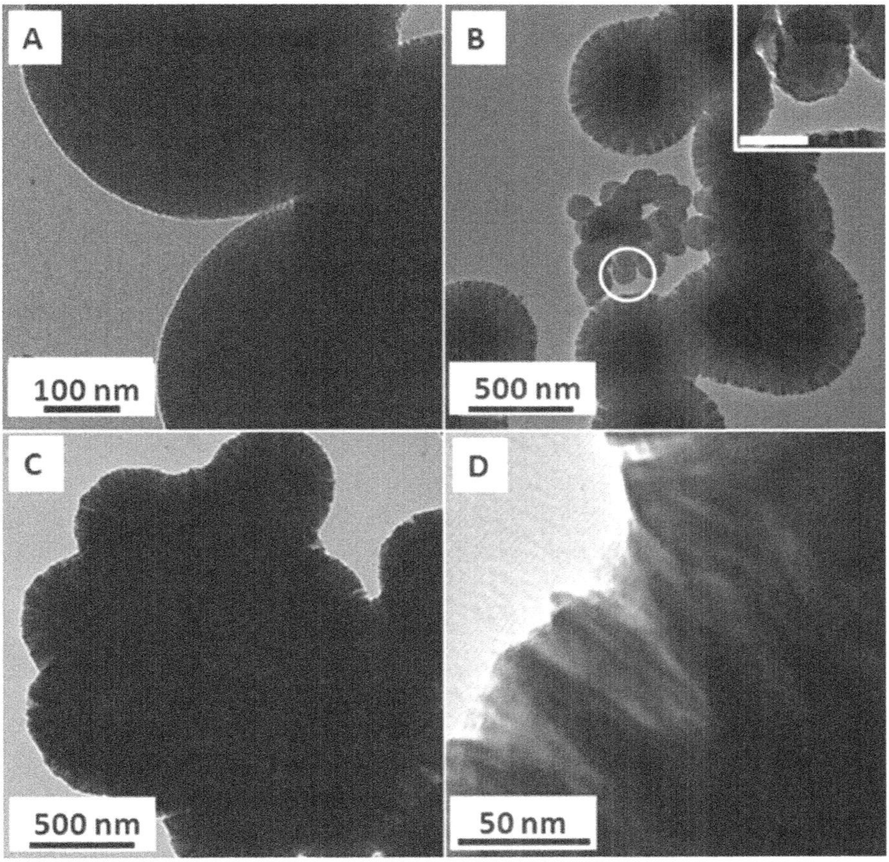

FIGURE 3.2 TEM images of hydrothermal reaction products: (A) HT–SiO_2–EtOH; (B) HT–SiO_2–iPrOH; and (C) HT–SiO_2–BuOH. The inset in (B) shows an enlargement of the circled region with a scale bar of 100 nm; (D) shows a higher magnification image of HT–SiO_2–iPrOH.

Source: Reproduced with permission from Ref. [19]. © 2011 American Chemical Society.

It has been observed that products obtained by different methods possess the same alignment of pores on the surface, but the texture of pore and the morphology of the surface are found to be different. The products obtained using different solvents after the hydrothermal reactions exhibit radial direction that has resulted from the combined effect of individual surfactant molecule that aligns on the surface of the sphere and thus the surface energy is reduced and also the tendency for assembling into micelles is reduced [20, 21].

3.3 SYNTHESIS OF TRANSPARENT MESOPOROUS SPHERES

By using emulsion-based chemistry, transparent marble-like mesoporous spheres is obtained. The silica spheres possess a uniform size ranging from 0.1 to 2 mm and size of mesoporous sphere of silica lie in the diameter range of 1–5 nm. The sorbets used for chromatographic separations depend upon the texture, porosity, size, and mechanical strength of the particles. The molecular selectivity and the space flow velocity are obtained because of the mono-dispersed pore size and high surface area. These particles are obtained through annealing and packing of the uniformly sized spheres at high temperatures.

Small spherical droplets of gelled silica are obtained through the process of spray drying. It can also be obtained by gelling the liquid through heating or by the chemical action in which the droplets have been sprayed [22]. Different size spheres are obtained by changing the size of the silica solution before it gets gelled. The silicon ester-like tetraethyl orthosilicate (TEOS) or Tetramethylorthoslicate (TMOS) acts as the source of silica. The obtained hydrogel spherical structures were first washed and then dried at a temperature of 120°C and at a high temperature of about 450°C. By controlling the silicon esters hydrolysis, spherical solid particles are obtained.

By partial hydrolysis of methyl or ethyl silicate or methyl silicate in the alcohol, an emulsion in water-alcohol solidifies into the gel beads. Silica beads with a diameter of 0.2–0.5 nm and having pore diameter with size 2 nm are obtained. The pore volume, distribution of pore size, pore shape and surface texture of silica sphere requires preparation conditions. Silica spheres prepared by these methods possess low surface area, and irregular shape [23]. Calcined spherical solid particles are also obtained by silicon ester hydrolysis. For example, the silica gel beads of size 0.2–0.5 mm diameter are obtained by partial hydrolysis of methyl or ethyl silicate in alcoholic solution containing less amount of water and in this HCl act as the catalyst and water-alcohol emulsion is formed which are solidified into the gel beads. Silica spheres obtained by this method possess irregular shape and have large pore size dispersion and low surface area.

3.3.1 EFFECT OF STIRRING ON SYNTHESIS OF MESOPOROUS SILICA SPHERES

The mesoporous silica spheres can also be prepared in single-step method. By varying the reaction volume and the stirring conditions, silica spheres of different sizes are obtained [24]. For synthesis of emulsions, sphere formation is influenced by the speed of stirring. Soft particles of gel are obtained when the speed of stirring is low, i.e., <200 rpm and hard spheres are obtained when the speed of stirring is medium, i.e., 200–400 rpm and fine powder is obtained when high speed of stirring is applied, i.e., >450 rpm [27]. The shape and size of the spheres prepared by using CTAB are shown in given Figure 3.3.

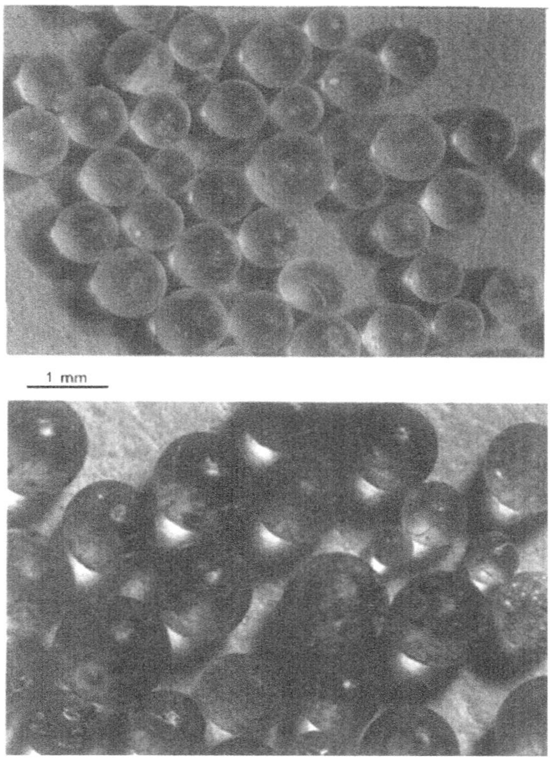

FIGURE 3.3 Size and shape of spheres using CTAB as surfactant: (a) as-made sample; (b) calcined at 500°C sample. The strength of the sphere depends upon the sphere's dryness.

Source: Reproduced with permission from Ref. [24]. © 1997 American Chemical Society.

3.4 SYNTHESIS OF SPHERES UNDER ACIDIC AND BASIC CONDITIONS

Efforts have been made for fabricating defined mesoporous spheres of silica. Under static conditions, various sizes of mesoporous silica sphere (MSS) are obtained by following the templating routes of cationic surfactant through stirring under the basic conditions. The sub micrometer-sized MSS's ranging from 0.4 to 1.1 μm and the millimeter sized particles of 0.1 to 2 mm are obtained [25].

Using the method of emulsion templating, under the acidic conditions, the MSS's having diameter within 1–100 μm are synthesized. After the removal of the organic phase, hollow microspheres are obtained and crushed readily [26]. Under the acidic conditions, mesoporous silica possessing various morphologies like disc like, spheroidal spiral and toroidal are produced. Templating route of cationic surfactant is followed to obtain the MSS's possessing the tunable mesopore size and the configuration. The nonionic surfactant interacts with the cationic surfactant and the mixed surfactant aggregates are obtained in the aqueous solution. Thus, it is expected that modified morphology is expected by particles of mesoporous silica which are obtained by mixing the cationic-non-ionic surfactant. Thus, the particles with the size 2–6 μm are synthesized under the static acidic conditions through mixing the cationic–non-ionic surfactant templating. Scanning electron microscopy (SEM) has been used to analyze the growth of MSS's the micrometer size, and it has been found that stirring is not suitable for the silica sphere formation, but the quiescent synthesis favors the spherical morphology. Stirring has great effect on the morphology of the obtained product. When stirring is applied for 30 minutes, then two morphologies are obtained. One of these consists of spheres of micrometer size and the other are the irregular large monoliths and when the sample is stirred for 960 minutes then only the irregular monoliths are obtained (Figure 3.4).

TEM images are also obtained at different stages during the formation of MSS's. Spherical particles showing the fluffy and spongy structure are obtained when the sample is aged for 5 hours under the static conditions. Tiny particles lying around the micrometer-sized particles aggregate together. The exterior surface of these particles is obtained by the subunit's aggregation. And these subunits possess the rough surface and consist of large number of nanometer pores and after aging for 16 hours the spongy and fluffy particles grow and dense and spherical particles with range 2 to 6 μm are developed, as shown in Figure 3.5.

Thus, the synthesized mesoporous product has not formed through drying and calcination but has resulted from growth in the mother solution. Thus, the process of unidirectional aggregation is followed to obtain these spheres. TEOS is hydrolyzed along with the cationic-non-ionic template to produce subunit of 0.1 μm in size. These subunit aggregates unidirectionally to obtain the fluffy and spongy spherical particles and these reconstructed into dense spheres possessing the smooth surface on aging through the internal cementation. Large monoliths are obtained when the soluble silica gets deposited on the surface of solid silica. The continuous stirring damages the particle morphology as unexpected aggregation occurs between the silica particles. This mesoporous sphere formation is an example of selective processing of this mesoporous silica into the macroporous form.

FIGURE 3.4 SEM micrographs of calcined mesoporous silica stirred for: (a) 0.5; and (b) 16 h.
Source: Reproduced with permission from Ref. [27]. ©1998 American Chemical Society.

FIGURE 3.5 TEM micrographs of mesoporous silica aged under static conditions for: (a–c) 5; and (d) 16 h.

Source: Reproduced with permission from Ref. [19]. ©1998 American Chemical Society.

3.5 SYNTHESIS OF SPHERES HAVING BIMODAL PORES

Dual templating approach or hydrothermal or chemical post-treatment of the mesoporous silica particles is used to obtain the porous materials possessing bimodal porosity. These bimodal pores do not have well defined structure and the size of the macropores lies in the range of 200 to 300 nm. Recently it has been discovered that different block co-polymers are being used as templates having large pores, and surfactants are being used as templates with small pores for synthesizing these bimodal mesopore material with tunable pore size and structure. The particles obtained possess irregular morphology and their particle size distribution is also broad. A simple method has been reported for synthesizing dual mesoporous silica sphere (DMSS) having core-shell structure. PS-b-PAA, an anionic block copolymer is being used as a template for forming large mesopores. By changing the lengths of the blocks of PS-b-PAA, the size of the pores and the thickness of the shell can be tuned. The dual MSS's prepared by this method possess uniform size of the particles and a defined structure of core-shell. The dual MSS's were prepared by hydrophobic nanoparticles encapsulation like iron oxide (IO) magnetic particles. It has been shown by the drug storing and releasing tests that approximately 500 mg of the Ibuprofen can be stored in one gram of DMSS. The following mechanism discussed has been proposed for the formation of the DMSS [28] (Scheme 3.1).

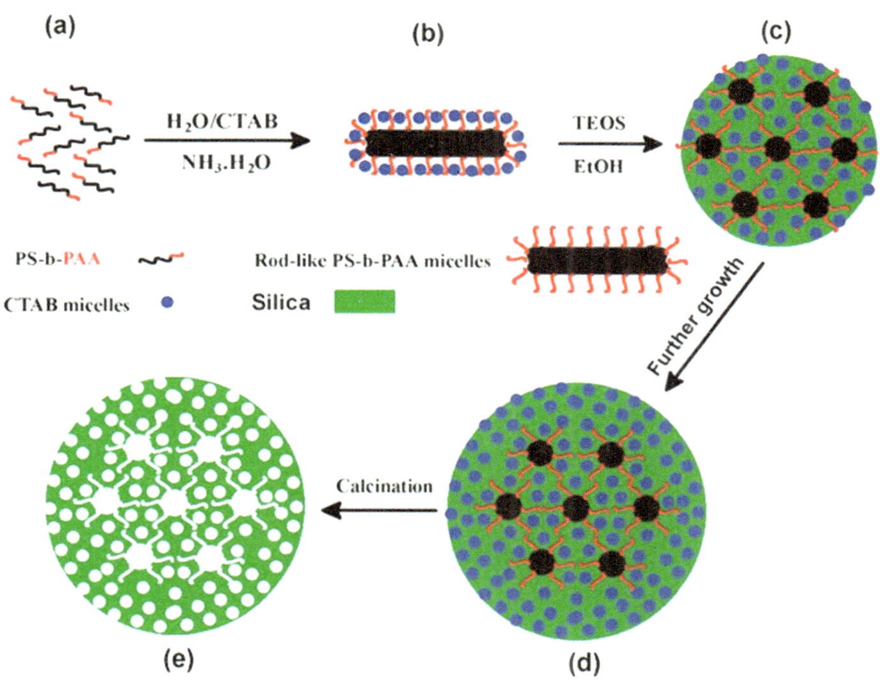

SCHEME 3.1 (a) THF solution of PS-b-PAA; (b) CTAB-coated PS-b-PAA aggregates formed by the electrostatic interaction between CTA+ and PAA–; (c) core part formed from the assembly between CTAB-coated PS-b-PAA aggregates and TEOS; (d) mesoporous shell formation via the self-assembly between the remaining CTAB and the additional TEOS; (e) final core-shell structured dual-mesoporous silica spheres after calcination.

Source: Reproduced with permission from Ref. [28]. © 2010 American Chemical Society.

By adding the small molecules of surfactants or the inorganic ions, PS-b-PAA micelles morphology can be changed from spheres to rod shape [29]. Rod-shaped aggregates were formed in the beginning of the experiment and it was due to the use of ammonia and the CTAB surfactant. The PAA hydrophilic block of aggregates of rod shape could couple via electrostatic interaction and the columbic forces with the micelles of CTAB and the composite micelles were formed also it leads to the CTAB coating on the rod-like particles. By using ethanol or TEOS, rod-like aggregates coated by CTAB, which have been made up of silicate oligomers were formed first. Core dual mesoporous structure has been obtained when composites of silica material packed in ordered fashion which was similar to the model that was proposed by Cai and co-workers [30]. These electrostatic interactions are responsible for the spherical morphology of the spherical silica. The remaining amount of the CTAB molecule was expelled out from the structure's core and when the structures core become large then the outer area of the core gets rich in the CTAB and it leads to the self-assembly between the TEOS and the CTAB molecules. The core gets surrounded with the outer shell consisting of small pores. CTAB has an important role in the structure's formation. Without CTAB, the coated aggregates could not have formed and also the self-assembly could not have been possible and the mesoporous structure would not have been formed. The solvent ethanol, also plays an important role in the formation of these aggregates. If no ethanol has been used or only a small amount of ethanol have been used then MSS's could not be obtained. This is so because usage of large amount of ethanol results in high CMC (critical micelle concentration) value. It has been found that metal nanoparticles could be incorporated into DMSS large pores so as to achieve the magnetic stabilization. By this encapsulation, the structure and monodispersity remains unaffected. By using the same procedure, different hydrophobic nanoparticles could be incorporated into spheres of DMSS and multifunctional spheres of silica could form having mesoporous dual structure.

3.6 DRUG RELEASE USING MESOPOROUS SILICA SPHERES

It has been found that drug release rate of the molecules can be tuned by adjusting the thickness of the DMSS shell. The three-step release of the drug has been observed; the first step involves fast drug release of the drug in 60 minutes. The second stage involves an increase in concentration of the drug in the next 6 to 24 h. Large stage involves the drug absorption in large pores. This release profile of DMSS is same with that of mesoporous structure of yolk-shell previously reported [31]. Mesoporous silica sphere possesses uniform characteristics as it has high specific surface area [32, 33].

Mesoporous silica can be synthesized under acidic and basic conditions. Spheres of spherical shape are obtained because of the minimum surface energy. The reaction rate can be reduced in the acidic media by carrying out the synthesis by using the non-ionic surfactant [34]. The self-assembly method induced by evaporation or the aerosol-assisted method is used for synthesizing the spherical MSS's [35, 36].

In basic conditions, surfactants like compounds of ammonia, salts of alkyl amines were used as surfactants. Modified Stöber method has been used to synthesize these spheres. Recently, spheres having hexagonal regularity have been synthesized by optimizing the concentration of TMOS (tetramethoxysilane) and C_nT_{MAX} (n-alkyl trimethylammonium) halide and these hexagonally ordered mesopores are aligned uniformly from center towards the spheres surface. Many researchers have investigated the mechanism for developing such homogeneous spheres. Mechanism of cooperative templating followed by templating mechanism of liquid crystal have been developed [32]. These reactions are conducted above surfactant's CMC value. The nucleation, particle growth, and spherical formation of the silica spheres cannot be interpreted by the single mechanism, as synthesis has been carried out under various reaction conditions. As reported by Ostafin and co-workers, the growth of particles has been attributed to oligomers and monomers in addition to the surface [37]. It has been found that at the silica sphere center, cubic structure was obtained, and from this structure aligned mesopores were obtained. Central core structure has governed alignment of the mesopore. It has been found that particle growth occurs rapidly during the reaction and is followed by propagation and termination and monodispersed particles formation. In Figure 3.6, the particle appearing small in size, i.e., of 140 nm at 144 seconds is found to grow bigger and obtain the size of 500 nm at 600 seconds 39.

Figure 3.6 confirms that small synthesized particles homogeneously grow into large particles and this leads to the generation of spherical monodispersed particles. From the given figure, it is confirmed that, during the reaction, small particles appeared in the reaction mixture along with the monodispersed particles. These particles can occur any time in the reaction mixture and thus it was suggested that new particles were kept on occurring during the reaction mixture. These small particles are found to grow homogeneously in big particles and finally leads to the formation of monodispersed particles.

It was observed that, under certain reaction conditions, small particles appeared suddenly when the particles have grown to their final size. Small particles were continuously observed during the synthesis. The result has been confirmed by the TEM observations (Figure 3.7). Mesoporous silica has a diameter within the range of 2–50 nm [32]. This size is comparable with the enzyme's diameter, and thus these materials can be used for the immobilization of enzyme.

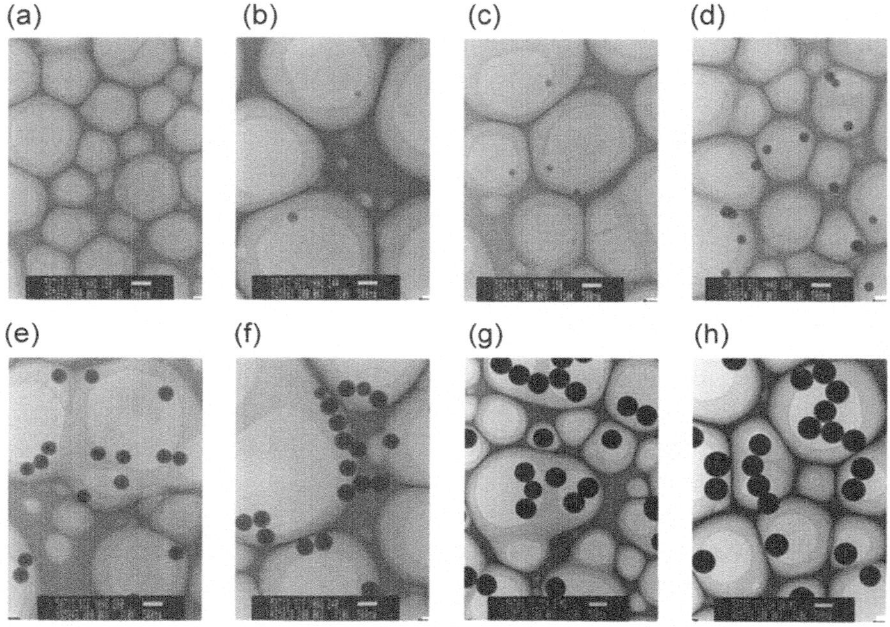

FIGURE 3.6 Transmission electron micrograph (TEM) images of samples obtained at different times: (a) 130; (b) 144; (c) 150; (d) 160; (e) 200; (f) 240; (g) 360; and (h) 600 s after the commencement of the experiment. The scale bar represents 500 nm.

Source: Reproduced with permission from Ref. [39]. © 2007 American Chemical Society.

3.7 SYNTHESIS OF MESOPOROUS SPHERES HAVING LARGE PORES

Recently, the simple, and useful method has been reported for preparing particles loaded with enzymes using mesoporous spheres of silica with large pore size. The enzyme can be immobilized on the surface of the sphere and a multilayer shell was made to assemble on the surface of the sphere via L-b-L, i.e., layer-by-layer assembly through electrostatic interactions [40].

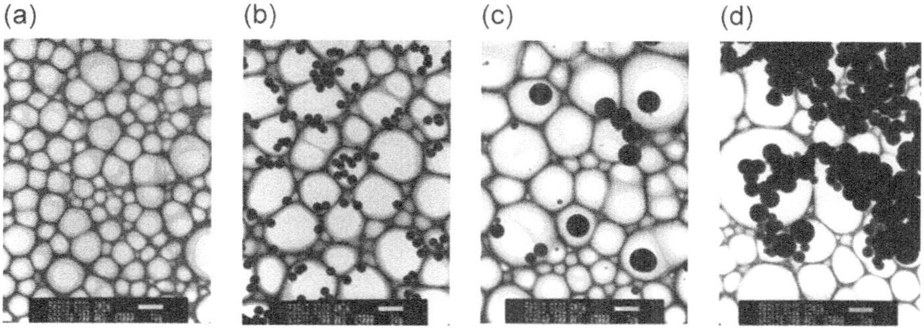

FIGURE 3.7 TEM images of samples obtained at different times: (a) 120; (b) 240; (c) 540; and (d) 720 s after the commencement of the experiment for a heterogeneous system. The scale bar represents 2 μm.

Source: Reproduced with permission from Ref. [39]. © 2007 American Chemical Society.

BMS particles enzyme immobilization capacity has been investigated for enzymes having different isoelectric points (pI) and different sizes. Hosts used for immobilization of enzymes are MS spheres and spheres of nonporous silica, constituting different pore sizes.

Following are the systems used for dense coatings on the BMS spheres loaded with enzymes: (i) a multilayer shell of PAA formed by deposition of poly(sodium-4-styrenesulfonate) (PSS) and poly(diallyl dimethylammonium chloride) (PDDA); (ii) formation of a composite shell via SiNP (Silica Nanoparticle) and PDDA (Figure 3.8).

3.8 ENCAPSULATION OF ENZYME

For encapsulation of enzyme (as shown in given Figure 3.9(a)) bimodal mesoporous spheres with size ranging between 2 and 4 μm were used. TEM studies (Figure 3.9) have confirmed their inner structure, shown in Figure 3.9(b) and (c). The homogeneous character of the particles has been shown in Figure 3.9(b). These bi-modal particles underwent slicing and resulted into large size distribution. Figure 3.9(c) shows the disordered structure of pores with size 10–40 nm.

The material is confirmed to possess a bimodal structure with small pores having a size of 2–3 nm and large pores having a size of 10–40 nm. Enzymes are the macromolecules and their encapsulation can occur in structures consisting of large pores. Figure 3.9(d) shows the nanoporous image of the

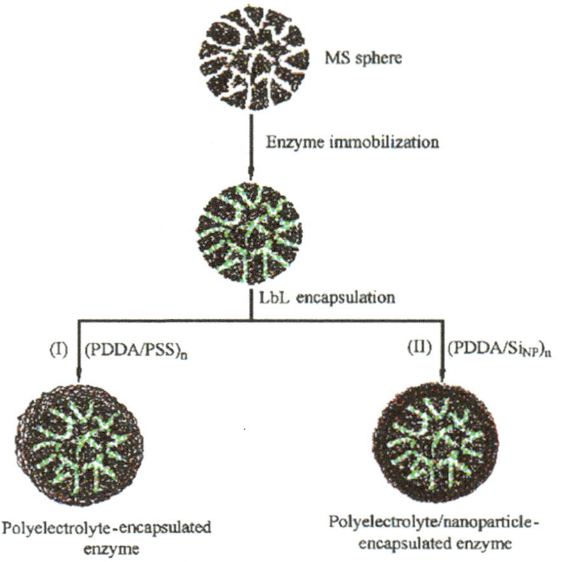

FIGURE 3.8 Schematic representation of enzyme encapsulation using MS spheres as supports. The enzyme is first immobilized and subsequently encapsulated by a (I) PE or (II) PE/nanoparticle multilayer shell.

Source: Reproduced with permission from Ref. [40]. © 2005 American Chemical Society.

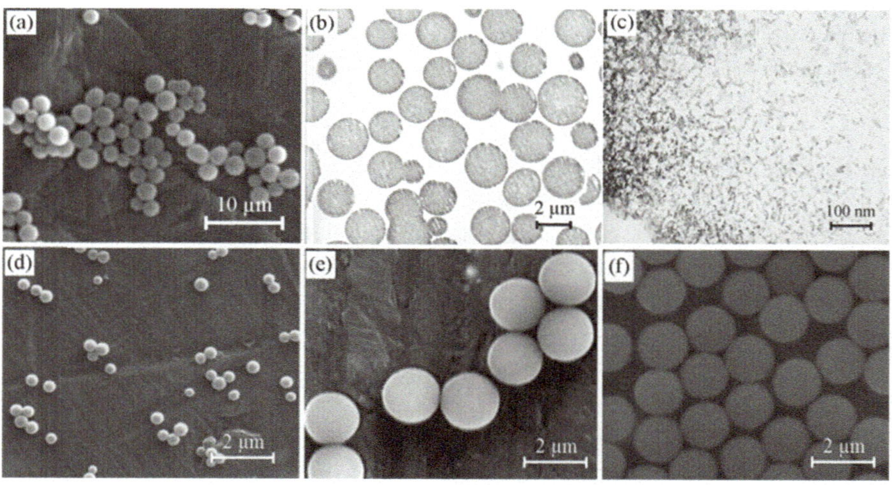

FIGURE 3.9 Electron microscopy images of the silica spheres employed: (a) SEM images of the BMS spheres; and (b, c) TEM of ultramicrotomed samples of the BMS spheres; SEM images of the (d) NS spheres; (e) MS-12 spheres; and (f) MS-16 spheres.

Source: Reproduced with permission from Ref. [40]. © 2005 American Chemical Society.

silica in which the size of the mesopores is 2–3 nm and Figures 3.9(e) and (f) depicts the image of particles of MS-12 and 16, which were synthesized in a homogeneous phase. SBA-15 with a pore size of 7 nm were also used as supports for enzyme immobilization. As compared with the BMS particles, the particle size of the MS-12 and MS-16 are found to be more homogeneous. Several enzymes with different pI values and having different molecular weights were used to determine the potential of enzyme encapsulation via BMS sphere, i.e., bimodal mesoporous silica sphere. It was observed that enzymes having low weight of molecule and high value of pI underwent high loading. Electrostatic forces have occurred for encapsulating the enzyme in the silica sphere because pI of silica is 3. For the encapsulation of larger enzymes having low value of pI, the tendency of loading is found to decrease. The size of pore and enzyme plays a significant role in encapsulation. The interactions between the enzyme and MSS's are so weak that enzymes undergo desorption when they are dispersed in the solution and to prevent this desorption, a protective barrier is provided for enzyme loading. For this purpose, a shell composed of multilayers was imposed on the surface of BMS sphere. Encapsulated catalase is being used in the food industry and helps in removing hydrogen peroxide (H_2O_2) from food. It is also used in the textile industry and analytical field. By measuring the enzyme activity, the amount of desorbed catalase can be measured by monitoring the H_2O_2 kinetics. After the deposition of the multilayer on the spheres, the leakage of enzyme was decreased to a great extent. The activity of the loaded sphere is dependent upon layer type and the number of layers deposited.

3.9 CONCLUSION

Spheres reactivity is affected depending upon the type of solvent used. As carbon atoms increase, the condensation decreases on moving from ethanol to propanol to butanol spheres prepared using ethanol are found to be mesoporous throughout. Under static conditions, Mesoporous silica with micrometer size dimensions have been synthesized using cationic–anionic surfactant templating. By varying the length of PS-*b*-PAA. The pore size of DMSS can be tuned. The thickness can be controlled by varying the TEOS concentration. These have potential applications in the field of separation, sorption, and biomedical field. The bimodal silica particles play a significant role in the immobilization of the enzyme. Because of the encapsulation, the properties of enzymes get enhanced.

3.10 CURRENT AND FUTURE PERSPECTIVE

Nanomedicines are found to have a great impact in disease therapy and are being used for diagnosis or conventional treatment. Research is being carried out for using this technology in treating multiple disease therapy. The mesoporous silica particles fabrication is costly but simple. These particles can also be used as nanocarriers for drugs placements and drug loading. The amount of these particles can be varied as in the same concentration drug cannot be loaded. FDA has approved the use of silica nanoparticles on the human trials for C dots. It is a very important step. Challenge lies because of the lack of genotoxicity, chronic toxicity, tissue compatibility, etc. Degradation mesoporous silica mechanism has not been discovered yet. Efforts are being made to cover the gap between clinical and preclinical use of mesoporous silica particles. The novel design of these materials can be used in diagnosing, imaging, and treating the patients need.

KEYWORDS

- **dual mesoporous silica sphere**
- **enzymes**
- **morphology**
- **silica spheres**
- **surfactant**
- **tetraethyl orthosilicate**

REFERENCES

1. Shokouhimehr, M., Piao, Y., Kim, J., Jang, Y., & Hyeon, T., (2007). A magnetically recyclable nanocomposite catalyst for olefin epoxidation. *Angew. Chem., Int. Ed., 46*(37), 7039–7043. doi: 10.1002/anie.200702386.
2. Zhang, Q., Lee, I., Ge, J., Zaera, F., & Yin, Y., (2010). Surface-protected etching of mesoporous oxide shells for the stabilization of metal nanocatalysts. *Adv. Funct. Mater., 20*, 2201–2214. /doi.org/10.1002/adfm.201000428.
3. Lai, C. Y., Trewyn, B. G., Jeftinija, D. M., Jeftinija, K., Xu, S., Jeftinija, S., & Lin, V. S. Y., (2003). A Mesoporous silica nanosphere-based carrier system with chemically removable CdS nanoparticle caps for stimuli-responsive controlled release of neurotransmitters and drug molecules. *J. Am. Chem. Soc., 125*(15), 4451–4459. doi: 10.1021/ja028650l.

4. Lu, J., Liong, M., Zink, J. I., & Tamanoi, F., (2007). Mesoporous silica nanoparticles as a delivery system for hydrophobic anticancer drugs. *Small, 3*(8), 1341–1346. doi: 10.1002/smll.200700005.
5. Zhu, Y., Shi, J., Shen, W., Dong, X., Feng, J., Ruan, M., & Li, Y., (2005). Stimuli-responsive controlled drug release from a hollow mesoporous silica sphere/polyelectrolyte multilayer core-shell structure. *Angew. Chem., Int. Ed., 44,* 5083–5087. doi: 10.1002/anie.200501500.
6. Nooney, R. I., Thirunavukkarasu, D., Chen, Y., Josephs, R., & Ostafin, A. E., (2002). Synthesis of nanoscale mesoporous silica spheres with controlled particle size. *Chem. Mater., 14,* 4721–4728.doi.org/10.1021/cm0204371.
7. Wang, Y., & Caruso, F., (2005). Mesoporous silica spheres as supports for enzyme immobilization and encapsulation. *Chem. Mater., 17,* 953–961.doi.org/10.1021/cm0483137.
8. Wang, Y., & Caruso, F., (2006). Nanoporous protein particles through templating mesoporous silica spheres. *Adv. Mater., 18,* 795–800.doi.org/10.1002/adma.200501901.
9. Wang, Y., Yu, A., & Caruso, F., (2005). Nanoporous polyelectrolyte spheres prepared by sequential coating sacrificial mesoporous silica spheres *Angew. Chem., Int. Ed., 44,* 2888–2892.
10. Lu, Y., Fan, H., Stump, A., Ward, T. L., Rieker, T., & Brinker, C., (1999). Aerosol -assisted self assembly of mesostructured spherical nanoparticles. *J. Nature, 398,* 223–226.
11. Gr€un, M., Unger, K. K., Matsumoto, A., & Tsutsumi, K., (1999). Novel pathways for the preparation of mesoporous MCM-41 materials: Control of porosity and morphology. *Micro-Porous Mesoporous Mater., 27,* 207–216.doi.org/10.1016/S1387-1811(98)00255-8.
12. Yoon, S. B., Sohn, K., Kim, J. Y., Shin, C. H., Yu, J. S., & Hyeon, T., (2002). Fabrication of carbon capsules with hollow macroporous core/mesoporous shell structures. *Adv. Mater., 14,* 19–21.https://doi.org/10.1002/1521-4095.
13. Tan, B., Lehmler, H. J., Vyas, S. M., Knutson, B. L., & Rankin, S. E., (2005). Controlling nanopore size and shape by fluorosurfactant templating of silica. *Adv. Mater., 17,* 2368–2371. doi.org/10.1021/cm048991t.
14. Ganguly, A., Ahmad, T., & Ganguli, A. K., (2010). Silica mesostructures: Control of pore size and surface area using a surfactant-templated hydrothermal process. *Langmuir, 26,* 14901–14908. DOI: 10.1021/la102510c.
15. Kim, T. W., Chung, P. W., & Lin, V. S. Y., (2010). Facile synthesis of monodisperse spherical MCM-48 mesoporous silica nanoparticles with controlled particle size. *Chem. Mater.* 2010,doi.org/10.1021/cm1017344.
16. St€ober, W., Fink, A., & Bohn, E., (1968). Controlled growth of monodisperse silica spheres in the micron size range. *J. Colloid Interface Sci.*, 62–69.
17. Martin, T., Galarneau, A., Renzo, F. D., Fajula, F., & Plee, D., (2002). Morphological control of MCM-41 by pseudomorphic synthesis. *Angew. Chem., Int. Ed., 41,* 2590–2592.10.1002/1521-3773(20020715).
18. Botella, P., Corma, A., & Navarro, M. T., (2007). Single gold nanoparticles encapsulated in monodispersed regular spheres of mesostructured silica produced by pseudomorphic transformation. *Chem. Mater., 19,* 1979–1983.doi.org/10.1021/cm0629457.
19. Yoo, W., & Stein, A., (2011). Solvent effects on morphologies of mesoporous silica spheres prepared by pseudomorphic transformations. *Chem. Mater., 23,* 1761–1767.doi.org/10.1021/cm102829m |.
20. Yoon, S. B., Kim, J. Y., Kim, J. H., Park, Y. J., Yoon, K. R., Park, S. K., & Yu, J. S., (2007). Synthesis of monodisperse spherical silica particles with solid core and mesoporous shell: Mesopore channels perpendicular to the surface J. Mater. Chem., 17, 1758–1761.

21. Deng, Y., Qi, D., Deng, C., Zhang, X., & Zhao, D., (2008). Superparamagnetic high-magnetization microspheres with an Fe_3O_4@SiO_2 Core and perpendicularly aligned mesoporous SiO_2 Shell for removal of microcystins. *J. Am. Chem. Soc., 130*, 28–29.doi.org/10.1021/ja0777584.
22. Von, R. K., (1979). The chemistry of silica. Solubility, polymerization, colloid and surface properties, and biochemistry. *J. Angewandte. Chichester, XXIV*, 886. John Wiley and Sons.
23. Van, D., Grift, C. J. G., Geus, J. W., Barten, H., Leferink, R. G. I., Van, M. J. C., & Den, O. A. T., (1988). In: Unger, K. K., Rouquuerol, J., Sing, K. S. W., & Kral, H., (eds.), *Characterization of Porous Solids, Stud. Surf. Sci. Catal.* (Vol. 39, p. 619). Elsevier: Amsterdam.
24. Huo, Q., Feng, J., Schu, F., & Stucky, G., (1997). Preparation of hard mesoporous silica spheres. *Chem. Mater., 9,* 14–17.
25. Grun, M., Lauer, I., & Unger, K. K., (1997). The synthesis of micrometer and sub-micrometer sized spheres of ordered mesoporous oxide MCM-41*Adv. Mater., 9*, 254–257.
26. Schacht, S., Huo, Q., Voigt-Martin, I. G., Stucky, G. D., & Schuth, F., (1996). Oil-water interface templating of mesoporous macroscale structures. *Science, 273,* 768. DOI: 10.1126/science.273.5276.768.
27. Limin, Q., Jiming, M., Humin, C., & Zhenguo, Z., (1998). Micrometer-sized mesoporous silica spheres grown under static conditions. *Chem. Mater., 10*, 1623–1626.
28. Dechao, N., Zhi, M., Yongsheng, L., & Jianlin, S., (2010). Synthesis of core-shell structured dual-mesoporous silica spheres with tunable pore size and controllable shell thickness. *J. AM. Chem. Soc., 132*, 15144–15147.
29. Yu, Y., & Zhang, L., (1998). Eisenberg, morphogenic effect of solvent on crew-cut aggregates of amphiphilic diblock copolymers.doi.org/10.1021/ma971254gA. *Macromolecules, 31*, 1144–1154.
30. Cai, Q., Luo, Z., Pang, W., Fan, Y., Chen, X., & Cui, F., (2001). Dilute solution routes to various controllable morphologies of MCM-41 silica with a basic medium *Chem. Mater., 13*, 258–263.doi.org/10.1021/cm990661z.
31. Liu, J., Qiao, S. Z., Hartono, S. B., & Lu, G. Q., (2010). Monodisperse yolk-shell nanoparticles with a hierarchical porous structure for delivery vehicles and nanoreactors. *Angew. Chem., Int. Ed., 49*, 4981–4985.
32. Kresge, C. T., Leonowicz, M. E., Roth, W. J., Vartuli, J. C., & Beck, J. S., (1992). Ordered mesoporous molecular sieves synthesized by a liquid-crystal template mechanism. *Nature, 359*, 710.
33. Beck, J. S., Vartuli, J. C., Roth, W. J., Leonowicz, M. E., Kresge, C. T., Schmitt, K. D., Chu, C. T. W., et al., (1992). Synthesis and characterization of mesoporous silica nanoparticles (MSNp) MCM 41 from natural waste rice husk. *J. Am. Chem. Soc., 114*, 10834.
34. Qi, L., Ma, J., Cheng, H., & Zhao, Z., (1998). Micrometer-sized mesoporous silica spheres grown under static conditions *Chem. Mater., 10*, 1623.doi.org/10.1021/cm970811a.
35. Lu, Y., Fan, H., Stump, A., Ward, T. L., Rieker, T., & Brinker, C. J., (1999). Aerosol assisted self assembly of meso- structured spherical nanoparticles. *Nature, 398*, 223.
36. Bore, M. T., Rathod, S. B., Ward, T. L., & Datye, A. K., (2003). Hexagonal mesostructures in powders produced by evaporation induced self-assembly of aerosol from aqueous tetra ethoxy silane solution. *Langmuir, 19*, 256.
37. Nooney, R. I., Thirunavukkarasu, D., Chen, Y., Josephs, R., & Ostafin, A. E., (2002). Synthesis of nanoscale mesoporous silica spheres with controlled particle size. *Chem. Mater., 14*, 4721. doi.org/10.1021/cm0204371.

38. Pauwels, B., Tendeloo, G. V., Thoelen, C., Rhijn, W. V., & Jacobs, P. A., (2001). Structure determination of spherical MCM-41 particles. *Adv. Mater., 13*, 1317.
39. Nakamura, T., Mizutani, M., Nozaki, H., Suzuki, N., & Yano, K., (2007). Formation mechanism for monodispersed mesoporous silica spheres and its application to the synthesis of core/shell particles. *J. Phys. Chem. C, 111*. doi.org/10.1021/jp0648240.
40. Wang, W., & Caruso, F., (2005). Mesoporous silica spheres as supports for enzyme immobilization and encapsulation *Chem. Mater., 17*, 953–961. 10.1021/cm0483137.

CHAPTER 4

TRENDS IN SMART BIOPOLYMER COMPOSITES AND POLYMER SOLIDS WITH MULTIDISCIPLINARY APPLICATIONS IN NANOSCIENCE AND NANOTECHNOLOGY

AMIR GULL[1], RAEES UL HAQ[2], MOHD. AAQIB SHEIKH[3], TARIQ AHMAD GANAIE[4], MUSHTAQ AHMAD LONE[5], and SHAHEEN KHURSID[6]

[1]*Department of Food Science and Technology, University of Kashmir, Srinagar, Jammu and Kashmir, India*

[2]*Department of Food Engineering and Technology, SLIET, Sangrur, Punjab, India*

[3]*Department of Food Technology, IUST, Awantipora, Jammu and Kashmir, India*

[4]*Department of Food Science and Technology, Government Degree College Shopian, Jammu and Kashmir, India*

[5]*Directorate of Planning and Monitoring, SKUAST-K, Srinagar, Jammu and Kashmir, India*

[6]*Department of Zoology, Government Degree College Tral, Jammu and Kashmir, India*

Nanotechnology-Enhanced Solid Materials: Design, Synthesis, Properties, Applications, and Perspectives. Lionello Pogliani, Ann Rose Abraham, A. K. Haghi, & Prabhat Ranjan (Eds.)

ABSTRACT

Living systems respond to external stimuli by adapting themselves to changing conditions. Smart polymers or intelligent materials are materials that can exhibit stimuli-sensitive properties and are becoming important in many commercial applications. These polymers undergo large reversible changes, either physical or chemical, in their properties as a consequence of small environmental variations such as temperature, pH, stress, UV irradiation, magnetic or electric field, ionic factors, etc., and respond in an anticipated and desired way. During recent years, research on stimuli-responsive polymers has been extensively performed, especially in bio-separation, affinity precipitation, bio-catalysis, waste minimization, glucose sensors, drug delivery, and protein purification. The versatility and potential of these materials have emerged with great potential and enabled the development of various types of substances that are biologically inspired. This chapter is aimed at creating a multidisciplinary forum of discussion on the most recent research advances in the application of smart polymer, providing a potential guided outlook for future studies.

4.1 INTRODUCTION

Polymers are mostly carbon-based chains of monomers, occasionally in amalgamation with metals and non-metals. Properties of polymers enabling them to respond to different external stimuli (electromagnetic field, the intensity of light, pH, temperature, and so on) are designated as smart polymers. Changes in different environmental conditions result in reversible or large physical and chemical changes in the properties of smart polymers. The polymers change either by folding into various conformations or change the color in response to environmental stimuli. To live up to the tag of 'smart,' polymers must possess the ability to detect environmental alterations and respond in an anticipated and desired way. The polymers may respond either to single or to multiple stimuli causing transitions from solution to gel, swelling, and collapsing of molecules and so on, therefore the term stimuli-responsive polymers is interchangeable for smart polymers [1]. The response of polymers to the environmental conditions depends upon the physical state of molecular chains with reversible solution to gel states arising from solubilized linear macromolecules while as the cross-linked polymers are associated with the transition of an expanded state from a collapsed state by virtue of reorganizing the polymeric chain. Another mechanism involves the development of responsive interfaces by changing the hydrophilic nature of smart surfaces with response to different stimuli. Application of the above-mentioned responses can be employed for designing smart materials for different practical applications in tissue engineering, pharmaceutical delivery systems, etc. [2]. The uniqueness of these polymers has gained considerable attention from last two decades and have witnessed exponential growth in the field of biotechnology, medicine, and engineering. Smart or intelligent polymers have also been used for bio-separation, drug delivery, waste minimization, and other biotechnological applications such as in purification techniques, affinity precipitation, controlled permeation membranes, thermo-sensitive chromatography, development of biocatalysts, biomimetic actuators, aqueous polymer two-phase partitioning, modulation of catalytic processes and as surfaces with switchable hydrophobic-hydrophilic properties [3]. The use of smart polymers, in combination with non-chromatographic technologies, represents a useful approach to the development of new separation techniques. These can be physically or chemically conjugated to biomolecules to yield polymer-biomolecule systems that can respond to biological, physical, or chemical stimuli. Smart polymers could be prepared by either by membrane processing and surface modification and the commonly used smart polymers are alginates, Eudragit S-100, Eudragit L-100, chitosan, poly(N-isopropyl acrylamide), poly(methacrylic acids), poly(acrylic acids), sulfobetaine methacrylate, etc. The structure of polymers can be altered by integrating the chain with different functional molecules like biological receptors by pre- and post-polymerization methods [4, 5]. Controlled living radical polymerization techniques are used in the management of nanoparticles and nano-assemblies by forming graft, star, and block copolymers, gels, and surface modifications [6, 7]. Nitroxide-mediated radical polymerization (NMP), one amongst three processing techniques involves the use of dormant and active moieties via temperature regulation. Proliferation of radicals in reversible addition-fragmentation chain transfer (RAFT) polymerization takes place by inclusion of thiocarbonylthio group within polymerization system and lastly atom-transfer radical polymerization (ATRP) transition uses oxidation-reduction reaction. Composite supported shape memory alloys (SMAs) are considered to be promising candidates for the supposed "smart composite materials" [8]. Shape memory composites with shape memory material systems give potential for making new paradigms in material-structural interactions [9]. Shape memory polymers (SMPs) are a category of stimuli-responsive materials which will change their shapes on command, from a short-lived form to their permanent form upon exposure to associate degree external stimulus. This phenomenon is based on twin section system, during which one segment is very elastic and when the stimulant is applied other segment remarkably cut back its stiffness [10]. SMPs have various benefits, which includes higher capability for elastic strain (more than 100% in most cases), lower density, tunable transformation temperature, and the most important is low cost which allows the employment of SMPs for extensive scope of applications [10]. The demand of shape memory composites for engineering and technical applications is increasing with time [11].

4.2 APPLICATION OF BIOPOLYMER COMPOSITES

4.2.1 CHITOSAN

Chitosan, polycationic derivatives of poly-*N*-acetyl-d-glucosamine (chitin) is most abundant natural polymer extracted from the algae, fungi cell walls as well as in the exoskeleton of invertebrates (crab, lobster, and shrimp) [12, 13]. Chitosan physicochemical and biological properties are determined by the distribution and composition of acetylated monomers (GlcNAc; A-unit) and deacetylated monomers (GlcN; D-unit) [14]. The degree of acetylation (DA) in chitosan can vary from DA-0 (fully deacetylated) to DA-1 (partially acetylated) depends on whether *the* *N*-acetyl group is either fully or partially replaced by NH_2 group. Being biocompatible, biodegradable, and nontoxic, it has gained great attention in food, cosmetic, pharmaceutical, and biomedical field [14] Also, the antibacterial property of this polymer makes it a suitable material for tissue engineering and wound healing. Wound healing properties of chitosan hydrogel/nano zinc oxide (ZnO) composite was evaluated by Sudheesh et al. [15]. Results indicated that chitosan composites exhibited excellent antibacterial activity, hemostatic potential and activated platelets attached on the composite bandage surface. Similarly, Madhumathi et al. [16]; Sudheesh et al. [17]; Chen et al. [18] reported high swelling ratio, antibacterial potential, controlled biodegradation, and excellent blood clotting ability of silver-incorporated chitosan hydrogel composite bandages. Studies of Choi et al. [19]; Salomäki & Kankare [20] showed chitosan-hyaluronate sponges exhibited enhanced wound healing potential than corresponding controls.

4.2.2 PECTIN

Pectin is a high-molecular-weight polymer consisting mainly of esterified D-galacturonic acid residues linked by α-(1-4) chain. Pectin is soluble in water, but it is insoluble in most organic solvents. Its solubility rate is associated with the degree of polymerization, distribution, and number of methyl-ester groups. Being made up of several structural units, homogalacturonan (HG) composed of (1→4)-linked α-d-Gal*p*A residues and type I rhamnogalacturonan (RG-I) consists of disaccharide-repeating unit [→4)-α-d-Gal*p*A-(1→2)-α-l-Rha*p*-(1→]n with a variety of side chains consisting of l-arbinosyl and d-galactosyl residues are the important structural units. A number of moles of methanol or acetic acid/100 mol GalA are defined as DA and degree of methylation (DA). Native pectin is having DA in order of 70–80 and it can range from 0–100%. High methoxyl pectin are having DA >50 and pectin with DA <50 is known as low methoxyl pectin. The most significant attribute that affects solubility, gelling, and film forming properties of pectin is the degree of esterification, and it varies according to plant source, plant harvesting, processing, and storage conditions [21, 22]. Nowadays, there is an increased interest in commercial use of pectin hydrogels for biomedical applications such as drug delivery, gene delivery, tissue engineering, and wound healing. As reported by Munarin, Tanzi, & Petrini [21]; Mishra et al. [22]; Coimbra et al. [23], pectin exhibits several benefits as it can be easily tailored into hydrogels, films, scaffolds, and nanoparticles. Due to its excellent swelling and erosion properties, it is useful for colon targeting systems. Besides this, it has been successfully investigated for constructing the colon-biodegradable drug delivery systems apart from being used as coating material for tablets, pellets, microparticles, and beads [24, 25].

4.2.3 HYALURONAN

Hyaluronan is a linear polysaccharide composed of repeating β-1,4-linked D-glucuronic acid (GlcA) and β-1,3-linked N-acetyl-D-glucosamine (GlcNAc) disaccharide units. Hyaluronan being the simplest glycosaminoglycan (GAG) polymer is commonly found in the extracellular matrix of all vertebrate epithelial, neural, and connective tissues. Apart from this, it is also present in bacteria capsules (e.g., Streptococci strains), but is lacking in fungi, plants, and insects. Hyaluronan highest content in the human body is found in synovial fluid, umbilical cords and in the eye vitreous humor. Due to its special viscoelastic, rheological properties and high molar mass of 10^7 DA make it an interesting biomaterial for medical applications including drug delivery, ophthalmic surgery, osteoarthritis treatment and tissue engineering [26]. Studies of Zacchi et al. [27] showed that the pattern of expression of particular dermal and epidermal molecules was similar to that found in normal skin, while demonstrated the use of HA benzyl ester in the coculture human fibroblasts and keratinocytes. Collagen-HA scaffold matrix produced by cross-linking formyl group-modified HA polymers with collagen fibers, possessed excellent biocompatibility and increased osteoconductivity [28]. N-succinyl chitosan/aldehyde HA composite hydrogel promoted the attachment and encapsulation of chondrocytes *in vitro*, thus demonstrating the potential applications of these hydrogels in cartilage tissue engineering [29].

4.2.4 GELLAN GUM

Gellan gum is a linear, anionic exopolysaccharide, consisting of a-L-r hamnose, β-D-glucose, and β-D-glucuronate units. Gellan is having molecular weight of 5×10^5 DA, produced by *Sphingomonas elodea* bacteria. Due to its versatile properties such as thermo-responsive, nontoxic, biocompatibility, potential to remain stable over extensive heat make it an interesting biomaterial for tissue engineering and pharmaceutical fields [30]. Elastic and soft gels are produced by gellan gum composites, while as pure gellan gum produces hard and transparent gels. Gelrite TM (acyl GG or acylated GG) and Kelcogel TM (low acyl GG or deacylated GG) are the commercially available gellan gum under different trade names. Besides, it is widely used for gene transfection, gene therapy, wound healing, cell adhesion, guided bone regeneration, dental care, ophthalmic formulations, biological signaling and delivery agents [31, 32]. In tissue engineering field it has been used as a material for cartilage reconstruction. Implants consisting of gellan microparticles immersed in gellan matrix were designed by Pereira et al. [33]. Although storage and loss moduli differ from those of human intervertebral disc. However, it turned out that implant swelling and degradation behavior can be regulated by adjusting the concentration of low- and high-acyl gellan. Application of Hesperidin (heteropolysaccharide) along with gellan gum have been successfully used in tissue engineering applications and confirmed the cartilage regeneration, cell adhesion, and differentiation ability of the prepared scaffold, using 3-(4,5-dimethylthiazol-2-yl)-2,5-Diphenyltetrazolium Bromide (MTT), SEM, and RT-PCR studies [34]. Chondrocyte encapsulated gellan gum-based HA hydrogel for cartilage regeneration have been successfully prepared by Kim et al. [35] results showed enhanced cell adhesion, viability, proliferation, and gene expression in an *in vitro* and in an *in vivo* model through hydrogel.

4.2.5 ALGINATES

Alginates belong to a group of natural polymers found widely in brown seaweeds, that are formed by the linear copolymers of d-mannuronic acid and l-guluronic acid units joined together by β-1-4 linkages. The different species of brown algae (*Macrocystis pyrifera, Ascophyllum nodosum, Laminaria japonica, Laminaria digitata,* and *Laminaria hyperborean*) used for commercial extraction of alginates are treated with alkaline aqueous solutions preferably sodium hydroxide solution. Calcium or sodium chloride is added to the extract for precipitating alginate that is further treated with hydrochloric acid for transforming alginate into alginic acid [36]. The application of alginates in biomedical science and various engineering fields has been widely explored by researchers and the reason for its success involves its biocompatibility and gelation properties. The presence of the carboxylic group within the structural chain permits the polymer to form gels whenever there is the presence of bipolar ions. The ease of developing an ionic gel from alginate polymers has led to its use in encapsulating cells aimed towards drug delivery systems and artificial organs [37]. The high-water absorption property of alginate polymers along with its porous structure makes it as an ideal material for developing wound dressings [38]. In addition to maintaining a moist microenvironment around the injury, alginate wound dressings reduce the chances of bacterial infection and enhance the wound healing.

4.2.6 CELLULOSE

Structurally cellulose is made up of repeated glucose monomers joined together by β-1-4 glycosidic linkages in such a way that one out of every two consecutive glucose molecules are rotated by 180°C with respect to other. Cellulose forms microfibrils in plants that are regarded as the basic structural framework of plants. Protofibrils, a chain of around 30 cellulose molecules, assemble together into much larger units known as microfibrils that ultimately form the cellulose fibers. The higher membrane permeability of cellulose is the reason for its consideration as a hemodialysis membrane [39]. Cellulose in plants is mainly associated with lignocellulose, hemicellulose, pectin, lignin, etc., whereas cellulose from microbial origin is devoid of any contaminants. The application of cellulose ranges from cutting-edge techniques in biotechnology to modern procedures of papermaking. Application of cellulose from the plant origin is much more complex than the pure bacterial cellulose (BC) due to the presence of different impurities in the former. Different

Gram-negative bacterial cultures used for the production of BC include *Acetobacter, Achromobacter, Agrobacter, Azotobacter, Rhizobium, Sarcina*, etc. [40, 41]. Both the static as well as shaking culture techniques are employed for the production of BC with the former process resulting in a comparatively lower production yield. The existence of hydroxyl groups in addition to better mechanical properties allows cellulose polymers for purification of proteins. Carboxymethyl cellulose fibers have been developed into dressings under the trade name of Aquacel that upon hydration forms a strong cohesive gel. Water solubility of cellulose fibers has enabled researchers to study these polymers as components for drug delivery systems [42].

4.2.7 LIGNIN

Lignin along with hemicellulose, cellulose, and other different extractives are among the widely available natural polymers found on the planet earth forming the bulk of biomass. The occurrence of lignin as lignin-carbohydrate complex (LCC) in plants poses challenges in separating lignin from cellulose. Lignin known as redox polymer of aryl propane units is a complex naturally occurring polymer made up of phenyl propane structural units without a distinctive primary structure. The quantity and type of aromatic ring substitutes in the phenyl propane units are the basis of its differentiation involving units of guaiacyl, hydroxyphenyl, and syringyl groups designated as G, H, and S, respectively [43]. Absence of glucose molecules within the polymeric structure is regarded as the unique characteristic feature of lignin. The most common use of lignin involves production of dispersants and carbon fibers. The low heat capacity and a higher glass transition temperature of lignin are associated with small amounts of methoxyl groups in the structure [44]. Recent advancements in the field of biotechnological sector have enhanced the role and use of lignin for: (i) modifying plants in order to improve their transformation into fuels; (ii) enhancing the potential yield of biomass; (iii) optimization of overall biomass productivity; (iv) creation of promising and improved enzyme systems for effective downstream processing of biomass into biofuels.

4.3 APPLICATIONS OF SMART POLYMERS

Application of smart polymers along with stimuli response of controlled drug release are shown in Figure 4.1.

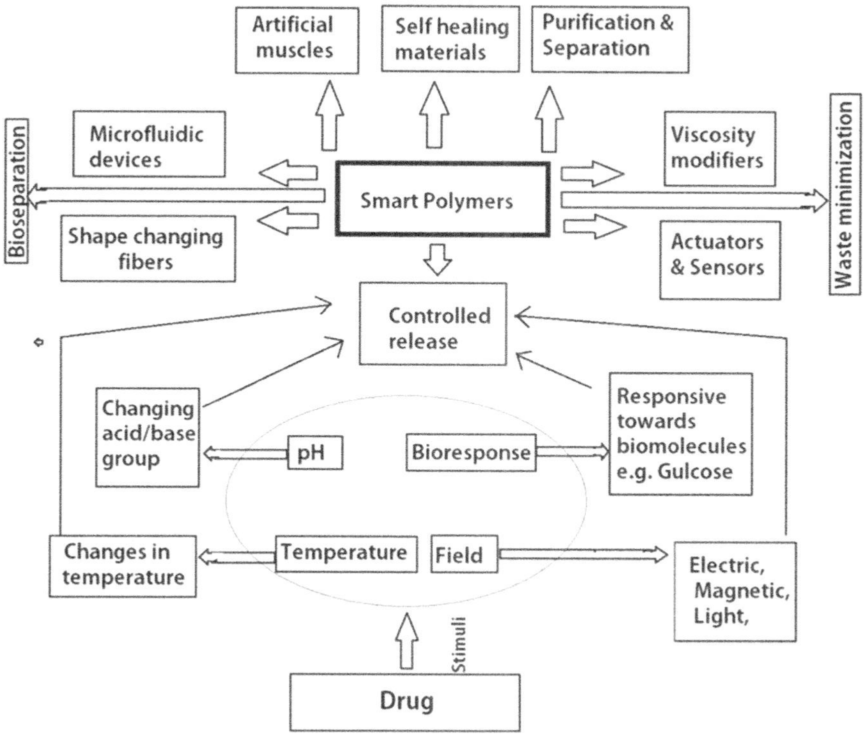

FIGURE 4.1 Smart polymer application combined with stimulus-response-controlled medication release.

4.3.1 BIO-SEPARATION

Advancement in biotechnology has led to the enlargement of a number of products using bio-separation processes and other techniques. Bio-separation process generally includes three steps:

- Favorable partitioning of the desired component and contamination;
- Phase separation by mechanical methods; and
- Recovery of the desired component.

Cost purification is a critical factor in determining the overall production cost of the desired product (biomolecule). Smart polymers have the capability to respond non-linearly by undergoing change in one or more of their inherent properties as a consequence of slight variations in their surrounding environment, such as pH, light, temperature, humidity, ionic strength, electrical potential, mechanical stress, and polarity, etc. [45]. Smart polymers show phase transition property in response to changes in the environmental stimuli and changes are translated to avalanche changes in the whole system, such as phase separation and facilitate the precipitation process. The potential of smart polymers to undergo phase separation makes it possible to use them in bio-separation process, where the bio-responsive polymer secures the formation of a new phase, including polymer precipitate enriched with desired component [46]. In an aqueous solution, smart polymers has the property to form a separate phase when the conditions are slightly changed, and the target component can be recovered easily by precipitation of the smart polymer. This property of smart polymers is utilized to isolate and purify target components such as protein, antibodies, etc. Smart polymeric materials versatility and untapped potentials are becoming increasingly attractive, which has led to some simple and economical strategies in bio-separation. Conjugates of smart polymers with ligands like (metal chelates, triazine dyes, carbohydrates, antibodies, antibiotics, nucleotides, protease inhibitors, etc., forms a complex with the desired component and phase separation of the complex is triggered by the slight change in the surrounding environment, transforming the polymer backbone into an insoluble state. Finally, the desired component is dissociated from the complex or precipitate, and remains in the supernatant in a purified form while as impurities (ligand-polymer conjugate) re-precipitates and remains in the solution. The ligands commonly used are p-aminobenamidene, biotin, aminophenyl-α-d-glucopyranoside, etc., have the affinity for given target component [47]. The use of bio-responsive polymers in the bio-separation process has developed into a broad area of application because it has been widely recognized as an efficient, scalable, and a single-step technique for the purification and recovery of desired products.

4.3.2 AFFINITY PRECIPITATION

The common merely used technique for separation is precipitation, purification, and recovery of the desired component. Affinity precipitation is a non-chromatographic method based on using a bio-responsive polymer, which selectively binds (electrostatic or hydrophobic interactions) with the target component and facilitates the precipitation process. The choice of the smart polymer carrier for affinity precipitation is decisive for the success of the purification process. Smart polymers have the ability to sense the surrounding environmental changes and respond to these changes in a pre-programed and marked way [48]. Phase separation of the complex is triggered by small changes of medium property (pH, temperature, ionic strength, electrical potential, etc.). The bio-responsive polymer changes its confirmation and precipitates in the complex with the target component, transforming the polymer backbone into an insoluble form. For separation, the desired component can be recovered from the precipitate by dissolution and dissociation, however, all the impurities remain in solution. The smart polymers facilitate phase separation as a consequence of slight variation in ionic strength and leads to target recovery with a high yield of the desired component. Attaching the target component with bio-responsive polymer makes the separation process simple, faster, and more efficient for purification, refolding, and recovery of the desired component. The affinity precipitation with smart polymers is one of the most successful and well-known biotechnological techniques usually characterized by reasonably high yields around 70–90% and purification folds up to 100 of desired components like proteins, antibodies, etc. [49]. Affinity precipitation is readily combined with other separation techniques like aqueous two-phase polymer systems (ATPSs). ATPSs is an important separation technique in biotechnology. It is a partition-based separation process usually composed of two incompatible polymers in water solution or a polymer and a high concentration of salt with a clear interface when the critical concentration of polymers is exceeded. The partitioning is a complex process and ATPSs plays a vital role in the selective partitioning of the target component between the two phases, which mainly depends on the surface properties of the target component. The possibility of easy recovery of target component from the reaction media when the conversion is completed as well as reuse of catalyst (smart polymer) makes it an efficient tool for the purification of desired component as it reduces whole purification process cost. The phase-forming components are selected depending upon the combination and application of biocatalysts with smart polymers, dispenses a novel opportunity for managing bio-catalytic reactions.

4.3.3 BIO-CATALYSIS

Over the last few decades, the conjugation of smart polymers with enzymes has achieved significant attention to establish more stable and competent enzymes. In an aqueous solution, smart polymers have the capability to form separate phases when there is a slight change in the surrounding environment. This property of smart polymer is used to generate reversibly soluble/

insoluble biocatalyst, which catalyze an enzyme reaction in their soluble state, provided that the enzyme molecule is covalently bounded and can be used in reactions with insoluble or poorly soluble substrates. As the reaction is complete, the conditions are altered to cause the catalyst to precipitate so that it can be separated from the product and reused. For waste minimization, catalyst recovery, and catalyst reuse polymer bound smart catalysts are very useful. With the use of smart polymers for immobilization of enzymes, the biocatalyst can be reused after homogenous catalysis [50]. The reversible biocatalyst inherits the advantages of immobilized enzymes, such as ease of separation and also overcomes the disadvantage of immobilized enzymes, such as difficulty in transport of substrate and products. The bio-conjugate can be separated simply by precipitation and in non-aqueous media, immobilization is used for enzyme stabilization against inactivation due to the media [51]. Reversibly soluble biocatalysts are used in various reactions which are impossible to carry out with the help of immobilized enzymes, for example, the use of reversibly soluble biocatalyst in the hydrolysis of high-molecular-mass substrates such as casein, dextrin, cellulose, and starch [52, 53].

4.3.4 WASTE MINIMIZATION

Non-biodegradable wastes have attracted widespread attention and require appropriate treatments before disposal. Reducing the amount of hazardous wastes, recycling polluted water and provide more freshwater is vital to environmental health. Polymer-bound smart adsorbents like nano-magnetic polymers, chitosan, cyclodextrins, etc., are extremely potent for adsorption of metal ions and effective in waste minimization [54]. They exhibit great potential in the elimination of contaminants (especially heavy metals) from wastewater due to their distinctive features like enhancing the adsorption capacity as well as selectivity for the targeted contaminants and also remarkably minimizes the pretreatment time [55]. Polymeric smart coatings have been developed for detecting, preventing, and remedying environmental damage due to hazardous contamination and play an important role in biological wastewater treatments like membrane bioreactors, activated sludge processes and fixed film processes by providing binding sites for other particles as well as metals. The use of polymeric smart materials to treat wastewater has grown exponentially due to their high sensitiveness, chemical strength under harsh conditions, remarkable selectivity for target components, swift, and accurate determination of trace environmental pollutants, etc.

4.3.5 GLUCOSE SENSORS

Diabetes mellitus, a disease characterized by high blood glucose levels, is a big threat to human health globally because untreated high sugar levels can damage the eyes, kidneys, heart, etc. Advances in smart polymers have promoted the fabrication of insulin (a hormone responsible for lowering blood sugar) delivery systems for the treatment of diabetic patients, which is of great significance in human healthcare and modern medicine [56]. Glucose sensors having smart polymers (conducting polymer hydrogels or pH-sensitive hydrogels) provides effective methods for carrying the exact amount of insulin at the time of need [57]. Due to quick responses, good sensitivity, precise detection at low sample volume, and versatile metabolic monitoring glucose sensors are widely used in monitoring the blood glucose level.

4.3.6 BIOMEDICAL FIELD

Utilization of smart polymers in the biomedical field is regarded vital and important aspect among the various developments in this field. Treatment of different diseases through advances in therapies along with the development of highly sophisticated devices that are programmed to react and function in response to an external stimulus or at the surrounding environment of tissues have been achieved. Different behavioral aspects of cells involving its proliferation, adhesion, and migration are accustomed to properties of surface substrate [58]. Polymers regulating the behavioral response with respect to different external stimuli are known as responsive substrates and are regarded to be of great significance in tissue engineering applications [59]. Attaining precise response towards smart surfaces in tissue engineering is achieved by functionalizing the surface of bioactive molecules, nanoscale 3D designing of hydrogels and tunable properties of surfaces. Application of different synthetic polymers in tissue engineering is tabulated in Table 4.1. Hydrogels, the 3D polymer network, are amongst the finest biomaterials used for scaffold preparation and to regulate their cellular alignment proper mechanical or electrical stimulus needs to be provided [60]. Enabling responses in smart hydrogels, physical (light, temperature, and electric), as well as chemical (ionic interactions, pH) stimuli, are necessary [61]. Polymers known for their shape memory are sensitive towards temperature and have been reported for developing medical devices employed in minimal invasive surgery procedures [62]. Introducing the medical device in its compressed form at the place of choice followed by its expansion by minimal invasive surgery is possible only through the unique properties of SMPs. In recent times developing the stents mainly used in urological and vascular procedures are among its most pertinent applications. In addition to the mechanical properties the polymeric stunts are regarded as more beneficial than metallic stunts due to the possible incorporation of a drug and its elution at the functional place; examples include reducing the infections in urological stents or reduction of thrombosis after vascular stent implantation [63].

TABLE 4.1 Uses of Some Smart Polymers in Tissue Engineering

Polymer	Cell/Tissue Type	References
Biomineralized polyurethane	Bone	[77]
Polyurethane + PEG hydrogel	Heart aortic valve	[78]
GelMA + reduced graphene oxide	Cardiac	[79]
Methacrylated gelatin (GelMA) + Gelatin	Bone marrow stem cells	[80]
GelMA + Chitin nanofibers	Vascularization	[81]
PEG-caprolactone (PEG-CAP) hydrogel end-capped with norbornene (PEG-CAP-NOR)	Cartilage	[82]
Poly(N-isopropylacrylamide) PNIPAAm – Gelatin	Cardiac	[83]
PPy with PLA	Peripheral nerve	[84]

Cancer therapy and its diagnosis have been greatly improved by the unique properties of smart polymers. Advances in magnetic resonance imaging diagnosis, cellular labeling, treating hyperthermia, immunoassay, and magnetic separation techniques are developed from magnetic nanoparticles (MNPs) [64]. Smart polymer-based biosensors find their applicability in forensics and clinical diagnosis because of the variations in analytic concentration (glucose in diabetes) or alterations in physical variables viz. pH and temperature (quantifying stomach CO_2 partial pressure by a pH sensor for diagnosing gastrointestinal ischemia) taking place in different diseases [65]. Pooling together biosensors with different medical devices has been employed with cochlear grafts and insulin-delivering devices as a few examples [66]. Identification of systematic concentration of analytes by combining with biosensors is used by medical devices based on microfluidics to discharge various bioactive materials in response to extreme or inadequate concentration of the above-mentioned analytes [67].

The role and use of smart polymers in permitting controlled drug delivery has been accredited to the selectively active treatments involving the transport of a particular drug to its precise place at the exact moment. Specific drug delivery procedure involves four mediums namely nano-fibers, hydrogels, nanoparticles, and assemblies, and lastly, bio-conjugates for successfully targeting the specific tissue [68, 69]. Mostly smart polymers made from poly N-isopropylacrylamide (PNIPAAm) along with its derivatives are used in biomedicine applications. Derivatives of the aforementioned polymers include molecules combining with hydroxyl or amino, ether or amide groups possessing hydrophobicity or in other words a very little affinity towards water. Laboratory synthesis of stimulus sensitive polymers along with well understanding of molecular biology aid in developing explicitly localized, effective, and personalized therapies, like the case of specific degradation of neutrophil elastase at inflammation areas acted upon by neutrophils [70, 71].

Minute variations in temperature cause solubility changes in thermo-sensitive polymers and this thermal-sensitive solubility successfully regulates the drug release rate without drastically compromising the polymer's biological activity and physicochemical stability inside the body [72]. Hydrophobicity of the alkyl groups necessary for solubility variation is important in instituting critical solution temperatures or more elaborately lower and upper critical temperatures. The normal range of human body temperature make upper critical solution temperature systems highly unlikely contenders for drug delivery due to requirements of considerably higher temperatures as the human body fails to provide the same. Changes in the conformation of polymers prominently occur above lower critical solution temperatures (LCST), making the system insoluble and hydrophobic, while as below the LCST, the polymer solubility increases in an aqueous solution [72]. PNIPAAm is an example of a thermo-sensitive polymer having a normal LCST of 32°C that can be successively enhanced towards the normal biological system of humans by addition of surfactants and additives. Possible role of PNIPAAm for *in situ* delivery of drugs is being currently explored by the researchers.

A drooping basic or an acidic group marks its presence in pH-sensitive or responsive polymers that behave in accordance with the pH of surroundings by accepting or donating the protons. Polyelectrolyte polymers possess various ionizable groups characterized further into weak poly bases (discharging protons at high pH and accepting them at neutral and high pH) and weak poly acids (take up protons at low pH and releasing them at high pH and neutral) [73]. High pH inside the human stomach paves the way for the practicality of pH-responsive polymers preventing denaturation of drugs administered orally that would otherwise breakdown in the presence of gastric juices. Oral administration of enteric drugs coated with a layer of polyacid remains in its intact form inside the high acidic environment of the stomach and gets released only inside the basic environment present within small intestines. Release of the drug is encouraged in small intestines where hydrogen bonds surrounding the material of choice break down and is taken up into the blood stream via absorption mechanism. Drugs coated within polybasic carriers are released at low pH environment, whereas at high pH conditions, the enteric coating gets more rigid [74].

Changes in the physical properties of smart polymers with respect to minute variations in electric current are termed as electric field responsive polymers. The majority of these polymers are ionizable polyelectrolytes functioning as acceptors as well as donors of electrons, thereby additionally acting as pH-responsive polymers. The significance of the relationship between electric current and pH change causes disturbances in hydrogen bonding of polymer chains resulting in degradation of polymers, thereby finding application in controlled drug delivery as a smart targeted mechanism for future application. Electrophoresis, diffusion, and forcing via syneresis water are commonly applicable methods of drug delivery involving the use of electric field responsive polymers [72]. Mechanical alteration of piezoelectric polymers develops surface charges without the application of any extra electrodes. Piezoelectric materials can be accustomed for the regeneration of tissues involving cartilage, bone, and neuronal tissue by using them as electroactive scaffolds producing electric signals in reaction to internal forces. Voltage-gated channels in biological cellular membranes receive the generated electric signals resulting in activation of biological events viz. tissue repair [75]. The use of polyhydroxybutyrate-co-valerate (PHBV) in bone tissue engineering is backed by its piezoelectric coefficient that is analogous with the coefficient of natural bone [76]. Light-responsive polymers are regarded as untraditional polymers that experience phase transition upon light exposure. Drugs coated with light-responsive polymers are transferred to the target, followed by the application of a light source enabling targeted delivery of the drug. Polymeric coatings sensitive towards light are regarded as novel along with reliably safe procedures as the active polymers are mainly water-soluble, biocompatible, and totally biodegradable [72]. Incorporation of chromophores or photosensitive molecules in light-responsive polymers is the basis for the functioning of such polymers.

4.4 CONCLUSION

Development of novel materials possessing a robust and more intricate structure from smart polymers needs scientists from multidisciplinary fields to form materials that are capable of undergoing rapid, reversible, physical, or chemical changes in their properties as a consequence of slight variations in their surrounding environment, such as light, temperature, humidity, ionic strength, electrical potential, mechanical stress, polarity, etc. Polymer-conjugated biomolecules, including proteins, polysaccharides, single or double-stranded oligonucleotides, simple lipids, phospholipids, ligands, synthetic drug molecules, etc., can be used in developing smart polymers and smart surfaces responsive to specific external stimuli. Different natural, as well as synthetic materials employed for the development of smart polymers, have both advantages and disadvantages. Therefore, their combination needs a very careful choice of selection. Researchers from polymer, organic, physics, pharmaceutical, biological, organic, and medical fields need to cooperate and work in tandem for effective responses to necessities of civilization related to advancement in quality of life in different fields of life.

KEYWORDS

- **atom-transfer radical polymerization**
- **biopolymer**
- **bio-separation**
- **drug delivery**
- **smart polymers**
- **waste minimization**

REFERENCES

1. Aguilar, C. M. R., Elvira, C., Gallardo, A., Vázquez, B., & San, R. J., (2007). Smart polymers and their applications as biomaterials. In: Ashammakhi, N., Reis, R. L., & Chiellini, E., (eds.), *Topics in Tissue Engineering*. E-book, Expert issues.
2. Arora, G., Singh, I., Nagpal, M., & Arora, S., (2011). Recent advances in stimuli induced pulsatile drug delivery system: A review. *Research Journal of Pharmacy and Technology, 4*, 691–703.
3. Ma, X., Tao, H., Yang, K., Feng, L., Cheng, L., Shi, X., Li, Y., et al., (2012). A functionalized graphene oxide-iron oxide nanocomposite for magnetically targeted drug delivery, photothermal therapy, and magnetic resonance imaging. *Nano Research, 5*, 199–212.

4. Arnold, R. M., Huddleston, N. E., & Locklin, J., (2012). Utilizing click chemistry to design functional interfaces through post-polymerization modification. *Journal of Materials Chemistry, 22*, 19357–19365.
5. Guillerm, B., Monge, S., Lapinte, V., & Robin, J. J., (2012). How to modulate the chemical structure of polyoxazolines by appropriate functionalization. *Macromolecular Rapid Communications, 33*, 1600–1612.
6. Ebara, M., Kotsuchibashi, Y., Narain, R., Idota, N., Kim, Y. J., Hoffman, J. M., et al., (2014). *Introductory Guide to Smart Biomaterials, Smart Biomaterials* (pp. 1–8). National Institute for Materials Science. Springer.
7. Barhoum, A., Bechelany, M., & Hamdy, M. A. S., (2019). *Handbook of Nanofibers: Fundamental Aspects and Experimental Setup, Synthesis, Properties, and Characterization* (Vol. I.). Springer. ISBN 978-3-319-53654-5.
8. Zheng, Y. J., Cui, L. S., & Schrooten, J., (2005). Basic design guidelines for SMA/epoxy smart composites. *Material Science Engineering, 390*, 139–143.
9. Jani, J. M., Leary, M., Subic, A., & Gibson, A. M. A., (2014). A review of shape memory alloy research, applications and opportunities. *Materials and Design, 56*, 1078–1113.
10. Leng, J. S., Zhang, D. W., Yu, K., & Lan, X., (2010). Study on the activation of styrene-based shape memory polymer by medium-infrared laser light. *Applied Physics Letters, 96*, 111–905.
11. Damanpack, A., Aghdam, M., & Shakeri, M., (2015). Micro-mechanics of composite with SMA fibers embedded in metallic/polymeric matrix under off-axial loadings. *European Journal of Mechanics a-Solids, 49*, 467–480.
12. Rinaudo, M., (2006). Chitin and chitosan: Properties and applications. *Progress in Polymer Science, 31*, 603–632.
13. Yen, M. T., & Mau, J. L., (2007). Physico-chemical characterization of fungal chitosan from shiitake stipes. *LWT -Food Science and Technology, 40*, 472–479.
14. Anitha, A., Sowmya, S., Sudheesh, K. P. T., Deepthi, S., Chennazhi, K. P., Ehrlich, H., Tsurkan, M., & Jayakumar, R., (2014). Chitin and chitosan in selected biomedical applications. *Progress in Polymer Science, 39*, 1644–1667.
15. Sudheesh, K. P. T., Lakshmanan, V. K., Biswas, R., Nair, S. V., & Jayakumar, R., (2012). Synthesis and biological evaluation of chitin hydrogel/nano ZnO composite bandage as antibacterial wound dressing. *Journal of Biomedical Nanotechnology, 8*, 1–10.
16. Madhumathi, K., Sudheesh, K. P. T., Abilash, S., Sreeja, V., Tamura, H., Manzoor, K., Nair, S. V., & Jayakumar, R., (2010). Development of novel chitin/nanosilver composite scaffolds for wound dressing applications. *Journal of Material Science. Materials in Medicine*, 807–813.
17. Sudheesh, K. P. T., Abhilash, S., Manzoor, K., Nair, S. V., Tamura, H., & Jayakumar, R., (2010). Preparation and characterization of novel β-chitin/nanosilver composite scaffolds for wound dressing applications. *Carbohydrate Polymers, 80*, 761–777.
18. Chen, M., Yang, Z., Wu, H., Pan, X., Xie, X., & Wu, C., (2011). Antimicrobial activity and the mechanism of silver nanoparticle thermosensitive gel. *International Journal of Nanomedicine, 6*, 2873–2877.
19. Choi, Y. S., Lee, S. B., Hong, S. R., Lee, Y. M., Song, K. W., & Park, M. H., (2001). Studies on gelatin-based sponges: Part III. A comparative study of cross-linked gelatin/alginate, gelatin/hyaluronate and chitosan/hyaluronate sponges and their application as a wound dressing in full-thickness skin defect of rat. *Journal of Material Science: Materials in Medicine, 12*, 67–73.

20. Salomäki, M., & Kankare, J., (2008). Influence of synthetic polyelectrolytes on the growth and properties of hyaluronan-chitosan multilayers. *Biomacromolecules, 10*, 294–301.
21. Munarin, F., Tanzi, M. C., & Petrini, P., (2012). Advances in biomedical applications of pectin gels. *International Journal of Biological Macromolecules, 51*, 681–689.
22. Mishra, R. K., Banthia, A. K., & Majeed, A. B. A., (2012). Pectin based formulations for biomedical applications: A review. *Asian Journal of Pharmaceutical and Clinical Research, 5*, 1–7.
23. Coimbra, P., Ferreira, P., De Sousa, H. C., Batista, P., Rodrigues, M. A., Correia, I. J., & Gil, M. H., (2011). Preparation and chemical and biological characterization of a pectin/chitosan polyelectrolyte complex scaffold for possible bone tissue engineering applications. *International Journal of Biological Macromolecules, 42*, 112–118.
24. Dafe, A., Etemadi, H., Dilmaghani, A., & Mahdavinia, G. R., (2017). Investigation of pectin/ starch hydrogel as a carrier for oral delivery of probiotic bacteria. *International Journal of Biological Macromolecules, 97*, 536–543.
25. Izadi, Z., Divsalar, A., Saboury, A. A., & Sawyer, L., (2016). β-Lactoglobulin-pectin nanoparticle- based oral drug delivery system for potential treatment of colon cancer. *Chemical Biology and Drug Design, 88*, 209–216.
26. Deangelis, P. L., (2012). Glycosaminoglycan polysaccharide biosynthesis and production: Today and tomorrow. *Applied Microbiology and Biotechnology, 94*, 295–305.
27. Zacchi, V., Soranzo, C., Cortivo, R., Radice, M., Brun, P., & Abatangelo, G., (1997). In vitro engineering of human skin-like tissue. *Journal of Biomedical Materials Research, 40*, 187–194.
28. Tan, H., Chu, C. R., Payne, K. A., & Marra, K. G., (2009). Injectable in situ forming biodegradable chitosan-hyaluronic acid based hydrogels for cartilage tissue engineering. *Biomaterials, 30*, 2499–2506.
29. Liu, L. S., Thompson, A. Y., Heidaran, M. A., Poser, J. W., & Spiro, R. C., (1999). An osteoconductive collagen/hyaluronate matrix for bone regeneration. *Biomaterials, 20*, 1097–1108.
30. Oliveira, J. T., Martins, L., Picciochi, R., Malafaya, P. B., Sousa, R. A., Neves, N. M., Mano, J. F., & Reis, R. L., (2010). Gellan gum: A new biomaterial for cartilage tissue engineering applications. *Journal of Biomedical Materials Research, 93*, 852–863.
31. Jeon, H. Y., Shin, E. Y., Choi, J. H., Song, J. E., Reis, R. L., & Khang, G., (2018). Evaluation of saponin loaded gellan gum hydrogel scaffold for cartilage regeneration. *Macromolecular Research, 26*, 724–729.
32. Kumar, S., Kaur, P., Bernela, M., Rani, R., & Thakur, R., (2016). Ketoconazole encapsulated in chitosan-gellan gum nanocomplexes exhibits prolonged antifungal activity. *International Journal of Biological Macromolecules, 93*, 988–994.
33. Pereira, D. R., Silva-Correia, J., Caridade, S. G., Oliveira, J. T., Sousa, R. A., Salgado, A. J., Oliveira, J. M., et al., (2011). Development of gellan gum-based microparticles/ hydrogel matrices for application in the intervertebral disc regeneration. *Tissue Engineering, Part C, 17*, 961–972.
34. Jeon, S. H., Lee, W. T., Song, J. E., Park, H., Choi, I. N., Kim, C. M., & Khang, G., (2017). Cartilage regeneration using hesperidin-containing gellan gum scaolds. *Polymer (Korea), 41*, 670–674.
35. Kim, W. K., Choi, J. H., Shin, M. E., Kim, J. W., Kim, P. Y., Kim, N., Song, J. E., & Khang, G., (2019). Evaluation of cartilage regeneration of chondrocyte encapsulated

gellan gum-based hyaluronic acid blended hydrogel. *International Journal of Biological Macromolecules, 141*, 51–59.
36. Clark, D. E., & Green, H. C., (1936). *Alginic Acid and Process of Making Same*. US Patent. 2036922.
37. Mumper, R. J., Huffman, A. S., Puolakkainen, P. A., Bouchard, L. S., & Gombotz, W. R., (1994). Calcium-alginate beads for the oral delivery of transforming growth factor-β1 (TGF-β1): Stabilization of TGF-β1 by the addition of polyacrylic acid within acid treated beads. *Journal of Controlled Release, 30*(3), 241–251.
38. Horncastle, J., (1945). Wound dressings. Past, present, and future. *Medical Device Technology, 6*(1), 30–34.
39. Von, B. H., Lajous-Petter, A., Debrandt, W., Hampl, H., Kochinke, F., & Herbst, R., (1988). Surface reactions on blood contact during haemodialysis and haemofiltration with various membrane types. *Journal of Membrane Science, 36*, 215–229.
40. Shoda, M., & Sugano, Y., (2005). Recent advances in bacterial cellulose production. *Biotechnology and Bioprocess Engineering, 10*, 1–8.
41. Mohite, B. V., & Patil, S. V., (2014). Physical, structural, mechanical and thermal characterization of bacterial cellulose by *G. hansenii* NCIM, 2529. *Carbohydrate Polymers, 106*, 132–141.
42. Thomas, S., Durand, D., Chassenieux, C., & Jyotishkumar, P., (2013). *Handbook of Biopolymer-Based Materials: From Blends and Composites to Gels and Complex Networks*. John Wiley and Sons.
43. Bogolitsin, K. G., Lunin, V. V., & Kosyakov, D. S., (2010). *Fizicheskaya Khimiya Lignina (Physical Chemistry of Lignin)*. Moscow: Akademkniga.
44. Li, H., & McDonald, A. G., (2014). Fractionation and characterization of industrial lignins. *Industrial Crops and Products, 62*, 67–76.
45. Wischerhoff, E., Badi, N., & Lutz, J. F., (2010). Smart polymer surfaces: Concepts and applications in biosciences. *Advances in Polymer Science, 240*, 1–33.
46. Galaev, I. Y., & Mattiasson, B., (2002). *Smart Polymers for Bioseparation and Bioprocessing*. London: Taylor and Francis.
47. Schneider, M., Guillot, C., & Lamy, B., (1981). The affinity precipitation techniques: Application to the isolation and purification of trypsin from bovine pancreas. *Annals of the New York Academy of Sciences, 369*, 257–263.
48. Gisser, K. R. C., geselbracht, M. J., Capellari, A., Hunsberger, L., Ellis, A. B., Perepezko, J., & Lisensky, G. C., (1994). Nickel-titanium memory metal. A smart material exhibiting a solid-state phase change and super elasticity. *Journal of Chemical Education, 71*, 334–340.
49. Hoffman, A. S., (2000). Bioconjugation of intelligent polymers and recognition proteins for use in diagnostics and affinity separation. *Clinical Chemistry, 46*, 1478–1486.
50. Roy, I., & Gupta, M. N., (2006). Design of smart biocatalysts: Immobilization of enzymes on smart polymers. In: Guisan, J. M., (ed), *Immobilization of Enzymes & Cells* (Vol. 22, pp. 87–95). Humana Press Inc., New York.
51. Gupta, M. N., (2000). *Methods in Non-Aqueous Enzymology*. Switzerland: Birkhauser Verlag.
52. Taniguchi, M., Kobayashi, M., & Fujii, M., (1989a). Properties of a reversible soluble-insoluble cellulose and its application to repeated hydrolysis of crystalline cellulose. *Biotechnology and Bioengineering, 34*, 1092–1097.
53. Hoshino, K., Taniguchi, M., Marumoto, H., & Fujii, M., (1989). Repeated batch conversion of raw starch to ethanol using amylase immobilized on reversible soluble- auto-precipitating carrier and flocculating yeast cell. *Agricultural and Biological Chemistry, 53*, 1961–1967.

54. Kakhki, M. R., (2015). Application of magnetic nanoparticles modified with cyclodextrins as efficient adsorbents in separation systems. *Journal of Inclusion Phenomena and Macrocyclic Chemistry, 82*, 301–310.
55. Euvrard, E., Morin-Crini, N., Druart, C., Bugnet, J., Martel, B., Cosentino, C., Moutarlier, V., & Crini, G., (2016). Cross-linked cyclodextrin-based material for treatment of metals and organic substances present in industrial discharge waters. *Beilstein Journal of Organic Chemistry, 12,* 1826.
56. Nambiar, S., & Yeow, J. T., (2011). Conductive polymer-based sensors for biomedical applications. *Biosensors and Bioelectronics, 26*, 1825–1832.
57. Zhao, Y., Cao, Y., Li, L., Cheng, W., Xu, L., Ping, X., Pan, L., & Shi, Y., (2016). Conducting polymers and their applications in diabetes management. *Sensors (Basel), 16*, 1787.
58. Alves, D. N. M., Pashkuleva, I., Reis, R. L., & Mano, J. F., (2010). Controlling cell behavior through the design of polymer surfaces. *Small, 6*, 2208–2220.
59. Perez, R. A., Won, J. E., Knowles, J. C., & Kim, H. W., (2013). Naturally and synthetic smart composite biomaterials for tissue regeneration. *Advanced Drug Delivery Reviews, 65*, 471–496.
60. Wu, Y., Wang, L., Guo, B., & Ma, P. X., (2017). Interwoven aligned conductive nanofiber yarn/hydrogel composite scaffolds for engineered 3D cardiac anisotropy. *ACS Nano, 11*(6), 5646–5659.
61. Eslahi, N., Abdorahim, M., & Simchi, A., (2016). Smart polymeric hydrogels for cartilage tissue engineering: A review on the chemistry and biological functions. *Biomacromolecules, 17*(11), 3441–3463.
62. Yakacki, C. M., & Gall, K., (2010). Shape-memory polymers for biomedical applications. *Advances in Polymer Science, 226*, 147–175.
63. Xue, L., Dai, S., & Li, Z., (2012). Synthesis and characterization of elastic star shape-memory polymers as self-expandable drug-eluting stents. *Journal of Materials Chemistry, 22*, 7403–7411.
64. Karimi, Z., Karimi, L., & Shokrollahi, H., (2013). Nano-magnetic particles used in biomedicine: Core and coating materials. *Materials Science Engineering, 33*, 2465–2475.
65. Herber, S., Bomer, J., Olthuis, W., Bergveld, P., & Van, D. B. A., (2005). A miniaturized carbon dioxide gas sensor based on sensing of pH-sensitive hydrogel swelling with a pressure sensor. *Biomedical Microdevices, 7*, 197–204.
66. Laursen, W., (2006). Breaking the sound barrier [cochlear implants]. *Engineering Technology, 1*, 38–41.
67. Do, J., Lee, S., Han, J., Kai, J., Hong, C. C., Gao, C., Nevin, J. H., Beaucage, G., & Ahn, C. H., (2008). Development of functional lab-on-a-chip on polymer for point-of-care testing of metabolic parameters. *Lab on a Chip, 8*, 2113–2120.
68. Abdal-hay, A., Agour, M., Kim, Y. K., Lee, M. H., Hassan, M. K., El-Ainin, H. A., et al., (2019). Magnesium-particle/polyurethane composite layer coating on titanium surfaces for orthopedic applications. *European Polymer Journal, 112*(3), 555–568.
69. Asadi, N., Del, B. A. R., Davaran, S., & Akbarzadeh, A., (2019). Common biocompatible polymeric materials for tissue engineering and regenerative medicine. *Materials Chemistry and Physics*, https://doi.org/10.1016/j.matchemphys.2019 122528.
70. Aimetti, C. A. A., Machen, A. J., & Anseth, K. S., (2009). Poly (ethylene glycol) hydrogels formed by thiol-ene photopolymerization for enzyme-responsive protein delivery. *Biomaterials, 30*, 6048–6054.

71. Fleige, E., Quadir, M. A., & Haag, R., (2012). Stimuli-responsive polymeric nanocarriers for the controlled transport of active compounds: Concepts and applications. *Advance Drug Delivery Reviews, 64*, 866–884.
72. Priya, J. H., Rijo, J., Anju, A., & Anoop, K. R., (2014). Smart polymers for the controlled delivery of drugs-a concise review. *Acta Pharmaceutica Sinica, 4*(2), 120–127.
73. Hamdy, M. A. S., & Abu-Thabit, N. Y., (2018). *Stimuli-Responsive Polymeric Nanocarriers for Drug Delivery Applications: Types and Triggers* (Vol. 1.). Elsevier. ISBN: 9780081019979.
74. Abu-Thabit, A. N. Y., & Hamdy, M. A. S., (2018). Historical development of drug delivery systems: From conventional macroscale to controlled, targeted and responsive nanoscale systems. In: Makhlouf, A. S. H., & Abu-Thabit, N. Y., (eds.), *Stimuli Responsive Polymeric Nanocarriers for Drug Delivery Applications: Types and Triggers* (Vol. 1, pp. 1–41). Elsevier.
75. Abzan, A. N., Kharaziha, M., & Labbaf, S., (2019). Development of three-dimensional piezoelectric polyvinylidene fluoride-graphene oxide scaffold by non-solvent induced phase separation method for nerve tissue engineering. *Materials and Design, 167*, 107636.
76. Gorodzha, S. N., Muslimov, A. R., Syromotina, D. S., Timin, A. S., Tcvetkov, N. Y., Lepik, K. V., et al., (2017). A comparison study between electrospun polycaprolactone and piezoelectric poly (3-hydroxybutyrate-co-3-hydroxyvalerate) scaffolds for bone tissue engineering. *Colloids and Surfaces B: Biointerfaces, 160*, 48–59.
77. Meskinfam, M., Bertoldi, S., Albanese, N., Cerri, A., Tanzi, M. C., Imani, R., et al., (2018). Polyurethane foam/nano-hydroxyapatite composite as a suitable scaffold for bone tissue regeneration. *Materials Science and Engineering, 82*, 130–140.
78. Puperi, D. S., Kishan, A., Punske, Z. E., Wu, Y., Cosgriff-Hernandez, E., West, J. L., et al., (2016). Electrospun polyurethane and hydrogel composite scaffolds as biomechanical mimics for aortic valve tissue engineering. *ACS Biomaterials Science and Engineering, 2*(9), 1546–1558.
79. Shin, S. R., Zihlmann, C., Akbari, M., Assawes, P., Cheung, L., Zhang, K., et al., (2016). Reduced graphene oxide-gelMA hybrid hydrogels as scaffolds for cardiac tissue engineering. *Small, 12*(27), 3677–3689.
80. Yin, J., Yan, M., Wang, Y., Fu, J., & Suo, H., (2018). 3D bioprinting of low-concentration cell-laden gelatin methacrylate (GelMA) bioinks with a two-step cross-linking strategy. *ACS Applied Materials and Interfaces, 10*(8), 6849–6857.
81. Hassanzadeh, P., Kazemzadeh-Narbat, M., Rosenzweig, R., Zhang, X., Khademhosseini, A., Annabi, N., et al., (2016). Ultrastrong and flexible hybrid hydrogels based on solution self-assembly of chitin nanofibers in gelatin methacryloyl (GelMA). *Journal of Materials Chemistry B, 4*(15), 2539–2543.
82. Neumann, A. J., Quinn, T., & Bryant, S. J., (2016). Nondestructive evaluation of a new hydrolytically degradable and photo-clickable PEG hydrogel for cartilage tissue engineering. *Acta Biomaterialia, 39*, 1–11.
83. Navaei, A., Truong, D., Heffernan, J., Cutts, J., Brafman, D., Sirianni, R. W., et al., (2016). PNIPAAm-based biohybrid injectable hydrogel for cardiac tissue engineering. *Acta Biomaterialia, 32*, 10–23.
84. Zhou, J. F., Wang, Y. G., Cheng, L., Wu, Z., Sun, X. D., & Peng, J., (2016). Preparation of polypyrrole-embedded electrospun poly (lactic acid) nanofibrous scaffolds for nerve tissue engineering. *Neural Regeneration Research, 11*(10), 1644.

CHAPTER 5

A COMPUTATIONAL STUDY OF METAL DOPED SILVER NANOPARTICLE-BASED CLUSTERS

SHAYERI DAS[1], PRABHAT RANJAN[1], and TANMOY CHAKRABORTY[2]

[1]*Department of Mechatronics Engineering, Manipal University Jaipur, Dehmi Kalan, Rajasthan, India*

[2]*Department of Chemistry and Biochemistry, School of Basic Sciences and Research, Sharda University, Greater Noida, Uttar Pradesh, India*

ABSTRACT

Transition metal-based nanoclusters have made a huge impact on applications in solid-state devices, solar cells, biotechnology, etc. In this report, the properties of nano alloy clusters Ag_nFe (n = 1–5) are studied using Gaussian 16 within the Density Functional Theory (DFT) framework. The ground state configurations have been used for the computation of the DFT-based global descriptors. All the clusters exhibit an odd-even oscillation of the computed descriptors. The bond length of these clusters is validated by their experimental values. The results show that the nanoclusters, due to the substantial highest occupied-lowest unoccupied molecular orbital gap, have good prospects in biomedical applications. Specifically, this study proposes that the nanoclusters Ag_nFe (n = 1–5) are appropriate for future

Nanotechnology-Enhanced Solid Materials: Design, Synthesis, Properties, Applications, and Perspectives. Lionello Pogliani, Ann Rose Abraham, A. K. Haghi, & Prabhat Ranjan (Eds.)

applications in biomedicine, since they could improve some contemporary difficulties of some nanoparticles, especially toxicity, as silver is non-toxic in nature. Similarly, Fe, also being non-toxic in nature, adds to the diversity in application.

5.1 INTRODUCTION

Among all transition metals, the noble metals, mainly copper, silver, and gold, have been minutely studied predominantly due to their significant deviation of nanoparticle properties from larger materials [1]. Both mono-metallic and bimetallic nanoclusters are the subjects of intense research, from mutually computational and experimental vantage point [2–4] and technical applications in an assortment of specializations have been projected due to the prevailing methodologies to manufacture functional nanoparticles of commercial importance [5–7]. It has been reported to be realistic to implement their physicochemical characteristics for explicit purposes, such as in medication [8], catalysis [9], food technology [10], and electronics, specifically in semiconductor devices [11, 12].

Specifically, silver nanoparticles have drawn huge interest for biomedical applications [13–15] as antimicrobial agents. Silver nanoclusters possess extremely less toxicity and are predominantly used in industries like health [16, 17], cosmetics and beauty products [18, 19], and other environmental applications [20–22].

Pereiro et al. [23] studied the biomedical application of pure silver clusters and observed its effects based on various-sized clusters. They reported the magnetic properties of those nanoparticles and their suitability in biomedicine, by eradicating difficulties like toxicity and opsonization. Roy Chowdhury et al. [16] synthesized silver nanoparticles and generated stable antimicrobial clusters. The researchers focused on plant-derived and biomolecule-based synthesis to avoid harmful chemicals. They reported that the cacao-based method generated stable Ag nanoparticles, which have applications in the biomedical field. Tavaf et al. [17] studied the antibacterial and other biomedicine-based activities of silver particles and reported that the silver nanoparticles with spherical and similar contours possess significant antibacterial, anti-biofilm, and anti-oxidant activities. Lin et al. [24] studied the synthesis of metallic nanoclusters. They reported that the nanoclusters are more bio-compatible and stable than organic dyes. The optical properties obtained render the clusters to be ideal for various applications in biomedical engineering and also molecular biology. Mathew et al. [25] stated that the novel properties which are a result of size quantization, for instance, luminescence open up new fields of application. They reviewed the possibilities that exclusive abilities and biocompatibility make the nanoclusters favorable contenders for the development of next-generation quantum medicine for disease diagnosis and treatment. Castro [26] studied the gold super atom clusters and reported their tunable properties like electronic, optical, and chiroptical properties. The authors reported the modulation or properties by different ligands for tunable clusters. Zhang et al. [27] performed a first principle investigation on noble metal clusters incorporating silver and gold. Based on the tunable optical properties, they reported that the outcome of the computed dielectric function indicated that the optical transition of Au clusters is reliant on concentration and the optical transition between HOMO and LUMO alters to the lower energy range as the number of Ag atom increases. These clusters are suitable for cell imaging, photothermal therapy as well as radiotherapy. Thakore et al. [28] efficiently synthesized silver and other nanoparticles using fruit latex. Comparative *in-vitro* toxicity has been observed by them, and it was deduced that the nanoclusters had good prospect in cellular imaging or photothermal therapy.

Kiruba et al. [29] reported an eco-friendly amalgamation of Au–Ag bimetallic nanoparticles using gripe water, where the synthesized Au–Ag nanoparticles have applications in surface-enhanced Raman spectroscopic and have been studied to heighten the Raman signals of rhodamine 6G. Sun et al. [30] reported a study of gold-doped iron nanoclusters, in order to utilize the magnetic properties in biomedical applications. They deduced that the stability of gold prevented oxidation and can be used for targeted drug delivery. Burdusel et al. [31] investigated silver nanoparticles and explored their nanostructures. They were motivated for their work on the most recent data concerning the biomedical application of Ag nanostructures, comprising of aspects associated to their potential toxicity, unique physiochemical properties, and biofunctional behaviors, and discussed the intrinsic biomedicine-based features of Ag-based nanostructures. It has been reported that modifications to contemporary biomaterials along with devices can be achieved using silver nanoparticles. Zhang et al. [32] reported that notable antimicrobial activity of Ag nanoparticles presents with a strong starting point for the design, development, and implementation of new and performance-biomedical products including target-based medical devices and supporting devices. Wei et al. [33] stated that the translation of Ag-based nanotechnology into clinical applications involves not only the improvement of viable methods for the synthesis of silver nanoparticles, but also a systematic comprehension of the related physicochemical particularities, *in vitro* and *in vivo* effects, biodistribution, safety control mechanisms, pharmacokinetics, and pharmacodynamics of silver nanoparticles. Yaqoob et al. [34] studied the biomedical applications of various transition metals like gold, silver, and palladium and reported that due to cytotoxicity further research is viable in this domain. The group suggested that metal-based nanoparticles can overcome the limitations of drug resistance as these clusters are applicable in the fabrication of more progressive binding affinity of biomolecules for detection of life-threatening diseases. Sanpo et al. [35] reported the biomedical applications of transition metal-substituted cobalt ferrite clusters and high-lighted the antibacterial properties of cobalt ferrite. They performed various analysis and indicated that transition metals strongly vary the microstructure and the crystal structure.

Chen et al. [36] reported the magnetic properties of Au-coated Fe nanoparticles. The group studied the morphology and magnetic properties using TEM and alternative gradient magnetometry. Gupta et al. [37] conducted a study on various iron oxide (IO) nanoparticles. The study suggests that the MNPs have the ability to bind to drugs, proteins, etc., and can be used in target-based cell therapy like tissue, tumor, etc. Figuerola et al. [38] reported the advancement of IO nanoalloys towards iron-based inorganic designs. They described the inorganic hybrid nanostructures of different materials as diagnostic and therapeutic agents. Lim et al. [39] stated the effects of iron and aluminum on graphene and reported about the enhanced catalytic reactivity of iron upon adsorption. Mahtout et al. [40] reported a study of iron-doped silicon clusters using DFT, where they found that system with spin-up electrons have high energy gap than the spin-down electrons. Huang et al. [41] fabricated iron-doped transition metal clusters and reported the variations in their structure and properties by using XRD, immersion tests and other methods. The cytotoxicity evaluation showed none of the materials exhibited significant toxicity. Janssens et al. [42] carried out DFT computation on single-doped Au_5X^+ systems (X=Au, Sc, Ti, Cr, Fe). The authors reported that the delocalized electrons depend on the constituting atoms as well as the equilibrium geometry of the cluster. The configurational, electronic, and magnetic features of Fe_n clusters (n=2–7) doped in icosahedral C_{60} and C_{80} fullerenes using first-principles calculations based on the DFT has been performed by Bezi Javan et al. [43] where they reported that magnetic moment of the larger clusters reduce with stronger hybridization.

Li et al. [44] studied the structure and electronic properties of Fe clusters where they showed that in the decomposition process the pentacarbonyl-iron molecule gives preference to remove CO and produce more stable molecules. They have reported that the HOMO-LUMO gap have higher stability than their neighboring atoms. Khanna et al. [45] reported about the geometry, with various features of Fe encapsulated Si_n (n= 9–11) atoms using the DFT and have global gradient approximation for exchange and correlation. The group studies the magnetic moment of all the clusters in the system. Singh et al. [46] conducted a study on magnetic moments of Fe, Co, and Ni and observed the total magnetic moment with respect to ground state structure using ab initio calculations. They stated that the atomic clusters are half-metallic as they exhibit very small HOMO-LUMO energy gap for minority spin in comparison to majority spin component. Kumar et al. [47] reported that metal-doped Si clusters, one of the dopants being iron depending upon the cluster size forms fullerene-like structures. Ma et al. [48] reported Fe-doped Si clusters Si_nFe (n = 1–14), they found that polarizability of system is enhanced after doing with iron. The outcomes of ab initio calculations for Ni_N (N = 2–16) and Fe_8Ni_5 clusters using DFT within the generalized gradient approximation have been reported by Sahoo et al. [49]. Postnikov et al. [50] stated that the largest clusters of Fe were relaxed, under a particular morphology type, the study of which they conducted using DFT. They found that the relaxation in cluster surface was accompanied by the enhancement of local magnetic moments. Ebert et al. [51] found a pronounced enhancement for the spin as well as the orbital moments of Fe and FePt clusters and obtained satisfactory results with available experimental counterpart. Košuth et al. [52] reported the electronic and magnetic properties of ferromagnetic semiconductor systems where they were able to check the magnetic moment with the magneto-crystalline anisotropy energy. Li et al. [53] investigated structures, electronic properties, magnetic moments, and growth strategies of Fe_nN clusters varying number of Fe atoms from 1 to 7, using DFT. The results indicated that nitrogen doping significantly alters the Fe_n clusters. Zhao et al. [54] reported a DFT approach of Fe-doped germanium clusters where they found that Fe dopants contribute to the reinforcement of the stability of the germanium framework. The HOMO-LUMO gaps were reduced due to the presence of Fe atoms in germanium clusters. Yu et al. [55] examined pure iron clusters, Fe_n (n = 1–8) by using DFT. They reported that the ground state structures of iron clusters prefer higher dimensional arrangement as well as coordination number.

These prospects of silver and iron in various fields have led to this computational study where Ag_nFe (n = 1–5) has been studied using DFT based descriptors where their electronic properties along with other DFT based descriptors have been computed and observed. This report has been presented in such a way that Section 5.2 provides a brief account of the computational methodology. Additionally, the subsequent sections provide the computational results along with the corresponding discussions. The entirety of the study has been briefed in Section 5.4.

5.2 COMPUTATIONAL DETAILS

Among several computational techniques, Density Functional Theory (DFT) has gained importance for being computational friendly. Various fields of study like alloy-based clusters, fluid mechanics, molecular, and nuclear physics, life sciences, etc., have adopted this technique [56–61]. DFT methods have been successfully used in various cases [62–65]. In the past few years, our group has performed various studies on metallic clusters [66–73] based on DFT.

In this report a theoretical investigation based on the transition metal-based clusters, Ag_nFe, where (n = 1–5) has been conducted. Optimization of the clusters have been instigated by means of Gaussian 16 [74] using DFT framework. For the optimization purpose, Local Spin Density Approximation, i.e., LSDA exchange-correlation with basis set LanL2dz is implemented. LSDA is a proven correlation for theoretical analysis of clusters [75, 76] whereas the basis set offers high accuracy in case of metallic clusters [75, 76].

Ionization energy (I) and electron affinity (A) of all the alloys are determined via the following equations [77]:

$$I = -\varepsilon_{HOMO} \tag{1}$$

$$A = -\varepsilon_{LUMO} \tag{2}$$

By means of values of 'I' and 'A,' which are the conceptual descriptors, molecular hardness (η), molecular softness (S), electronegativity (χ) and electrophilicity index (ω) have been computed. The equations based on which computation have been performed have been presented in Eqns. (3) to (6):

$$\chi = -\mu = \frac{I + A}{2} \tag{3}$$

here; μ symbolizes the chemical potential of the cluster:

$$\eta = \frac{I - A}{2} \tag{4}$$

$$S = \frac{1}{2\eta} \quad (5)$$

$$\omega = \frac{\mu^2}{2\eta} \quad (6)$$

The bond order parameter of Ag_nFe (n = 1–5) cluster is defined as [78]:

$$\sigma = (N_{Ag\text{-}Ag} + N_{Fe\text{-}Fe} - N_{Ag\text{-}Fe})/(N_{Ag\text{-}Ag} + N_{Fe\text{-}Fe} + N_{Ag\text{-}Fe}) \quad (7)$$

where; $N_{Ag\text{-}Ag}$, $N_{Fe\text{-}Fe}$, and $N_{Ag\text{-}Fe}$ are the number of nearest-neighbor bonds between Ag-Ag, Fe-Fe, and Ag-Fe, respectively.

5.3 RESULTS AND DISCUSSION

5.3.1 EQUILIBRIUM GEOMETRY

The ground state or lowest energy structures of Ag_nFe (n = 1–5) has been presented in Figure 5.1, along with the point groups and spin multiplicity. Figure 5.1 also shows the bond lengths between all the atoms in every structure. Bond length is elucidated as the distance between the centers of two atoms that are covalently bonded atoms. The length of the bond is confirmed using the total number of bonded electrons. The increase in the bond order, leads to the stronger attraction between the two atoms and results in shorter magnitude of bond length. In general, the length of the bond between two atoms is almost equal to the sum of the covalent radii of the atoms. Bond length is represented in Angstrom (Å).

AgFe displays ground state configuration is a linear structure presented in Figure 5.1(a) with Ag atom linked in a line with Fe atom. The orientation has the symmetry group $C_{\infty V}$ and doublet spin multiplicity. The bond length of Ag-Fe is 2.378 Å.

Ag_2Fe has straight line ground state structure, presented in Figure 5.1(b), which has Fe placed at one of the ends. The structure comprises of point group of $C_{\infty Vs}$ and triplet spin multiplicity. The bond length of Ag-Ag is 2.680 Å and that of Ag-Fe is 2.320 Å.

Ag_3Fe in Figure 5.1(c) represents structural modification of a quadrilateral with the diagonals connected to each other, exhibiting point group C_{2v} and doublet spin multiplicity. In this structure, the bond lengths vary diagonally as well as at the vertices. The Ag-Ag bonds connected directly has bond length 2.619 Å and that connected diagonally has bond length 3.471 Å. Similarly, the Ag-Fe atoms which are directly placed at corresponding vertices possess a bond length of 2.463 Å and the diagonal bond length is 3.312 Å.

The structure possessing the lowest energy for Ag_4Fe is presented in Figure 5.1(d), which is the most stable structure, point group C_{2v} and quintet spin multiplicity. The structure resembles a trapezium-like shape. The Ag-Ag bond length at the non-parallel sides is both 2.674 Å. The Ag-Ag bond length at corresponding side is 2.713 Å. The Fe atom in placed in the center of the base and the four Ag atoms are linked with it. The Ag atoms on the same side create Ag-Fe bonds of the length 2.609 Å and that on the opposite arm have bond lengths 2.459 Å.

Figure 5.1(e) shows the ground state structure of Ag_5Fe which has point group C_{2v} and multiplicity detect Å. The structure is hexagonal in shape with Fe atom being centrally located. All the Ag atoms are directly connected with the Fe atom. The Ag-Ag bonds created with the Ag atom at the topmost vertex has bond length 2.528 Å. The Ag-Ag bonds on the parallel sides have bond length 2.578 Å. Ag-Fe bond length with Ag atoms at the base side is 2.703 Å, with the top vertex is 2.664 Å and the remaining atoms on the parallel side is both 2.598 Å.

The reported experimental bond length of Ag-Ag atoms is 2.530 Å [79]. The computed data obtained is in range with the experimental value as the shortest bond length is approximately equal to the experimental value with a mere difference of 0.002 Å. The bond length between silver atoms follows odd-even oscillations where for structures having an odd number of total atoms, the bond lengths are larger in value. The shortest bond length of Ag-Ag bond is 2.528 Å which is exhibited by Ag_5Fe and the longest bond length for the pair is 3.471 Å which is possessed by Ag_3Fe. In case of Ag-Fe the bond length remains same for both AgFe and Ag_2Fe, which is also the shortest bond length of 2.378 Å. The highest bond length for Ag-Fe is 3.312 Å. The bond lengths then keep on increasing with the total number of atoms. In the entire series of clusters, the Ag_3Fe cluster is an exception. This cluster breaks the pattern of change in bond length and exhibits highest bond length for both Ag-Ag and Ag-Fe atom sets. The reported bond length of Ag with Au is 2.500 Å [80]. So, it can be inferred that by substituting Au with Fe, the bond length with Ag does not vary to a huge extent. The transition metals thus exhibit similar bond lengths even when substituted with another of its kind. The change in bond lengths with an increase in cluster size has been presented in Table 5.1.

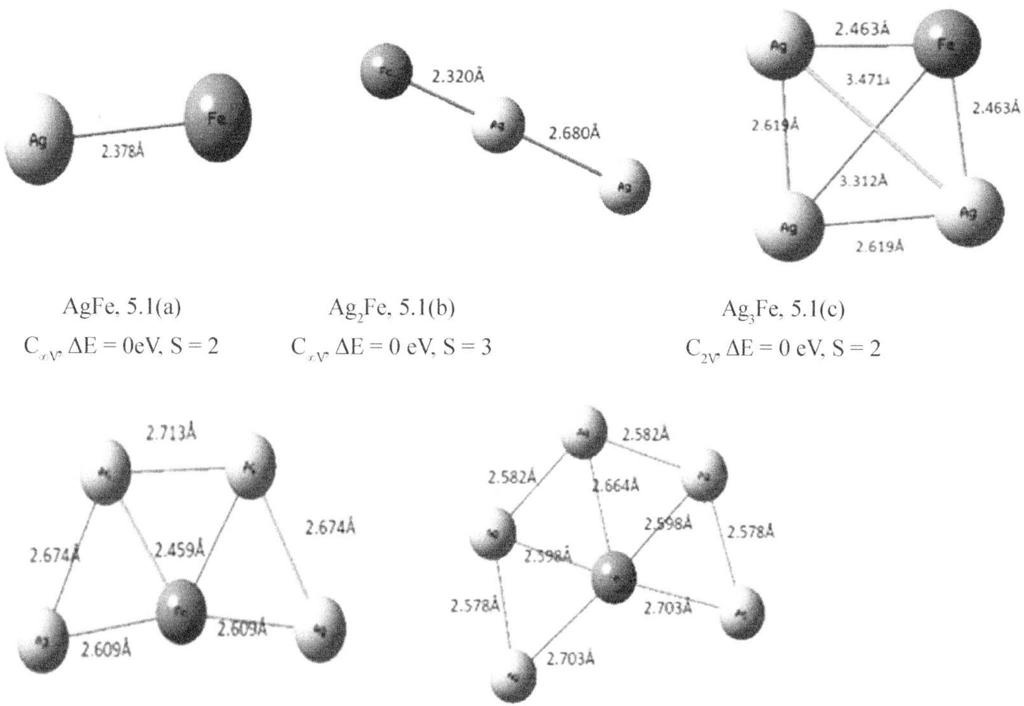

AgFe, 5.1(a)
$C_{\infty V}$, $\Delta E = 0eV$, $S = 2$

Ag_2Fe, 5.1(b)
$C_{\infty V}$, $\Delta E = 0$ eV, $S = 3$

Ag_3Fe, 5.1(c)
C_{2V}, $\Delta E = 0$ eV, $S = 2$

Ag_4Fe, 5.1(d)
C_{2V}, $\Delta E = 0$ eV, $S = 5$

Ag5Fe, 5.1(e)
C2V, $\Delta E = 0$ eV, $S = 10$

FIGURE 5.1 Lowest energy structures of Ag_nFe (n = 1–5).

TABLE 5.1 Bond Lengths of Ag-Ag Atoms in Ag_nFe (n = 1–5) Clusters

Cluster	Ag-Ag Bond Lengths in Structure
AgFe	None
Ag_2Fe	2.680 Å
Ag_3Fe	2.619 Å, 3.471 Å
Ag_4Fe	2.674 Å, 2.713 Å
Ag_5Fe	2.528 Å, 2.578 Å

Bond order is another integral part of the atomic structure. Bond order determines the stability of a bond between a pair of electrons. Both bond length and bond order influence the type and strength of the bond formed. They vary inversely to each other, i.e., when bond order increases, the bond length will decrease. Bond order provides a quantitative measure of segregation in nanoalloy clusters [81–85]. The values of bond order can be expressed in positive, zero or negative values. When the value of bond order is positive, the structure possesses segregation of atoms in their respective clusters. The bond order zero signifies that the concerned structure is a disorderly mixed cluster and if the value is negative in magnitude, then the species has mixed phases of clusters in it.

The bond order of the clusters has been computed using Eqn. (7). The tabulated representation of bond order of all the clusters under study is in Table 5.2 of this report.

TABLE 5.2 Bond Orders of Ag_nFe (n = 1–5) Clusters

Species	Bond Order
AgFe	–1
Ag_2Fe	0.5
Ag_3Fe	0.5
Ag_4Fe	0.43
Ag_5Fe	0.44

Comparing the computed data with criteria of bond order variations, we can state that AgFe is in a mixed phase of clusters and the rest of the clusters contain segregation of atoms in them.

5.3.2 *ELECTRONIC PROPERTIES AND DFT-BASED DESCRIPTORS*

Using electronic structure theory, the computational study of transition metallic silver-iron cluster is performed. The global descriptors which are based on density functional theory like molecular hardness, molecular softness, electronegativity, and electrophilicity are evaluated using Eqns. (3) to (6). The same is presented in Table 5.3.

TABLE 5.3 DFT-based Global Descriptors of Ag_nFe (n = 1–5) Clusters

Species	HOMO-LUMO Gap (in eV)	Hardness (eV)	Softness (eV)	Electronegativity (in eV)	Electrophilicity Index (in eV)	Dipole Moment (in Debye)
AgFe	0.396	0.198	2.526	4.507	51.306	0.791
Ag_2Fe	1.653	0.827	0.605	4.725	13.504	1.025
Ag_3Fe	0.519	0.259	1.928	5.474	57.774	0.216
Ag_4Fe	1.649	0.824	0.606	4.259	11.003	0.378
Ag_5Fe	0.131	0.066	7.609	3.769	108.087	0.409

It has been stated that molecular orbitals of donor and acceptor have an imperative role in the transfer of charge and bond creation during the establishment of the donor-acceptor system [86–92]. The energy alteration between orbitals HOMO and LUMO is an essential factor to compute the electronic properties like chemical hardness, softness, and other values depending on the energy gap of HOMO and LUMO [93, 94]. Kinetic stability of a cluster decreases with the decrease of the HOMO-LUMO gap. It states the minimum energy requirement of an electron to move from occupied orbital to unoccupied orbital [95]. In this study, Ag_2Fe exhibits the highest HOMO-LUMO gap and Ag_5Fe, the least value. Reported data [96] shows that the best HOMO-LUMO gap suitable for biomedical application in 0.39 eV. In that respect presented range of HOMO-LUMO gap of clusters Ag_nFe (n = 1–4) are suitable for biomedical application, only Ag_5Fe falls below the reported value. The computed HOMO-LUMO gaps also present with an odd-even oscillation where the clusters bearing an odd number of atoms exhibit a higher energy gap than those with an even number of electrons. Correspondingly the gap increases in case of even number of total atoms ($AgFe<Ag_3Fe$) and decreases in case of odd number of total atoms ($Ag_2Fe > Ag_4Fe$). Ag_5Fe remains an exception even in this aspect as the HOMO-LUMO gap instead of increasing in value, diminishes by a huge range.

In order to comprehend molecular structure, stability, binding, and dynamics of chemical species, molecular hardness is a vital factor [97]. It is the natural tendency of every molecule to configure themselves with maximum hardness [98]. Molecular hardness is an electronic parameter of clusters which depict the stability of the molecular system. Movement of a molecule from stable to unstable state reduces its hardness value and vice versa [88]. According to computed date, Ag_2Fe exhibits the highest value and Ag_5Fe, the least, the range varying from 0.066 eV to 0.827 eV. The cluster with the maximum HOMO-LUMO gap also possesses the highest value of hardness and vice versa.

The clusters that possess the peak HOMO-LUMO gap, are the least prone to respond against any external agitation. Hence the molecular softness shares an inverse relation with the HOMO-LUMO gap of the cluster. Correspondingly the cluster with the maximum hardness gap exhibits minimum softness and vice versa. The molecular softness ranges from 0.605 eV–7.609 eV, with Ag_5Fe being highest and Ag_2Fe presenting with the least value. The change is values of molecular softness is abrupt as for Ag_nFe (n = 1–4) the values are increasing in a linear rate whereas for cluster Ag_5Fe, the value increases at an extremely steep rate and in more than triple of the second highest value of softness. The variation of softness with the HOMO-LUMO gap is plotted in Figure 5.2.

Another most vital factor to comprehend the transfer of charge from donor to acceptor is electronegativity [99, 100]. In the computed data the electronegativity varies from 3.769 eV to 5.474 eV. Maximum value is found for Ag_3Fe and the minimum for Ag_5Fe. According to the computed data, the lowest value of electronegativity is exhibited by the structure with the lowest HOMO-LUMO gap. In the case of electronegativity all the computed values are close in magnitude with no huge variations in value.

The electrophilicity index is the constraint for computation of the expanse of energy reduction due to additional flow of electrons for the duration of donor-acceptor interaction, which is influenced by both ionization energy and electron affinity [101]. The cluster with lowest HOMO-LUMO gap exhibits the maximum electrophilicity index. The margin of the highest value is almost double to that of the second highest one. Ag_5Fe cluster has the highest electrophilicity index and the least belongs to Ag_4Fe. The entire range of computed data shows that electrophilicity index has an odd-even oscillating relation. The clusters with an even sum of atoms possess much higher values of electrophilicity index than the ones with an odd number of electrons. Ag_5Fe being the only exception in change of magnitude, the odd and even pairs show small change in computed values.

The tabulation also contains the dipole moment of the cluster. The optimized structures presented with most values to be less than 1 Debye. Dipole moment is related to the structural symmetry of a cluster, among the clusters Ag_2Fe presents with the highest value. The range of dipole moment is

0.216–1.025 Debye. The cluster Ag_2Fe which has the highest dipole moment also retains the highest HOMO-LUMO gap. The variation between numbers of Ag atoms with dipole moment is expressed in Figure 5.3.

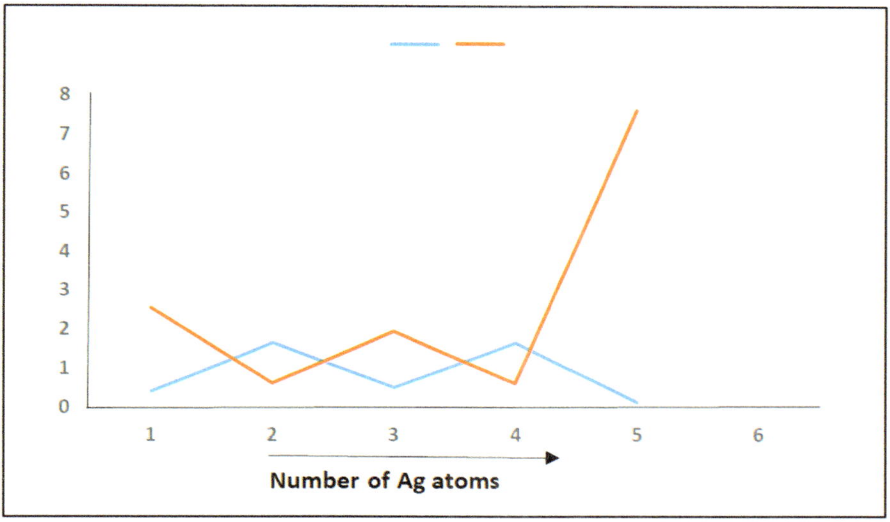

FIGURE 5.2 Relationship among HOMO-LUMO gap with molecular softness.

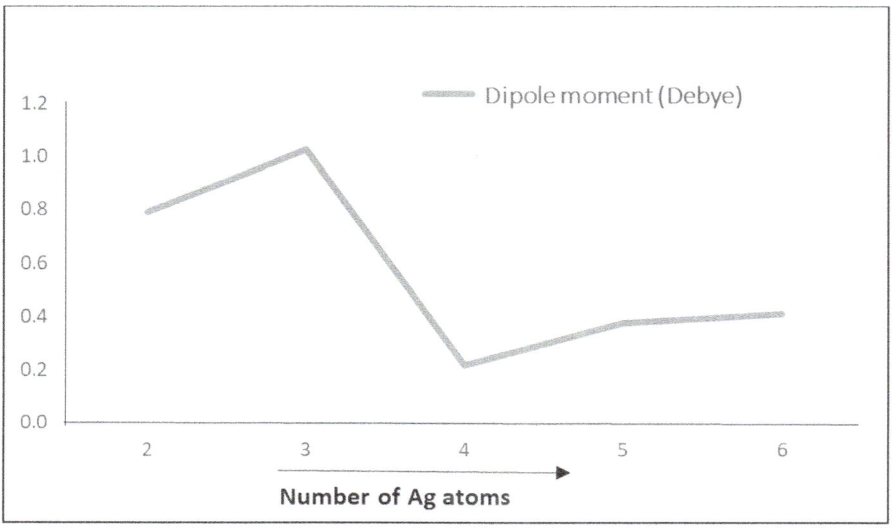

FIGURE 5.3 Relationship between number of Ag atoms with dipole moment.

5.4 CONCLUSION

A computational study of iron-doped silver nanoparticle-based clusters has been performed. DFT-based global descriptors have been computed based on the clusters varying 1 to 5 silver atoms with a single Fe atom, possessing the most stable configuration. The obtained bond length of the ground state configuration of the clusters validates their reported experimental values. The bond order of the clusters AgFe is in a mixed phase of clusters and the rest of the clusters contain segregation of atoms in them. The computed data present with HOMO-LUMO gap which is validated by reported values of metallic clusters. The nanoalloy clusters, excluding Ag_5Fe, present with an energy gap that is suitable for biomedical applications, thus opening new avenues of research incorporating organic compounds due to their non-toxicity. All the clusters present a unique odd-even oscillating character in terms of molecular hardness, softness, electronegativity, and electrophilicity index. The values of hardness are limited within a small range varying from 0.066 eV to 0.827 eV. On the contrary, the values of molecular softness exhibit a higher range of 0.605 eV to 1.928 eV. Similar to molecular hardness, electronegativity also exhibits a small range in terms of computed magnitude. Only the cluster Ag_5Fe shows unnatural progression in magnitude of the computed data, especially in the case of electrophilicity.

KEYWORDS

- **density functional theory**
- **electronegativity**
- **pharmacokinetics**
- **photothermal therapy**
- **radiotherapy**
- **silver nanoparticles**

REFERENCES

1. Wang, J., Wang, G., & Zhao, J., (2003). *Chem. Phys. Lett., 380*, 716.
2. Nørskov, J. K., Bligaard, T., Rossmaisl, J., & Christensen, C. H., (2009). *Nat. Chem., 1*, 37.

3. Zhang, Q., Xie, J., Yu, Y., & Lee, J. Y., (2010). *Nanoscale, 2*, 1962.
4. Ferrando, R., Jellinek, J., & Johnston, R. L., (2008). *Chem. Rev., 108*, 845.
5. Harding, C., Habibpour, V., Kunz, S., Farnbacher, A. N., Heiz, U., Yoon, B., & Landman, U., (2009). *J. Am. Chem. Soc., 131*, 538.
6. Fang, B., Wanjala, B. N., Hu, X., Last, J., Loukrakpam, R., Yin, J., Luo, J., & Zhong, C. J., (2011). J. *Power Sources, 196*, 659.
7. Wang, W., Wang, S., Ma, X., & Gong, J., (2011). *Chem. Soc. Rev., 40*, 3703.
8. Huh, Y. M., Lee, E. S., Lee, J. H., Jun, Y. W., Kim, P. H., Yun, C. O., Kim, J. H., Suh, J. S., & Cheon, J., (2007). *Adv. Mater., 19*, 3109.
9. Campelo, J. M., Luna, D., Luque, R., Marinas, J. M., & Romero, A. A., (2009). *Chem. Sus. Chem., 2*, 18.
10. Sozer, J., & Kokini, L., (2009). *Trends Biotech., 27*, 82.
11. Steiner, D., Mokari, T., Banin, U., & Millo, O., (2005). *Phys. Rev. Lett., 95*, 056805.
12. Saraf, D., & Kshirsagar, A., (2014). *Phys. Chem. Chem. Phys., 16*, 10823.
13. Alexander, J. W., (2009). *Surg. Infect., 10*, 289–292.
14. Ioan-Avram, N., Anton, F., Maria, S., Denisa, F., Ovidiu, O., & Ecaterina, A., (2014). *Curr. Org. Chem., 18*, 173–184.
15. Geraldo, D. A., Needham, P., Chandia, N., Arratia-Perez, R., Mora, G. C., & Villagra, N. A., (2016). *Biointerface Res. Appl. Chem., 6*, 1263–1271.
16. Chowdhury, N. R., MacGregor-Ramiasa, M., Zilm, P., Majewski, P., & Vasilev, K., (2016). *J. Colloid Interface Sci., 482*, 151–158.
17. Tavaf, Z., Tabatabaei, M., Khalafi-Nezhad, A., & Panahi, F., (2017). *Eur. J. Integr. Med., 12*, 163–171.
18. Domeradzka-Gajda, K., Nocun, M., Roszak, J., Janasik, B., Quarles, C. D. Jr., Wasowicz, W., Grobelny, J., Tomaszewska, E., Celichowski, G., Ranoszek-Soliwoda, K., et al., (2017). *Toxicol. Lett., 272*, 38–48.
19. Kraeling, M. E. K., Topping, V. D., Keltner, Z. M., Belgrave, K. R., Bailey, K. D., Gao, X., & Yourick, J. J., (2018). *Regul. Toxicol. Pharm., 95*, 314–322.
20. Kejlová, K., Kašpárková, V., Krsek, D., Jírová, D., Kolářová, H., Dvořáková, M., Tománková, K., & Mikulcová, V., (2015). *Int. J. Pharm., 496*, 878–885.
21. Zhang, L., Zeng, G., Dong, H., Chen, Y., Zhang, J., Yan, M., Zhu, Y., Yuan, Y., Xie, Y., & Huang, Z., (2017). *Bioresour. Technol., 230*, 132–139.
22. Gupta, S. D., Agarwal, A., & Pradhan, S., (2018). *Ecotoxicol. Environ. Saf., 161*, 624–633.
23. Pereiro, M., Baldomir, D., Botana, J., Arias, J. E., Warda, K., & Wojtczak, L., (2008). *J. Appl. Phys., 103*, 07A315.
24. Lin, C. A. J., Lee, C. H., Hsieh, J. T., Wang, H. H., Li, J. K., Shen, J. L., Chan, W. H., Yeh, H. I., & Chang, W. H., (2009). *Jpn. J. Med. Electron. Biol., 29*(6), 276–283.
25. Mathew, A., & Pradeep, T., (2014). *Part. Part. Syst. Charact., 31*, 1017–1053.
26. Muñoz-Castro, A., (2019). *Inorg. Chem. Front.*, **6**, 2349–2358.
27. Zhang, X. D., Guo, M. L., Wu, D., Liu, P. X., Sun, Y. M., Zhang, L. A., She, Y., et al., (2011). *Int. J. Mol. Sci., 12*, 2972–2981.
28. Thakore, S. I., Nagar, P. S., Jadeja, R. N., Thounaojam, M., Devkarc, R. V., & Rathore, P. S., (2019). *Arabian J. Chem., 12*, 694–700.
29. Kirubha, E., & Palanisamy, P. K., (2014). *Adv. Nat. Sci.: Nanosci. Nanotechnol., 5*(4), 6.
30. Sun, Q., Kandalam, A. K., Wang, Q., Jena, P., Kawazoe, Y., & Marquez, M., (2006). *Phys. Rev. B, 73*, 134409.

31. Burdușel, A. C., Gherasim, O., Grumezescu, A. M., Mogoantă, L., Ficai, A., & Andronescu, E., (2018). *Nanomaterials, 8*, 681.
32. Zhang, X. F., Liu, Z. G., Shen, W., & Gurunathan, S., (2016). *Int. J. Mol. Sci., 17*, 1534.
33. Wei, L., Lu, J., Xu, H., Patel, A., Chen, Z. S., & Chen, G., (2015). *Drug Discov. Today, 20*, 595–601.
34. Yaqoob, S. B., Adnan, R., Rameez, K. R. M., & Rashid, M., (2020). *Front. Chem., 8*, 376.
35. Sanpo, N., Berndt, C. C., Wen, C., & Wang, J., (2013). *Acta Biomater., 9*, 5830–5837.
36. Chen, M., Yamamuro, S., Farrell, D., & Majetich, S. A., (2003). *J. Appl. Phys., 93*, 7551.
37. Gupta, A. K., & Gupta, M., (2005). *Biomaterials, 26*, 3995–4021.
38. Figuerola, A., Di Corato, R., Manna, L., & Pellegrino, T., (2010). *Pharmacol. Res., 62*, 126–143.
39. Lim, D., Negreira, A. S., & Wilcox, J., (2011). *J. Phys. Chem. C., 115*, 8961–8970.
40. Mahtout, S., & Belkhir, M. A., (2006). *Phys. Lett. A, 360*, 384–389.
41. Huang, T., Cheng, J., Bian, D., & Zheng, Y., (2015). *J. Biomed. Mater. Res., Part B, 104*, 225–240.
42. Janssens, E., Tanaka, H., Neukermans, S., Silverans, R. E., & Lieven, P., (2004). *Phys. Rev. B., 69*, 085402.
43. Bezi, J. M., & Tajabor, N., (2012). *J. Magn. Magn. Mater., 324*, 52–59.
44. Li, Z., Shao, T., & Zhao, Z., (2021). *Phase Transitions, 94*(5), 298–307.
45. Khanna, S. N., Rao, B. K., Jena, P., & Nayak, S. K., (2003). *Chem. Phys. Lett., 373*, 433–438.
46. Singh, R., & Kroll, P., (2008). *Phys. Rev. B, 78*, 245404.
47. Kumar, V., & Kawazoe, Y., (2003). *Phys. Rev. Lett., 91*, 199901.
48. Ma, L., Wang, J., & Wang, G., (2013). *J. Chem. Phys., 138*, 094304 (1–9).
49. Sahoo, S., Rollmann, G., & Entel, P., (2005). *Phase Transitions, 78*, 723–731.
50. Postnikov, A. V., & Entel, P., (2007). *Phase Transitions, 77*, 149–159.
51. Ebert, H., Bornemann, S., Minár, J., Kosbreveuth, M., Sbreveipr, O., Dederichs, P. H., Zeller, R., & Cabria, I., (2005). *Phase Transitions, 78*(1–3), 71–83.
52. Košuth, M., Minár, J., Cabria, I., Perlov, A., Crisan, V., Ebert, H., & Akai, H. (2002). *Phase Transitions, 75*(1, 2), 113–123.
53. Li, Z., & Zhao, Z., (2017). *Mater. Chem. Phys., 187*, 54–59.
54. Zhao, W. J., & Wang, Y. X., (2008). *Chem. Phys., 352*, 291–296.
55. Yu, S., Chen, S., Zhang, W., Yu, L., & Yin, Y., (2007). *Chem. Phys. Lett., 446*, 217–222.
56. Illas, F., & Martin, R. L., (1998). *J. Chem. Phys., 108*, 2519.
57. Gyorffy, B., Staunton, J., Stocks, G., Gross, E., & Dreizler, R., (1995). *Plenum* (p. 461). NY, 461.
58. Kummel, S., & Brack, M., (2001). *Phys. Rev. A, 64*, 022506.
59. Car, R., & Parrinello, M., (1985). *Phys. Rev. Lett., 55*, 2471.
60. Koskinen, M., Lipas, P., & Manninen, M., (1995). *Nucl. Phys. A, 591*, 421.
61. Schmid, R. N., Engel, E., & Dreizler, R. M., (1995). *Phys. Rev. C, 52*, 164.
62. Chen, P., Qin, M., Chen, H., Yang, C., Wang, Y., & Huang, F., (2013). *Phys. Status Solidi A, 210*, 1098.
63. Palacios, P., Sanchez, K., Conesa, J. C., & Wahnon, P., (2006). *Phys. Stat. Sol., 203*, 1395–1401.
64. Palacios, P., Sanchez, K., Conesa, J. C., Fernandez, J. J., & Wahnon, P., (2007). *Thin Solid Films, 515*, 6280–6284.
65. Palacios, P., Wahnon, P., Pizzinato, S., & Conesa, J. C., (2006). *J. Chem. Phys., 124*, 014711.

66. Ranjan, P., & Chakraborty, T., (2020). *J. Nanopart. Res., 22*, 35-1-11.
67. Ranjan, P., & Chakraborty, T., (2020). *J. Nanopart. Res., 22*, 280-1-11.
68. Ranjan, P., Dhail, S., Venigalla, S., Kumar, A., Ledwani, L., & Chakraborty, T., (2015). *Mater. Sci. -Pol., 33*, 719–724.
69. Ranjan, P., Venigalla, S., Kumar, A., & Chakraborty, T., (2014). *New Front. Chem., 23*, 111–122.
70. Ranjan, P., Kumar, A., & Chakraborty, T., (2016). *Mat. Today Proc., 3*, 1563–1568.
71. Ranjan, P., Chakraborty, T., & Kumar, A., (2020). *Mater. Sci. Pol.,38*, 97–107.
72. Ranjan, P., & Chakraborty, T., (2018). *Key Eng Mater., 777*, 183–189.
73. Ranjan, P., Das, S., Yadav, P., Tandon, H., Chaudhary, S., Malik, B., Rajak, S. K., et al., (2021). *Theor. Chem. Acc. 140*. 59 (1–12).
74. Gaussian 16, Revision C.01, Frisch, M. J., Trucks, G. W., Schlegel, H. B., Scuseria, G. E., Robb, M. A., Cheeseman, J. R., et al., Gaussian, Inc., Wallingford CT, (2016).
75. Chen, F. Y., & Johnston, R. L., (2007). *Apply. Phys. Lett., 90*, 153123–1-3.
76. Chen, F. Y., & Johnston, R. L., (2008). *Acta Mater., 56*, 2374–2380.
77. Parr, R. G., & Yang, W., (1989). *Density Functional Theory of Atoms and Molecules*. Oxford University Press, New York, Oxford.
78. Li, W., & Chen, F., (2013). *J. Nanopart. Res., 15*, 1809.
79. Beutel, V., Kramer, H. G., Bhale, G. L., Kuhn, M., Weyers, K., & Demtroder, W., (1993). *J. Chem. Phys., 98*, 2699.
80. Fabbi, J. C., Langenberg, J. D., Costello, Q. D., Morse, M. D., & Karlsson, L., (2001). *J. Chem. Phys., 115*, 7543.
81. Ranjan, P., Chakraborty, T., & Kumar, A., (2018). Theoretical analysis: Electronic, Raman, Vibrational, and magnetic properties of CunAg (n= 1-12) nanoalloy clusters. In: *Theoretical and Quantum Chemistry at the Dawn of the 21st Century* (pp. 25–58). Apple Academic Press.
82. Mayer, I., (1983). Charge, bond order and valence in the AB Initio SCF theory. *Chem. Phys. Lett., 97*(3), 270–274.
83. Cordy, W., (1947). *J. Chem. Phys., 15*, 305.
84. Brenner, D. W., Shenderova, O. A., Harrison, J. A., Stuart, S. J., Ni, B., & Sinnott, S. B., (2002). *J. Phys.: Condens. Matter., 14*, 783.
85. Horsfield, P., Bratkovsky, A. M., Fearn, M., Pettifor, D. G., & Aoki, M., (1996). *Phys. Rev. B, 53*, 12694.
86. Fujimoto, H., Kato, S., Yamabe, S., & Fukui, S., (1974). *J. Chem. Phys., 60*, 572.
87. Fujimoto, H., Kato, S., Yamabe, S., & Fukui, S., (1974). *J. Am. Chem. Soc., 96*, 2024.
88. Ghosh, D. C., & Bhattacharyya, S., (2004). *Int. J. Mol. Sci., 5*, 239.
89. Ghosh, D. C., (1984). *Indian J. Pure Appl. Phys., 22*, 346.
90. Ghosh, D. C., (1989). *Indian J. Pure Appl. Phys., 27*, 160.
91. Xiao, H., Kheli, J. T., & Goddard, III. W. A., (2011). *J. Phys. Chem. 2*, 212–217.
92. Saravanan, S., & Balachandran, V., (2014). *Spectrochim. Acta, Part A, 120*, 351–364.
93. Azam, F., Alabdullah, N. H., Ehmedat, H. M., Abulifa, A. R., Taban, I., et al., (2018). *J. Biomol. Struct. Dyn., 36*, 2099–2117.
94. Parr, R. G., & Zhou, Z., (1993). *Acc. Chem. Res., 26*, 256–258.
95. Xiao, H., Kheli, J. T., & Goddard, III. W. A., (2011). *J. Phys. Chem. Lett., 2*, 212.
96. Gowri, S., Gopinath, K., & Arumugam, A., (2018). *J. Photochem. Photobiol. B, 180*, 166–174.

97. Chattaraj, P. K., & Sengupta, S., (1999). *J. Phys. Chem., 103*, 6122–6126.
98. Pearson, R. G., (1987). *J. Chem. Ed., 64*, 561–567.
99. Sanderson, R. T., (1951). *Science, 114*, 670–672.
100. Sanderson, R. T., (1952). *Science, 116*, 41, 42.
101. Parr, R. G., Szentpaly, L. V., & Liu, S., (1999). *J. Am. Chem. Soc., 121*, 1922–1924.

CHAPTER 6

GRAPHENE OXIDE AS AN EMERGING SOLID-BASED NANOCOMPOSITE MATERIAL FOR HIGHLY EFFICIENT SUPERCAPACITORS

RANJIT DEBNATH and MITALI SAHA

Department of Chemistry, National Institute of Technology, Agartala, Tripura, India

ABSTRACT

Nanomaterials have great potential for electronic conductivity and ionic transport with respect to conventional batteries or supercapacitors (SCs). They have the capability of occupation of each intercalation site available in their particle volume, which enables them for fast ion diffusion and high specific capacities. Over the last two decades, various allotropic forms of carbon, such as fullerenes, graphene, carbon nanotubes (CNTs), etc., have been investigated as sources of energy generation, storage, and ease of production. Since the development of the first graphene-based supercapacitor in 2008, a tremendous amount of research work was reported on the fabrication and modification of graphene-based SCs. Although the energy storage capacity of pristine graphene is very poor, its energy storage capacity can be modified by using nanocomposites of graphene with other materials. Interestingly, it has been found that functionalized graphene, i.e., graphene oxide (GO) exhibits higher capacitance than graphene due to *pseudo*-capacitance effect of oxygen-containing functional groups present on its basal planes.

Nanotechnology-Enhanced Solid Materials: Design, Synthesis, Properties, Applications, and Perspectives. Lionello Pogliani, Ann Rose Abraham, A. K. Haghi, & Prabhat Ranjan (Eds.)

In view of easy synthesis, shorter processing time, lower cost, and higher capacitance, GO may be a better choice and more efficient material than graphene for SCs. This chapter represents the recent research progress on GO-based materials for application in SCs.

6.1 INTRODUCTION

Carbon-based supercapacitors (SCs) have proven efficient power storage systems which can store electrical energy between the adsorbed electrolyte layer and carbonaceous electrode surface. Currently, the real challenge to develop SCs is probably low cost, good electrochemical stability, environmental safety, excellent conductivity, and large specific surface area. Carbon nanostructures with different spatial dimensions show unique properties during their usage as electrode materials for SCs. It has been found that commercial SCs containing activated carbons (ACs) suffer from serious problem of limited energy densities due to restricted charge accumulation and sluggish mass diffusion. To overcome the issues, enormous efforts are being carried out to design and develop SCs with different forms of nanocarbons. This chapter highlights the significance and recent advances of the graphene-based SCs. Figure 6.1 shows the schematic diagram of a supercapacitor.

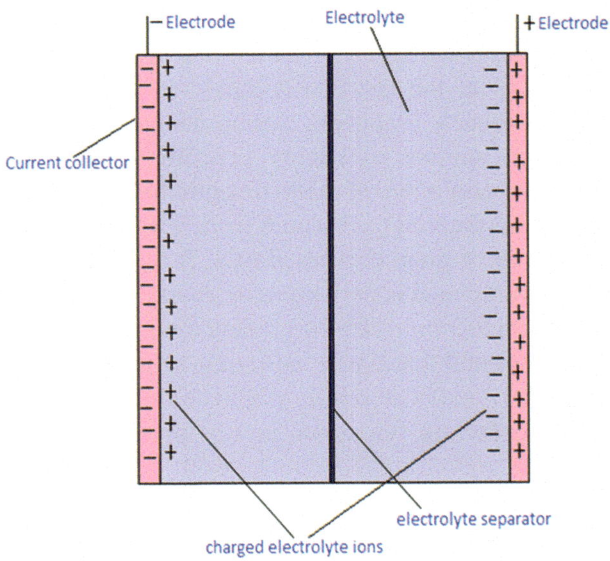

FIGURE 6.1 The schematic diagram of supercapacitor.

6.2 CARBON MATERIALS FOR SUPERCAPACITORS (SCS)

Various materials like templated carbon, carbon nanotubes (CNTs), carbon nanofiber, AC, carbon nanocomposites, graphene, and nanocomposites are currently being used as important materials for supercapacitor applications. The reason of great importance of carbon materials for supercapacitor lies on the fact that carbon is present in different forms, naturally abundant, its thermal and chemical stability as well as good electrical conductivity which enabled the researchers to design low-priced SCs. An efficient supercapacitor requires a favorable material containing large active surface area, suitable pore size, excellent conductivity, surface wettability, rapid charge/discharge rate, high cycle stability, etc. To design a relevant storage system, it is necessary to give importance to the lifetime, reliability, storage capacity, cost, and environmental impact. In this context, carbon materials not only exhibit lightweight, high-power density, long cyclic stability, low cost but also provides several designs for energy storage purpose.

6.2.1 ZERO-DIMENSIONAL CARBON NANOMATERIALS FOR SUPERCAPACITORS (SCS)

0-dimensional carbon nanomaterials refer to materials which are usually composed of a small number of atoms and molecules. Carbon nanoparticles without any defined crystalline structures were exploited due to their broad optical absorptions, comparable to CNTs or graphene, etc. Due to higher electron affinity, fullerenes are considered to be one of the best electric charge acceptors which make them promising materials for SCs. The redox activity of C_{60} has been utilized in molecular electronics, solar cells, and energy storage devices and when researchers find difficult to control layers in graphene, a number of walls and length diameter in CNTs, fullerenes possess no problems and efficiently worked for reversible and reproducible electroreductions. Carbon or graphene quantum dots (QDs) also exhibits optical properties and photoinduced redox characteristics that easily rival those of fullerenes. In fact, in addition to fullerenes, carbon dots can be considered as another nanoscale allotrope of carbon with zero-dimension, which will represent a rapid and advance research field in the near future. Capacitance is not only determined by electrical conductivity or surface area, but also depends on pore size distribution, pore shape, structure, etc., and in this context, ACs play an important role having a wide range of pore size distribution such as micropores with <2 nm, mesopores of 2–50 nm and macropores of >50 nm.

6.2.2 ONE-DIMENSIONAL CARBON NANOMATERIALS FOR SUPERCAPACITORS (SCS)

1-dimensional nanostructures such as nanotubes, nanofibers, nanobelts, nanopillars, nanoribbons, nanowires (NWs), etc., are being given considerable attention due to the dependence of their functional properties on dimensions. They include NWs, nanopillars, nanoribbons, and nanobelts. For high performance flexible SCs, 1-D nanostructured materials have gained increasing interest owing to their high specific area, efficient electron transport, and excellent mechanical strength, thus enabling them to be flexible SCs with some suitable properties, such as high energy densities, superb power densities, and great flexibilities. 1D nanodomain of carbon materials particularly CNTs has shown tremendous impact as an effective electrode material to construct supercapacitor devices as it imparts high thermal and electrical conductivity, high tensile strength, and high surface area. Although 1-D carbon nanostructures can be excellent supercapacitor electrode materials, the main challenges are to grasp the physiochemical properties along with the electrochemical behavior of these nanomaterials within multicomponent-fabricated charge storage systems.

6.2.3 TWO-DIMENSIONAL CARBON NANOMATERIALS FOR SUPERCAPACITORS (SCS)

Two-dimensional (2D) carbon materials were greatly explored as potential and promising candidates for the development of flexible supercapacitor devices with improved electrochemical performance. Although, these 2D nanomaterials can be assembled into a dense film to be used for flexible SCs, yet the high packing density creates restacking problems along with poor volumetric capacitance. Since all the atoms are present on the surface of 2D carbon materials, the design of surface structure becomes highly critical for determination of bulk electrochemical properties. However, such 2D structures are well fitted to store the surface charges in a SCs. Since the first isolation of graphene, several 2D nanomaterials have been exploited for energy storage systems with respect to their bulk counterparts. Interestingly, graphene and other porous carbon materials were found to store charge through electrostatic interaction and also found to be capable of achieving rapid charging and high power with high stability.

6.2.4 THREE-DIMENSIONAL CARBON NANOMATERIALS FOR SUPERCAPACITORS (SCS)

During the last few years, a growing interest has been observed in the development of natural and synthetic bulky carbon with 3D structures rather than two-dimensional structures due to the increase of active surface area throughout the entire 3D carbon structures. As compared to 1D and 2D structures, 3D morphology offers more opportunities in rational design and development of hierarchical porous carbon nanostructures with 3D configuration, which are assumed to enhance the performances of energy storage devices. From the perspective of supercapacitor applications, three-dimensional porous carbon materials are highly demanding as they generally possess highly ordered channels along with large specific surface area, high electronic and ionic conductivity, etc. They also possess a large amount of open pores and low tortuosity, which is why as an electrode material, it not only provides a fast electron transport path, shortening the ion diffusion length significantly to achieve fast kinetic performances, but also suppresses the agglomeration of nanoparticles. Furthermore, the porous interconnected structure of bulky carbon along with the high surface area can provide a large contact area with the electrolyte, therefore continuous ion transportation pathways always occur in the entire space, thereby enhancing the rate performance. The as-grown 3-D carbon foams (CFs) are also being used as electrode materials for energy storage devices like SCs and batteries.

6.3 GRAPHENE OXIDE (GO)-BASED COMPOSITES FOR SUPERCAPACITORS (SCS)

Due to aggregation of graphene nanosheets, the available surface area of graphene is usually smaller than the theoretical one and its capacitance value is also low. Although, it has been theoretically claimed that graphene oxide (GO) is not suitable for SCs, but GO-based nanocomposites with conducting polymers, metal oxides, etc., exhibits lower cost, shorter processing time, higher capacitances, good cycling performances, etc. Figure 6.2 represents the schematic diagram of GO.

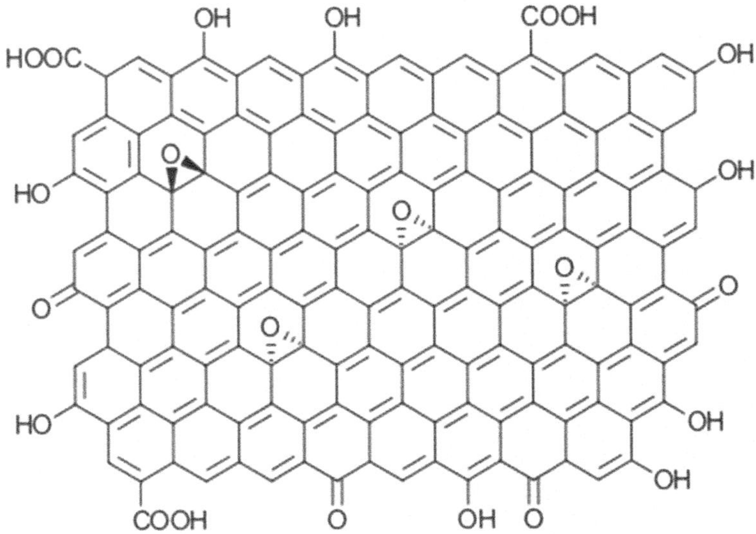

FIGURE 6.2 The schematic diagram of graphene oxide.

In 2010, Xin et al. fabricated a composite of needle-like MnO_2 nanocrystals supported by GO through a simple soft chemical route and showed that the electrochemical performance of nanocomposites could be modified by the chemical interaction between MnO_2 and GO, and this method provided a direct approach to deposit MnO_2 nanoparticles onto the GO sheets [1]. Zhixiang Wei and coworkers constructed nanocomposites electrode by combining one dimensional polyaniline (PANI) NWs with two-dimensional GO nanosheets. The composite showed specific capacitance of 555 $F.g^{-1}$ at a discharge current of 0.2 $A.g^{-1}$ and the capacitance retention of 92% after 2000 consecutive cycles [2]. In 2011, Konstantin et al. reported the synthesis of nanocomposites-based electrode by GO and multi-walled carbon nanotube (MWCNT) and compared its performance with pristine GO and multi-walled CNT. They found that, in the potential range of 0.1 to 0.5 V, the specific capacitances of GO/MWCNT, MWCNTs, and GO were 251, 85, and 60 $F.g^{-1}$, respectively [3]. Hu Zhongai and coworkers prepared a nanocomposites electrode using cobalt nitrate and aluminum nitrate as the raw material, where graphite oxide was employed as carrier. Al-substituted α-$Co(OH)_2$/GO showed specific capacitance of 1,137 $F.g^{-1}$ in 6 M KOH solution at 1 $A.g^{-1}$ current density within the potential range of 0–0.5 V and only 12% of the initial specific capacitance were losses after 500 cycles [4].

Bin et al. compared graphene and GO both as an electrode material in SCs and they found that GO exhibited higher capacitance up to 189 $F.g^{-1}$ than graphene due to a pseudo-capacitance effect of attached oxygen-containing functional groups on its planes [5].

In 2012, Jipeng et al. synthesized CoAl-layered double hydroxide (CoAl-LDH) with GO and a maximum specific capacitance of 772 F/g was obtained at 1 A/g in 6 mol/L KOH solution for 12.9% containing GO and the composite also exhibited excellent cyclic stability of 73% specific capacitance remains even after 10,000 cycles [6]. In 2013, Jun et al. fabricated nanocomposite electrode using MnO_2, CNT, and graphene oxide (MCGO), where GO acted as a surfactant to disperse CNTs for the subsequent deposition of MnO_2. The specific capacitance of this composite was 4.7 times more than that of pure MnO_2. The composite also showed high energy density of 30.4–14.2 $Wh.kg^{-1}$, large power density of 2.6–50.5 $kWkg^{-1}$, maintaining 94% of the initial specific capacitance value after 1,000 cycles [7]. Bao-Hang et al. synthesized nanocomposite material by GO-based benzimidazole cross-linked network (GOBIN). The composite showed the highest specific capacitance value up to 370 F/g at a current density of 0.1 A/g and maintained cyclic stability about 90% of the original capacitance after 5,000 cycles [8]. Chandrakant D. Lokhande and coworkers improved the capacitance of supercapacitor electrode by carbon-based GO and manganese oxide (Mn_3O_4), and the composite achieved specific capacitance of 344 F/g at a scan rate of 5 mV/s [9]. Wei et al. prepared a material by graphene oxide and polyaniline (GO/PANI) using one-step electrochemical co-deposition method. The composites achieved the highest specific capacitance of 1136.4 F/g with GO concentration of 10 mg/L, at a scan rate of 1 mV/s, which was almost two times higher than that of PANI (484.5 F/g). The GO/PANI composite also showed 89% cycling stability of its initial capacitance after 1,000 cycles [10]. In the year 2014, Liming et al. presented a new type of electrochemical SCs based on GO NWs and polypyrrole (PPy) NWs into a microporous Al_2O_3 template. The composite electrode achieved high capacitance of 960 F/g due to the large surface area of the vertically aligned (VA) NWs and the superficial contact between the NWs and the substrate electrode. It also retained high stability after 300 charging and discharging cycles [11]. Hyoung-Joon Jin and coworkers prepared carbon aerogels from silk proteins and GO by a flash lyophilization process followed by carbonization, the nitrogen-enriched carbon aerogels show a high capacitance of 298 $F.g^{-1}$ because of the pseudocapacitive effects and also achieved a specific energy of 63 $Wh.kg^{-1}$, specific power of 20 kW.kg^{-1}, and high cyclic stability even after 5,000 cycles [12]. Chunnian Chen and coworkers synthesized nanocomposite electrodes using GO, nano-sulfur, and PPy. In 1M Li_2SO_4 electrolytic solution, the composite showed excellent conductivity because of the higher surface area of GO and porous structure of PPy. The GO/nano-S/PPy composite exhibited a better specific capacitance of 892.5 F.g^{-1} at scan rates of 5 mV.s^{-1} and the capacitance retained about 81.2% after 500 cycles [13]. Hyung et al. studied SCs containing GO electrodes using molecular dynamics (MD) simulations. The full range of electrode oxidation from 0% to 100% (fully oxidized GO) was investigated by decorating the graphene surface with hydroxyl groups and ionic liquid 1-ethyl-3-methylimidazolium tetrafluoroborate ($EMI^+BF_4^-$) as an electrolyte. Their study suggested that capacitance has a tendency to decrease with increasing electrode oxidation [14]. Chunliang et al. prepared nanocomposite from aniline (ANI), GO, and ammonium persulfate (APS) by polymerization with the mass ratio (mANI:mGO) 1000:1, 100:1, and 10:1, respectively. The PANI/GO composite with the mass ratio 1000:1 possessed excellent capacitive behavior with a specific capacitance as high as 355.2 F.g^{-1} at 0.5 A.g^{-1} in 1 M H_2SO_4 electrolyte [15]. Jian Xie and coworkers developed covalently-grafted PANI and GO nanocomposites. The covalently-grafted (PANI/GO) composites showed unique morphology of much larger specific surface area and pore volume, and this maximized the synergistic effect between PANI and GO, resulting in excellent electrochemical performances. The specific capacitance of PANI/GO (6:1) was found to be 442 F/g vs. 226 F/g of pure PANI with improved cyclic stability of 83% after 2,000 cycles, whereas for PANI/GO (6:1) the cyclic stability was 54.3% after 1,000 cycles [16]. De Yan et al. synthesized hybrid electrode based on GO flakes with manganese dioxide (MnO_2) nanosheets. The MnO_2-GO hybrid exhibited enhanced specific capacitance of 213 F/g at a current density of 0.1 A/g, and specific capacitance retention of 98.1% at 1 A/g after 1,000 cycles compared with the MnO_2 and GO [17]. Yanfeng et al. designed electrode via an *in situ* oxidative polymerization from GO and conducting polymers (PANI and PPy). The obtained specific capacitance of APGO/PPy and APGO/PANI are 404 and 522 F.g^{-1}, respectively, and the composites exhibited good electrical conductivity, low internal resistance, and stable cyclic performance [18].

In 2015 Haihan Zhou and coworkers introduced a simple and low-cost electrochemical co-deposition method to fabricate nanocomposites by polypyrrole and graphene oxide (PPy/GO) and the nanocomposite showed a high areal capacitance of 152 m F.cm^{-2} at 10 mV.s^{-1} with good cyclic stability [19]. Yufeng et al. developed a composite electrode using MnO_2, GO, and 3D hierarchical porous structure (HPC) in 1 M Na_2SO_4 aqueous electrolyte.

The optimized MnO_2/GO/HPC asymmetric supercapacitor showed a highest specific capacitance of 84 F/g, energy density of 46.7 $Wh.kg^{-1}$ at a power density of 100 $W.kg^{-1}$ and also showed superior long cyclic stability of 100% capacitance retention after 1,000 cycles and 93% after 4,000 cycles [20]. Yaming Wang and coworkers fabricated a 3D composite electrode from GO and PPy in an aqueous solution of $LiClO_4$. The PPy/GO composite electrode with a mass loading of 0.26 mg cm^{-2} exhibited excellent specific capacitance value of 481.1 F/g, whereas the electrode containing larger mass loading of 1.02 mg cm^{-2} exhibited the best area capacitance of 387.6 mF cm^{-2}, at a current density of 0.2 mA cm^{-2} along with capacitance retention of 80% as the current density load increased from 0.2 to 10 mA cm^{-2} [21]. Jianhui Qiu and coworkers reported a novel method for the preparation of composites material by GO, PPy, and MWCNTs. The composite showed high capacitance of 406.7 F/g at current density of 0.5 A/g and along with the cyclic stability of 92% after 1,000 cycles [22]. Lim H. N and coworkers prepared a composite material with polypyrrole, graphene oxide and zinc oxide (PPy/GO/ZnO) through potentiostatic electrochemical polymerization. The composite showed a specific capacitance of 94.6 $F.g^{-1}$ at a current density of 1 $A.g^{-1}$ and the capacitance retention of the supercapacitor was 74% of its initial value after 1,000 charge/discharge cycles [23]. Weihua et al. developed an electrode using GO with bacterial cellulose (BC) fibers through one-step esterification having a 3D network structure. The composite electrode showed a good specific capacitance of 160 F/g at a current density of 0.4 A/g and also demonstrated an outstanding capacitance retention of 90.3% after 2,000 recycles [24]. Kaushik Pal and coworkers prepared a GO-based nanocomposites through *in-situ* polymerization of pyrrole and aniline to study the interaction of GO with PPy and PANI. The specific capacitance of PPy-coated GO increased by ~122.73% compared to pristine GO and due to the presence of oxygen-containing functional group, the binding energy of polypyrrole-GO was found to be better than polyaniline-GO [25]. Chang et al. reported a novel strategy for self-assembly of GO-induced monolithic NiCo–carbonate hydroxide (NiCo–CH) NWs-based composite films (G–CH). The GO surface having oxygen-containing functional groups helped effectively to induce the formation of monodisperse NiCo–CH NWs. The obtained monolithic G–CH architectures showed extraordinary supercapacitor performance as binder- and conductive agent-free electrode and also showed superior volumetric capacitance of 2,936 F cm^{-3} with good cyclic stability [26].

In 2016, Yusran Sulaiman and coworkers developed an electrode material using poly(3,4-ethylene dioxythiophene) and graphene oxide (PEDOT/GO) potentiostatically at 1.2 V with different electrodeposition times (1–30 min) and different concentrations of GO. The composite was found to have high power density and energy of 496.64 Wh/kg and 18.24 W/kg, respectively and showed excellent specific capacitance of 157.17 $F.g^{-1}$ at a current density of 1.0 $A.g^{-1}$ [27]. Huaihe et al. synthesized electrode by carbonization of GO, resorcinol-formaldehyde aerogels using inverse emulsion method. The composite showed a stabilized specific capacitance value of 123.6 $F.g^{-1}$ at a current density of 0.1 $A.g^{-1}$ and 113.9 $F.g^{-1}$ at a current density of 1 $A.g^{-1}$ after 2,000 cycles [28]. Mei et al. developed a homogeneous electrode material using PANI and GO with a redox system in which iron dichloride ($FeCl_2$) and cumene hydroperoxide (CHP) acted as reductant and oxidant, respectively. The nanocomposites showed much higher specific capacitance of 308.3 $F.g^{-1}$ than that of the pure PANI at a current density of 1 $A.g^{-1}$ in 2 M H_2SO_4 electrolyte [29]. Zheng et al. developed a honeycomb shape composites electrode using GO, polyaniline, and metal hydroxide with cobalt ions or nickel ions. The GO/PANI/$Ni(OH)_2$ nanocomposite showed highest specific capacitance of 743 $F.g^{-1}$ whereas GO/PANI/$Co(OH)_2$ showed specific capacitance of 723 $F.g^{-1}$ and cyclic stability of specific capacitance was found to be 15.6% after 2,000 cycles [30]. Seong Chan Jun and coworkers developed an electrode using GO and layered molybdenum disulfide (MoS_2). The MoS_2/GO hybrid electrode showed a high specific capacitance of 613 $F.g^{-1}$ at low scan rate. Moreover, the MoS_2/GO electrode-based asymmetric supercapacitor device exhibited ultra-high energy (23 Wh kg^{-1}) with a power density of 17 kW kg^{-1} [31]. Sumit et al. investigated low-cost electrode material for supercapacitor based on polyvinyl alcohol (PVA) matrix by integrating GO as filler. The nanocomposite (GOPVA) prepared by adding 0.5% (w/w) of GO in PVA showed significantly higher specific capacitance of 400 $F.g^{-1}$ [32]. Yang et al. synthesized electrode with GO with Mn_3O_4 nanocrystals via a facile solvothermal self-assembly method. The composites showed the specific capacitance of 225 F/g at the scan rate of 5 mV/s and also showed high electrochemical stability that retained about 99.5% capacitance after 6,000 cycles and the energy density was found to be as high as 34.1 Wh/kg at the power density of 251 W/kg [33]. Haihan Zhou and coworkers fabricated composite films of heterogeneous conducting polymers-coated GO with carbon nanotubes (CPs-GO/CNTs) via one-step electrochemical co-deposition. The as-prepared PPy-GO/CNTs and PEDOT-GO/CNTs ternary composites achieved a high specific capacitance of 142.2 $mF.cm^{-2}$ and 99.0 $mF.cm^{-2}$ at 1.0 mA cm^{-2}, respectively. The composite showed excellent cycle stability of 97.3% and 99.2% of initial capacitance after 5,000 cycles, respectively [34]. Haihan et al. fabricated electrode material by GO nanosheets loaded with poly(3,4-ethylene dioxythiophene) (GO/PEDOT). This nanocomposite showed the improved capacitive behaviors and achieved a high capacitance of 66.3 mF cm^{-2} at 10 mV/s [35]. Lixian et al. prepared a composite of PPy, Prussian blue and GO which showed a specific capacitance of 525.4 F/g at a current density of 5 A/g in the potential range from 0.2 to 0.6 V using 1 M KNO_3 as electrolyte, and the as-prepared supercapacitor retain a specific capacitance of 96% after 2,000 cycles [36].

In 2017, Saeed Shahrokhian and coworkers developed an electrode based on vanadium pentoxide (V_2O_5), PPy, and GO via one-step electrochemical deposition process and this nanocomposite showed a specific capacitance value of 750 $F.g^{-1}$ at a current density of 5 $A.g^{-1}$, maintaining 83% of its initial value after 3,000 cycles. It showed better electrochemical stability of the electrode and device assembled by V_2O_5/PPy/GO yielded a maximum power density of 13,680 W kg^{-1} at an energy density of 22.8 W h kg^{-1} [37]. N. Manyala and coworkers developed prepared material in the form of gel, based on GO and carbon nanorods showing specific capacitance of 719.5 $F.g^{-1}$ corresponding to specific capacities of 80.8 mAh g^{-1} at a current density of 0.5 $A.g^{-1}$ [38]. Nasser Arsalani and coworkers developed an electrode based on polyaniline, GO, vanadium pentoxide and ammonium persulfate, and the resultant composite showed significantly higher specific capacitance of 712 $F.g^{-1}$ at the current density of 0.5 $A.g^{-1}$ in 1 M H_2SO_4 solution. The composite retained 83% of its original specific capacitance after 6,000 charge-discharge cycles [39]. Yusuke et al. reported the fabrication of a new composite, combining GO sheets with Prussian blue nanoparticles used a precursor for iron oxide (IO). The GO/IO hybrid (prepared from GO: PB =75: 25) showed a better specific capacitance of 91 $F.g^{-1}$ at a scan rate of 20 mV s^{-1}, compared to pure GO (81 $F.g^{-1}$) and pure IO (47 $F.g^{-1}$) [40]. Li et al. developed flexible electrode from GO, carbon nano-onion (OLC) and Ni_3S_2 ($GCNi_3S_2$) and it showed a high specific capacitance value of 1064.3 $F.g^{-1}$, energy density of 3.63 mWh cm^{-3} and retaining long cycle stability of 97.2% capacitance after 5,000 cycles [41]. Chenghua et al. synthesized materials based on ultrathin α-FeOOH nanorods and GO, the resultant α-FeOOH nanorods/GO composite with 20% iron content has the largest specific capacitance of 127 $F.g^{-1}$ at 10 $A.g^{-1}$, excellent rate capability of 100 $F.g^{-1}$ at current density of 20 $A.g^{-1}$ and good cyclic performance by retaining 85% after 2,000 cycles [42]. Lin Zhang and coworkers developed nanocomposite based on nickel aluminum layered double hydroxides (NiAl-LDHs) with GO, containing 12 wt.% GO, showing excellent specific capacitance of 959 $F.g^{-1}$ at current density of 1 $A.g^{-1}$ and the better capacitance retention of 67% even at 10 $A.g^{-1}$ [43].

In 2018, Ignacio Martin-Gullon et al. synthesized electrode using carbon xerogel and GO (concentrations range from 1.2 to 2.5%), which showed specific capacitance of 120 $F.g^{-1}$ and power densities above 30 kW/kg, almost 180% more as compared to pristine xerogels [44]. Himadri Acharya and coworkers synthesized a hybrid electrode by GO and copper terephthalate metal-organic framework (Cu-(BDC)-MOF) which delivered high specific capacitance of 606 F/g at scan rate of 2 mV/s with better stability over 93% retention of initial capacitance after 1,000 cycles [45]. Qiang Ma and coworkers synthesized an electrode based on PPy and GO, the specific capacitance of which remained at 389 F/g under a current density of 2 A/g after 1,000 cycles. The highest conductivity of the composite achieved, was 27 times better than that of Ppy [46]. Innocenzo et al. developed material using PVA, GO coupled with Ni-Co through electrodeposition method, where the nanocomposite electrode showed a specific capacitance value of about 2,100 ± 200 $F.g^{-1}$ by maintaining the potential scan rates in the range of 10 mV s^{-1} to 250 mV s^{-1} [47].

3D CF has been largely used for preparing self-supporting supercapacitor electrodes. To improve the performance of such electrodes, Xiaodong Hong and coworkers prepared material by spin-coating method using GO and CF, the CF/GO-4,500 (4,500 rpm, 1.2 wt.% GO) composite showed the specific capacitance of 159 Fg^{-1} (258 mF cm^{-2}) at 0.5 Ag^{-1} [48]. Nasser et al. developed an electrode from polyaniline with ammonium persulfate in presence of acidic solution of GO, and the electrode exhibited a better current density in cyclic voltammetry studies with specific capacitance of 727 $F.g^{-1}$ and excellent cyclic stability by controlling only 4.3% loss after 5,000 cycles [49]. Yaoming Xiao and coworkers developed an electrode in which two sides of porous polypropylene films (PPF) was covered with the composite of PPy-GO via galvanostatic method to establish the sandwich structured PPy-GO//PPF//PPy-GO composite films, presenting specific capacitance value of 108.0 mF cm^{-2} at scan rate of 2 mV s^{-1} along with excellent flexibility and cyclic stability of 84.9% capacitance retention after 10,000 cycles [50]. Tong et al. fabricated electrode in which PANI nanofibers placed into GO architecture by the adsorption of aniline onto stacked GO sheets, performed a high capacitance of 780 F/g at 0.5 A/g and excellent rate performance of 521 F/g at 50 A/g compared to pristine PANI of 323 $F.g^{-1}$ at 0.5 $A.g^{-1}$. The assembled PGA50//PGA50 supercapacitor achieved a high energy density of 30 Wh kg^{-1} at a power density of 216 W/kg [51].

In 2019, Dan et al. reported a colloidal gel containing GO together with polyaniline nanofiber where the GO network helped in the gelation process and also served as an effective energy-transmission network to provide fast structural recovery after the gel is deformed due to high-rate shearing. This outstanding fast recovery property has made it possible to use in the technique of conventional air-brush spray to print the flexible surfaces and showed 46 times higher areal capacitance value than traditionally spray-printed ones [52]. Yifu et al. synthesized material by combining Co_2SiO_4 nanobelts with GO, and the Co_2SiO_4/GO composites showed high specific capacitance of 511 $F.g^{-1}$ at 0.5 $A.g^{-1}$ with excellent cyclic stability of 84% after 10,000 cycles [53]. Dongyu Zhao and coworkers developed an electrode using polyaniline, graphene oxide and copper (PANI/GO/Cu), which exhibited specific capacitance of 557.92 F/g with good cycling stability [54]. Haijun et al. developed a composite by polyoxometalate (POM), GO, and metallacalixarene viz. $[Ag_5 (C_2H_2N_3)_6][H_5\subset SiMo_{12}O_{40}]$ and $[Ag_5(C_2H_2N_3)_6][H_5\subset SiW_{12}O_{40}]$; ($C_2H_2N_3$ = 1H-1,2,4-triazole). Due to the synergetic effects of GO and POM-incorporated metallacalixarenes, the compound 1/15% GO-based electrode showed the highest specific capacitance of up to 230.2 $F.g^{-1}$ at current density equal to 0.5 $A.g^{-1}$ [55].

Xiankai et al. synthesized electrode by using GO and waterborne polyurethane (WPU) followed by treatment with ferric catalyst and boric acid to increase the specific surface area, the resultant materials provided high specific capacitance of 330 $F.g^{-1}$ at 0.5 $A.g^{-1}$ and also showed a good capacitance retention of about 89.9% after 5,000 cycles in 6 M KOH electrolyte [56]. Ill Min et al. developed highly effective nanocomposite using Ni nanoparticles with GO producing 3 wt.% NiO/GO(Ni/GO-1) and 8 wt.% NiO/GO(Ni/GO-2) where, Ni was present as $Ni(OH)_2$, NiO, Ni_2O_3, NiOOH, and NiO_2. The Ni/GO-2 as electrode material exhibited better specific capacitance of 461 $F.g^{-1}$ in 6 M KOH at a scan rate of 5 mV [57]. Nomxolisi et al. developed composite using GO decorated with copper, silver, or platinum (Pt) nanoparticles based on dodecylbenzene sulfonic acid (DBSA)-doped polyaniline (PANI). The composites GO/Cu/DBSA–PANI, GO/Ag/DBSA–PANI and GO/Pt/DBSA–PANI showed specific capacitance values of 192.8, 206.4, and 227.2 $F.g^{-1}$, respectively [58]. Yazhen et al. developed materials by blending 2-methylimidazole and cobalt nitrate hexahydrate together in aqueous solutions containing hexadecyltrimethylammonium bromide (CTAB), followed by GO wrapping to form nanocomposites which showed better specific capacitance of 100.41 F/g at a sweep rate of 5 mV/s [59]. Soo-Jin et al. developed a hybrid supercapacitor electrode using metal-organic framework with dual metal (Ni and Co) centers (Ni-Co-MOF) on graphene oxide sheets (Ni-Co-MOF/GO) through a one-pot solvothermal reaction. At a current density of 1 A/g, the maximum specific capacitance observed was 447.2 F/g with excellent rate capability. It also possessed outstanding life stability of 99.6% after 300 cycles [60].

Cobalt–aluminum layered double hydroxide (CoAl LDH) has been identified as a promising electrode material for pseudocapacitors. However, some properties of CoAl LDH, such as aggregation and poor conductivity limited its large-scale practical applications. To overcome such limitations, Yuxin et al. developed nanocomposites of GO with CoAl LDH and the product showed outstanding specific capacitance value of 1725.71 F/g at a current density of 1 A/g and optimal 87.73% capacitance retention at 7 A/g [61]. Paper-based SCs are more flexible and show advantages due to the adhesion between paper and active materials compared to other substrates like plastics. Yanwu et al. fabricated paper-based solid-state flexible SCs by inkjet printing using hybrid ink consisting of GO platelets and carbon quantum dots (CQDs), followed by casting of PVA/sulfuric acid (H_2SO_4) gel electrolyte. This showed a specific capacitance of ~1.0 mF cm^{-2} at a scan rate of 100 mV s^{-1}, an energy density of 0.078 mW h cm^{-3} at a power density of 0.28 mW cm^{-3} and also showed a retention of capacitance 98% after 1,000 cycles [62]. Xueping et al. synthesized composite electrode based on GO with polyaniline. They further added hexadecyltrimethylammonium bromide (CTAB) into the solution. As a result, GO was uniformly distributed in the solution and aniline monomers were easily adsorbed on the surface of GO, and when it was applied in SCs showed a specific capacitance value of 405 F.g^{-1} at 1 A.g^{-1} with 80% content of PANI [63].

In 2020, Xiankai Jiang and coworkers reported composite containing WPU, phenolic resin (PF), GO, ferric catalyst and boric acid (WPU/PF/GO/Fe/B). This supercapacitor electrode performed a high specific capacitance of 340.5 F/g at a current density of 0.5 A/g with excellent cycling stability [64]. Fabian et al. synthesized and compared different electrode materials using different metal oxides mixed with GO namely Co_3O_4@GO, MnO_2@GO, and Co_3O_4/MnO_2@GO. These sowed specific capacitance of 975, 1,358, and 1,718 F.g^{-1}, respectively at a current density of 1 A.g^{-1} [65]. Ali Maleki and coworkers presented electrode material, which contained melamine-functionalized graphene oxide nano-sheets (GO@melamine), zinc oxide nanoparticles (ZnO NPs) and IO nanoparticles (Fe_3O_4 NPs). The nanocomposite Fe_3O_4/GO@melamine-ZnO can sustain with scan rate up to 800 mV/s and showed specific volumetric capacitance of 0.00049889 F/g [66]. Elaheh et al. synthesized electrode using silsesquioxane containing graphene oxide (SSQ-GO) and tris[2-(2-methoxy ethoxy) ethyl] methyl ammonium iodide (TMEMAI) incorporated with conductive polymer, poly ortho amino phenol (SSQ-GO-TMEMAI-POAP). The specific capacitance calculated was 450 $F.g^{-1}$ at a current density of 0.5 $A.g^{-1}$ with excellent retention capacity of ~96% within 3,000 charge-discharge cycles [67]. Behzad et al. represented a simple method for the fabrication of nanocomposites based on graphene oxide and polychrysoidine (GO-PCHRYS), which performed better storage capacity and high specific capacitance of 715.3 $F.g^{-1}$ at the current density of 0.5 $A.g^{-1}$, while the electrode stability determined was 89.8% after 10,000 cycles [68]. Elaheh et al. prepared composite using triethyl amine (Et_3N) and $CrCl_3$ to host a Cr-containing molecule, grafted covalently on the surface of GO (Cr-GO). The Cr-GO composited electro-synthetically with poly ortho amino phenol (POAP) and then used as an efficient electrode material showing specific capacitance of 354 $F.g^{-1}$ at 0.5 $A.g^{-1}$ with excellent durability and cyclic stability of 94% after 5,000 cycles [69]. Lifang Liu and coworkers developed heteroatom-decorated hierarchical porous carbon nanofibers composites containing PANI, GO, phosphazene (N_3P_3(p-OC_6H_4-p-CHO)$_6$, HAPCP) and polymethyl methacrylate (PMMA). This novel free-standing flexible electrode exhibited excellent electrochemical performance of 680.8 F/g at 0.5 A/g with a good capacitance retention of 93.5% after 3,000 cycles [70].

Md. Mominul et al. synthesized material from 3D MWCNTs horizontally dispersed on 2D GO nanosheet framework, supported by MnO_2. The optimum contents of MWCNTs and MnO_2 in the composites showed a maximum specific capacitance of 1,380 $F.g^{-1}$ [71]. Haifeng et al. prepared materials for symmetric solid-state SCs by PPy/GO where with 2.5 wt.% GO, the PPy/GO nanocomposite showed a high specific capacitance of 97.3 mF cm^{-2} at current density of 1 mA cm^{-2} and the PPy/GO film supercapacitor also provided excellent cycling stability and good flexibility [72]. Xiaohong et al. prepared porous electrode materials by combination of carbon materials (GO) and active metal (Zn, Co, and Ni) (GO/Zn–Co–Ni). The GO/ZCN exhibits a specific capacitance of 843.0 C/g at a current density of 1 A/g and also exhibited capacitance retention of 97.71% at the end of 5,000 charge-discharge cycles [73]. Zhong-Shuai et al. introduced a new class of "all-inside-one" film SCs by encapsulating 2D interdigital microelectrodes in chemically cross-linked PVA-based hydrogel electrolyte containing GO which significantly improved the ionic conductivity and boosted robustness of hydrogel electrolyte. It showed high single-cell areal capacitance of 40.8 mF cm^{-2} from microelectrodes as thin as 5 μm [74]. Partha Pratim Das and coworkers developed electrode using vanadomanganate and graphene oxide ($K_7Mn^{IV}V_{13}O_{38}.18H_2O$-GO with 2:1 and 4:1 ratio), The MnV_{13}/GO composite of ratio 4:1 in 1M Na_2SO_4 electrolyte and that of ratio 2:1

in 1M LiCl electrolyte showed significant specific capacitance values of 387.02 $F.g^{-1}$ and 269.15 $F.g^{-1}$, respectively with energy density of 53.75 Wh/kg and 37.38 Wh/kg, respectively at a scan rate of 5 $mV.s^{-1}$ [75]. Juan et al. fabricated the nitrogen-doped hierarchical carbon nanosheets (i.e., HGPC-A) for electrode via one-step carbonization and activation strategy in the presence of GO using PVA/polyaniline hydrogels as precursors and they observed that when the current density increased from 0.5 $A.g^{-1}$ to 20 $A.g^{-1}$, it showed high specific capacitance of 311 $F.g^{-1}$ with superior rate capacity, excellent cycling stability of 88.5% over 10,000 cycles and better charge storage capacity of 7.0 Wh kg^{-1} [76].

In 2021, Wu et al. developed a sandwich-like composite electrode based on PANI, ZIF-67 and graphene oxide (GO/ZIF-67/PANI) which showed the specific capacitance up to 1,014 $F.g^{-1}$ at a current density of 1 $A.g^{-1}$. The assembled ASC based on GO/$Co(OH)_2$/PANI provided a high energy density of 53.42 $Wh{\cdot}kg^{-1}$ at a power density of 800 $W{\cdot}kg^{-1}$ and the cyclic capacitance remained 80% after 20,000 cycles [77]. Fouad Ghamouss and coworkers synthesized material by combination of polyaniline (PANI), protic ionic liquid (PIL) and GO, the PANI/PIL/GO containing 16 wt.% of GO material exhibited specific capacitance of 223 $F.g^{-1}$ at current density of 10 $A.g^{-1}$ in H_2SO_4 (1 mol L^{-1}) [78]. Somenath et al. developed a composite electrolyte using GO sheets, carboxylated carbon nanotubes (fCNTs), and the hybrid of fCNT/GO. The composite electrolyte fCNT/GO-PGE showed 50% higher thermal stability than original PGE. Finally, MnO_2-based SCs using PGE, GO-PGE, and fCNT/GO-PGE electrolytes were fabricated and the electrode showed specific capacitance as 39.5, 77.6, and 83.3 $F.g^{-1}$, respectively [79]. Nehal et al. developed electrode by polyindole (PIND), polyacrylonitrile (PAN) and GO. The PIND/PAN/GO nanofiber webs exhibited a better specific capacitance of 4,960 mF g^{-1} than that of PIND/PAN nanofiber webs (1,810 mF g^{-1}) in 1 M H_2SO_4 [80]. Mohammad Mazloum-Ardakani et al. synthesized material using arginine amino acid and GO, and the resultant electrode provided an exceptional specific capacitance of 295 F/g in 1 A/g with excellent cycling stability of 94% after 4,000 cycles [81]. Liang et al. developed nanocomposite with GO, PPy, and PANI. This composite when used as an electrode for supercapacitor, it exhibited specific capacitance of 372 F/g at current density of 0.4 A/g and also maintained cyclic stability of about 91% of the specific capacitance after 2,000 electrochemical cycles [82]. Debajyoti et al. developed a green approach for the synthesis of composite material from PANI and GO. The composite PANI-GO showed pseudocapacitive behavior with an enhanced specific capacitance of 510 F/g at a current density of 1 A/g and also maintained 92% cyclic stability even after completing 1,000 cycles [83]. Guoji Liu and coworkers synthesized electrode material based on graphene oxide, mesoporous silica, and polyaniline (GO/mSiO_2/PANI) where the composite showed a highest specific capacitance of 1218.4 F/g at 0.5 A/g and maintained capacitive stability of 92.5% after 1,000 cycles [84].

Yongjin et al. developed 3D composites using Co_3O_4 based on GO and MoS_2 (GO-MoS_2/Co_3O_4) which exhibited excellent specific capacitance of 135.3 $F.g^{-1}$ at current density of 1 $A.g^{-1}$ and also showed great cycling stability by retaining 87% after 3,000 cycles [85]. Eduardo et al. synthesized nanocomposites using GO, MnO_2 nanoparticles and PANI in presence of aqueous medium of $KMnO_4$ as oxidizing agent. The composites showed highest capacitances of 150 $F.g^{-1}$ with 68.2% cyclic stability after 500 charge-discharge cycles [86]. Wenfeng Qiu and coworkers developed a composite electrode with GO and lanthanum ntitanate ($La_2Ti_2O_7$, LTO) fibers. The composite showed a high specific capacitance of 900.6 F/g at a current density of 1 A/g in 1 M H_2SO_4 and 10 wt.% aqueous solution of sucrose as the electrolyte [87].

Shuihua et al. synthesized nanocomposite using Ni_2CoS_4 and GO (MH-Ni_2CoS_4/GO) by microwave (MW) heating and hydrothermal method. The MH-Ni_2CoS_4/GO electrode exhibited high specific capacitance of 2675.0 F/g at the current densities of 2 A/g, and showed cyclic stability of 95.0% even after 2,000 cycles [88]. Wanwan et al. developed electrode material by GO nanosheets and polyacrylonitrile (PAN-GO) and then nanofibers were coated on the surface of Ni-plated cotton yarn (NCY) through conjugate electrospinning process, followed by chemical deposition of PPy layer and the electrode exhibited a high specific capacitance of 28.34 mF cm^{-2} with high energy density of 3.98 μWh cm^{-2} and also showed high capacitance retention of 90.2% after 1,000 cycles [89]. The recent developments of GO-based SCs are mentioned in Table 6.1.

6.4 CONCLUSION

Graphene-based materials are extensively being investigated as promising materials for the development of low-cost and efficient SCs. As the capacitance is proportional to the surface area of the electrode materials, so chemically and electrochemically inert carbons such as graphene can be considered as good candidates for supercapacitor matrices. Scientists and researchers have suggested the incorporation of transition metals and metal oxides along with graphene and GO to improve the capacitance value of SCs

TABLE 6.1 Recent Developments of Graphene Oxide-based Supercapacitors

Composites	Specific Capacitance (F/g)	Current Density (A/g)	Cyclic Stability	References
NiO/GO	461	–	–	[57]
GO/Pt/DBSA–PANI	227.2	50.8	–	[58]
CTAB/GO	100.4	1	–	[59]
Ni-Co-MOF/GO	447.2	1	99.6%@300 cycles	[60]
CoAl LDH/GO	1725.7	1	85.56%@5000 cycles	[61]
GO/CQDs	–	–	98% @1000 cycles	[62]
GO/PANI	405	1	71% @700 cycles	[63]
WPU/PF/GO/Fe/B	340.5	0.5	–	[64]
Co_3O_4/MnO_2@GO	1,718	1	82.23%@5000 cycles	[65]
Fe_3O_4/GO@melamine-ZnO	0.00049889	–	72%@1000 cycles	[66]
SSQ-GO-TMEMAI-POAP	450	0.5	96%@3000 cycles	[67]
GO-PCHRYS	715.3	0.5	89.8%@10000 cycles	[68]
Cr-GO	354	0.5	94%@5000 cycles	[69]
PANI/GO/PMMA/HAPCP	680.8	0.5	93.5% @3000 cycles	[70]
MnO_2/MWCNT/GO	1,380	–	–	[71]
PPy/GO	–	–	94.1%@1000 cycles	[72]
GO/ZCN	843	1	97.7%@5000 cycles	[73]
GPAH/MP/GPAH	–	–	–	[74]
MnV_{13}/GO	387	0.5	–	[75]
HGPC-A	311	–	88.5%@10000 cycles	[76]
GO/ZIF-67/PANI	1,014	1	80%@20000 cycles	[77]

TABLE 6.1 *(Continued)*

Composites	Specific Capacitance (F/g)	Current Density (A/g)	Cyclic Stability	References
PANI/PIL/GO	223	10	–	[78]
fCNT/GO-PGE	83.3	–	–	[79]
PIND/PAN/GO	4.9	–	–	[80]
Arg/GO	295	1	94%@4000 cycles	[81]
GO/PPy/PANI	372	0.4	91%@2000 cycles	[82]
PANI/GO	510	1	92%@1000 cycles	[83]
GO/mSiO_2/PANI	1218.4	0.5	92.5%@1000 cycles	[84]
GO-MoS_2/Co_3O_4	135.3	1	87%@3000 cycles	[85]
GO/MnO_2/PANI	150	–	68.2%500 cycles	[86]
GO/LTO	900.6	1	–	[87]
MH-Ni_2CoS_4/GO	2,615	2	96%@2000 cycles	[88]
PGPG/NCY	–	–	92%@1000 cycles	[89]

due to additional redox reactions. This chapter highlights brief information on the requirement of graphene-based materials in different macrostructural complexity such as 0D, 1D, 2D, 3D, etc. The recent trends toward the rationalizing of the nanostructure and architecture of graphene-based nanocomposites to achieve better electrochemical performances in the field of SCs are discussed in detail.

KEYWORDS

- **carbon nanotubes**
- **GO-based benzimidazole cross-linked network**
- **graphene oxide**
- **multi-walled carbon nanotube**
- **polyaniline**
- **supercapacitors**

REFERENCES

1. Chen, S., Zhu, J., Wu, X., Han, Q., & Wang, X., (2010). Graphene oxide-MnO_2 nanocomposites for supercapacitors. *ACS Nano, 4*, 2822–2830.
2. Xu, J., Wang, K., Zu, S. Z., Han, B. H., & Wei, Z., (2010). Hierarchical nanocomposites of polyaniline nanowire arrays on graphene oxide sheets with synergistic effect for energy storage. *ACS Nano, 4*, 5019–5026.
3. Aboutalebi, S. H., Chidembo, A. T., Salari, M., Konstantinov, K., Wexler, D., Liu, H. K., & Dou, S. X., (2011). Comparison of GO, GO/MWCNTs composite and MWCNTs as potential electrode materials for supercapacitors. *Energy & Environmental Science, 4*, 1855–1865.
4. Chen, Y., Hu, Z., Chang, Y., Wang, H., Fu, G., Jin, X., & Xie, L., (2011). Layered Al-substituted cobalt hydroxides/GO composites for electrode materials of supercapacitors. *Chinese Journal of Chemistry, 29*, 2257–2262.
5. Xu, B., Yue, S., Sui, Z., Zhang, X., Hou, S., Cao, G., & Yang, Y., (2011). What is the choice for supercapacitors: Graphene or graphene oxide?. *Energy & Environmental Science, 4*, 2826–2830.
6. Fang, J., Li, M., Li, Q., Zhang, W., Shou, Q., Liu, F., & Cheng, J., (2012). Microwave-assisted synthesis of CoAl-layered double hydroxide/graphene oxide composite and its application in supercapacitors. *Electrochimica Acta, 85*, 248–255.
7. You, B., Li, N., Zhu, H., Zhu, X., & Yang, J., (2013). Graphene oxide-dispersed pristine CNTs support for MnO_2 nanorods as high performance supercapacitor electrodes. *ChemSusChem, 6*, 474–480.

8. Cui, Y., Cheng, Q. Y., Wu, H., Wei, Z., & Han, B. H., (2013). Graphene oxide-based benzimidazole-crosslinked networks for high-performance supercapacitors. *Nanoscale, 5*, 8367–8374.
9. Gund, G. S., Dubal, D. P., Patil, B. H., Shinde, S. S., & Lokhande, C. D., (2013). Enhanced activity of chemically synthesized hybrid graphene oxide/Mn3O4 composite for high performance supercapacitors. *Electrochimica Acta, 92*, 205–215.
10. Zhang, Q., Li, Y., Feng, Y., & Feng, W., (2013). Electropolymerization of graphene oxide/polyaniline composite for high-performance supercapacitor. *Electrochimica Acta, 90*, 95–100.
11. Chen, Z., Yu, D., Xiong, W., Liu, P., Liu, Y., & Dai, L., (2014). Graphene-based nanowire supercapacitors. *Langmuir, 30*, 3567–3571.
12. Yun, Y. S., Cho, S. Y., & Jin, H. J., (2014). Carbon aerogels based on regenerated silk proteins and graphene oxide for supercapacitors. *Macromolecular Research, 22*, 509–514.
13. Chen, C., Fu, X., Ma, T., Fan, W., Wang, Z., & Miao, S., (2014). Synthesis and electrochemical properties of graphene oxide/nanosulfur/polypyrrole ternary nanocomposite hydrogel for supercapacitors. *Journal of Applied Polymer Science, 131*.
14. DeYoung, A. D., Park, S. W., Dhumal, N. R., Shim, Y., Jung, Y., & Kim, H. J., (2014). Graphene oxide supercapacitors: A computer simulation study. *The Journal of Physical Chemistry C, 118*, 18472–18480.
15. Gui, D., Liu, C., Chen, F., & Liu, J., (2014). Preparation of polyaniline/graphene oxide nanocomposite for the application of supercapacitor. *Applied Surface Science, 307*, 172–177.
16. Li, Z. F., Zhang, H., Liu, Q., Liu, Y., Stanciu, L., & Xie, J., (2014). Covalently-grafted polyaniline on graphene oxide sheets for high performance electrochemical supercapacitors. *Carbon, 71*, 257–267.
17. Liu, Y., Yan, D., Li, Y., Wu, Z., Zhuo, R., Li, S., & Geng, Z., (2014). Manganese dioxide nanosheet arrays grown on graphene oxide as an advanced electrode material for supercapacitors. *Electrochimica Acta, 117*, 528–533.
18. Wu, W., Li, Y., Zhao, G., Yang, L., & Pan, D., (2014). Aldehyde–poly (ethylene glycol) modified graphene oxide/conducting polymers composite as high-performance electrochemical supercapacitors. *Journal of Materials Chemistry A, 2*, 18058–18069.
19. Zhou, H., Han, G., Xiao, Y., Chang, Y., & Zhai, H. J., (2014). Facile preparation of polypyrrole/graphene oxide nanocomposites with large areal capacitance using electrochemical code position for supercapacitors. *Journal of Power Sources, 263*, 259–267.
20. Zhao, Y., Ran, W., He, J., Huang, Y., Liu, Z., Liu, W., & Gao, F., (2015). High-performance asymmetric supercapacitors based on multilayer MnO_2/graphene oxide nanoflakes and hierarchical porous carbon with enhanced cycling stability. *Small, 11*, 1310–1319.
21. Cao, J., Wang, Y., Chen, J., Li, X., Walsh, F. C., Ouyang, J. H., & Zhou, Y., (2015). Three-dimensional graphene oxide/polypyrrole composite electrodes fabricated by one-step electrodeposition for high performance supercapacitors. *Journal of Materials Chemistry A, 3*, 14445–14457.
22. Wang, B., Qiu, J., Feng, H., & Sakai, E., (2015). Preparation of graphene oxide/polypyrrole/multi-walled carbon nanotube composite and its application in supercapacitors. *Electrochimica Acta, 151*, 230–239.
23. Chee, W. K., Lim, H. N., Harrison, I., Chong, K. F., Zainal, Z., Ng, C. H., & Huang, N. M., (2015). Performance of flexible and binderless polypyrrole/graphene oxide/

zinc oxide supercapacitor electrode in a symmetrical two-electrode configuration. *Electrochimica Acta, 157*, 88–94.

24. Liu, Y., Zhou, J., Zhu, E., Tang, J., Liu, X., & Tang, W., (2015). Facile synthesis of bacterial cellulose fibres covalently intercalated with graphene oxide by one-step cross-linking for robust supercapacitors. *Journal of Materials Chemistry C, 3*, 1011–1017.
25. Pal, K., Panwar, V., Bag, S., Manuel, J., Ahn, J. H., & Kim, J. K., (2015). Graphene oxide–polyaniline–polypyrrole nanocomposite for a supercapacitor electrode. *RSC Advances, 5*, 3005–3010.
26. Yang, J., Yu, C., Fan, X., Zhao, C., & Qiu, J., (2015). Ultrafast self-assembly of graphene oxide-induced monolithic NiCo–carbonate hydroxide nanowire architectures with a superior volumetric capacitance for supercapacitors. *Advanced Functional Materials, 25*, 2109–2116.
27. Azman, N. H. N., Lim, H. N., & Sulaiman, Y., (2016). Influence of concentration and electrodeposition time on the electrochemical supercapacitor performance of poly (3, 4-ethylene dioxythiophene)/graphene oxide hybrid material. *Journal of Nanomaterials*.
28. Lei, Q., Song, H., Chen, X., Li, M., Li, A., Tang, B., & Zhou, D., (2016). Effects of graphene oxide addition on the synthesis and supercapacitor performance of carbon aerogel particles. *RSC Advances, 6*, 40683–40690.
29. Li, M., Yin, W., Han, X., & Chang, X., (2016). Hierarchical nanocomposites of polyaniline scales coated on graphene oxide sheets for enhanced supercapacitors. *Journal of Solid State Electrochemistry, 20*, 1941–1948.
30. Ma, L., Su, L., Zhang, J., Zhao, D., Qin, C., Jin, Z., & Zhao, K., (2016). A controllable morphology GO/PANI/metal hydroxide composite for supercapacitor. *Journal of Electro-analytical Chemistry, 777*, 75–84.
31. Patil, U. M., Nam, M. S., Kang, S., Sohn, J. S., Sim, H. B., Kang, S., & Jun, S. C., (2016). Fabrication of ultra-high energy and power asymmetric supercapacitors based on hybrid 2D MoS 2/graphene oxide composite electrodes: A binder-free approach. *RSC Advances, 6*, 43261–43271.
32. Pawar, P. B., Shukla, S., & Saxena, S., (2016). Graphene oxide–polyvinyl alcohol nanocomposite based electrode material for supercapacitors. *Journal of Power Sources, 321*, 102–105.
33. Zhang, C., Wang, L., Zhao, Y., Tian, Y., & Liang, J., (2016). Self-assembly synthesis of graphene oxide double-shell hollow-spheres decorated with Mn_3O_4 for electrochemical supercapacitors. *Carbon, 107*, 100–108.
34. Zhou, H., & Han, G., (2016). One-step fabrication of heterogeneous conducting polymers-coated graphene oxide/carbon nanotubes composite films for high-performance supercapacitors. *Electrochimica Acta, 192*, 448–455.
35. Zhou, H., Zhai, H. J., & Han, G., (2016). Adjust the electrochemical performances of graphene oxide nanosheets-loaded poly (3, 4-ethylene dioxythiophene) composites for supercapacitors with ultralong cycle life. *Journal of Materials Science: Materials in Electronics, 27*, 2773–2782.
36. Zou, Y., Wang, Q., Xiang, C., She, Z., Chu, H., Qiu, S., & Sun, L., (2016). One-pot synthesis of ternary polypyrrole-Prussian-blue-graphene-oxide hybrid composite as electrode material for high-performance supercapacitors. *Electrochimica Acta, 188*, 126–134.
37. Asen, P., & Shahrokhian, S., (2017). One step electrodeposition of V_2O_5/polypyrrole/graphene oxide ternary nanocomposite for preparation of a high performance supercapacitor. *International Journal of Hydrogen Energy, 42*, 21073–21085.

38. Oyedotun, K. O., Madito, M. J., Bello, A., Momodu, D. Y., Mirghni, A. A., & Manyala, N., (2017). Investigation of graphene oxide nanogel and carbon nanorods as electrode for electrochemical supercapacitor. *Electrochimica Acta, 245*, 268–278.
39. Tabrizi, A. G., Arsalani, N., Namazi, H., & Ahadzadeh, I., (2017). Vanadium oxide assisted synthesis of polyaniline nanoarrays on graphene oxide sheets and its application in supercapacitors. *Journal of Electroanalytical Chemistry, 798*, 34–41.
40. Tanaka, S., Salunkhe, R. R., Kaneti, Y. V., Malgras, V., Alshehri, S. M., Ahamad, T., & Hossain, M. S., (2017). A. Prussian blue derived iron oxide nanoparticles wrapped in graphene oxide sheets for electrochemical supercapacitors. *RSC Advances, 7*, 33994–33999.
41. Wang, C., Wu, C., Chen, S., Zhao, X., Xu, W., Niu, Z., & Song, L., (2017). Membrane-assisted assembly strategy of flexible electrodes for multifunctional supercapacitors. *Carbon, 125*, 419–428.
42. Wei, Y., Ding, R., Zhang, C., Lv, B., Wang, Y., Chen, C., & Li, Y., (2017). Facile synthesis of self-assembled ultrathin α-FeOOH nanorod/graphene oxide composites for supercapacitors. *Journal of Colloid and Interface Science, 504*, 593–602.
43. Zhang, L., Yao, H., Li, Z., Sun, P., Liu, F., Dong, C., & Zhao, B., (2017). Synthesis of delaminated layered double hydroxides and their assembly with graphene oxide for supercapacitor application. *Journal of Alloys and Compounds, 711*, 31–41.
44. Ramos-Fernández, G., Canal-Rodríguez, M., Arenillas, A., Menéndez, J. A., Rodríguez-Pastor, I., & Martin-Gullon, I., (2018). Determinant influence of the electrical conductivity versus surface area on the performance of graphene oxide-doped carbon xerogel supercapacitors. *Carbon, 126*, 456–463.
45. Islam, D. A., Chakraborty, A., Roy, A., Das, S., & Acharya, H., (2018). Fabrication of graphene oxide (GO)-supported sheet-like CuO nanostructures derived from a metal-organic-framework template for high-performance hybrid supercapacitors. *Chemistry Select*, 11816–11823.
46. Fu, C., Ma, Q., Liu, H., & Tang, W., (2018). Effects of graphene oxide on the conductivity and capacitance of polypyrrole. *Int J. Electrochem Sci., 13*, 4267–4275.
47. Coviello, D., & Casella, I. G., (2018). Electrochemical performance of binary Ni-Co particles deposited on graphene oxide/polyvinyl alcohol substrate in alkaline medium. *Electrochimica Acta, 261*, 104–112.
48. Hong, X., Lu, Y., Li, S., Wang, R., & Liu, Y., (2018). Extremely low fractions of graphene oxide in carbon foam prepared by a spin-coating method as freestanding supercapacitor electrodes. *Journal of Materials Science, 53*, 16476–16483.
49. Tabrizi, A. G., Arsalani, N., Mohammadi, A., Ghadimi, L. S., Ahadzadeh, I., & Namazi, H., (2018). A new route for the synthesis of polyaniline nanoarrays on graphene oxide for high-performance supercapacitors. *Electrochimica Acta, 265*, 379–390.
50. Wang, N., Han, G., Xiao, Y., Li, Y., Song, H., & Zhang, Y., (2018). Polypyrrole/graphene oxide deposited on two metalized surfaces of porous polypropylene films as all-in-one flexible supercapacitors. *Electrochimica Acta, 270*, 490–500.
51. Zhou, Q., Wei, T., Yue, J., Sheng, L., & Fan, Z., (2018). Polyaniline nanofibers confined into graphene oxide architecture for high-performance supercapacitors. *Electrochimica Acta, 291*, 234–241.
52. Xiong, Z., Yun, X., Qiu, L., Sun, Y., Tang, B., He, Z., & Li, D., (2019). A dynamic graphene oxide network enables spray printing of colloidal gels for high-performance micro-supercapacitors. *Advanced Materials, 31*, 1804434.

53. Cheng, Y., Zhang, Y., & Meng, C., (2019). Template fabrication of amorphous Co_2SiO_4 nanobelts/graphene oxide composites with enhanced electrochemical performances for hybrid supercapacitors. *ACS Applied Energy Materials, 2*, 3830–3839.
54. Ma, Y., Zhao, D., Chen, Y., Huang, J., Zhang, Z., Zhang, X., & Zhang, B., (2019). A novel core-shell polyaniline/graphene oxide/copper nanocomposite for high performance and low-cost supercapacitors. *Chemical Papers, 73*, 119–129.
55. Hou, Y., Chai, D., Li, B., Pang, H., Ma, H., Wang, X., & Tan, L., (2019). Polyoxometalate-incorporated metallacalixarene@ graphene composite electrodes for high-performance supercapacitors. *ACS Applied Materials & Interfaces, 11*, 20845–20853.
56. Li, R., Qin, C., Zhang, X., Lin, Z., Lv, S., & Jiang, X., (2019). Boron/nitrogen co-doped carbon synthesized from waterborne polyurethane and graphene oxide composite for supercapacitors. *RSC Advances, 9*, 1679–1689.
57. Gopiraman, M., Saravanamoorthy, S., Deng, D., Ilangovan, A., Kim, I. S., & Chung, I. M., (2019). Facile mechanochemical synthesis of nickel/graphene oxide nanocomposites with unique and tunable morphology: Applications in heterogeneous catalysis and supercapacitors. *Catalysts, 9*, 486.
58. Dywili, N. R., Ntziouni, A., Ikpo, C., Ndipingwi, M., Hlongwa, N. W., Yonkeu, A. L., & Iwuoha, E. I., (2019). Graphene oxide decorated nanometal-poly (anilino-dodecylbenzene sulfonic acid) for application in high performance supercapacitors. *Micromachines, 10*, 115.
59. Cao, W., Han, M., Qin, L., Jiang, Q., Xu, J., Lu, Z., & Wang, Y., (2019). Synthesis of zeolitic imidazolate framework-67 nanocube wrapped by graphene oxide and its application for supercapacitors. *Journal of Solid State Electrochemistry, 23*, 325–334.
60. Hong, J., Park, S. J., & Kim, S., (2019). Synthesis and electrochemical characterization of nanostructured Ni-Co-MOF/graphene oxide composites as capacitor electrodes. *Electrochimica Acta, 311*, 62–71.
61. Jing, C., Huang, Y., Xia, L., Chen, Y., Wang, X., Liu, X., & Zhang, Y., (2019). Growth of cobalt-aluminum layered double hydroxide nanosheets on graphene oxide towards high performance supercapacitors: The important role of layer structure. *Applied Surface Science, 496*, 143700.
62. Liu, J., Ye, J., Pan, F., Wang, X., & Zhu, Y., (2019). Solid-state yet flexible supercapacitors made by inkjet-printing hybrid ink of carbon quantum dots/graphene oxide platelets on paper. *Science China Materials, 62*, 545–554.
63. Wang, Y., Wang, Y., Tian, Y., Ma, L., Wang, C., & Gao, X., (2019). Fabrication and characterization of graphene oxide/polyaniline electrode composite for high performance supercapacitors. *ECS Journal of Solid State Science and Technology, 8*, M103.
64. Zhang, X., Lin, Z., Qin, C., Guo, X., Ma, Y., & Jiang, X., (2020). One-step carbonization production of B/N co-doped carbon from polyurethane/phenolic/GO composite for supercapacitors. *Journal of Materials Science: Materials in Electronics, 31*, 715–727.
65. Obodo, R. M., Onah, E. O., Nsude, H. E., Agbogu, A., Nwanya, A. C., Ahmad, I., & Ezema, F. I., (2020). Performance evaluation of graphene oxide based Co_3O_4@ GO, MnO_2@ GO and Co_3O_4/MnO_2@ GO electrodes for supercapacitors. *Electroanalysis, 32*, 2786–2794.
66. Eivazzadeh-Keihan, R., Taheri-Ledari, R., Khosropour, N., Dalvand, S., Maleki, A., Mousavi-Khoshdel, S. M., & Sohrabi, H., (2020). Fe_3O_4/GO@ melamine-ZnO nanocomposite: A promising versatile tool for organic catalysis and electrical capacitance. *Colloids and Surfaces A: Physicochemical and Engineering Aspects, 587*, 124335.

67. Ajdari, F. B., Kowsari, E., Nadri, H. R., Maghsoodi, M., Ehsani, A., Mahmoudi, H., & Ramakrishna, S., (2020). Electrochemical performance of silsesquioxane-GO loaded with alkoxy substituted ammonium-based ionic liquid and POAP for supercapacitor. *Electrochimica Acta, 354*, 136663.
68. Zarandi, R. F., Rezaei, B., Ghaziaskar, H. S., & Ensafi, A. A., (2020). Synthesis of graphene oxide-polychrysoidine nanocomposite for supercapacitor applications. *Journal of Energy Storage, 29*, 101334.
69. Ashtiani, A. A., Kowsari, E., Haddadi-Asl, V., Yousefi, M., Naderi, H. R., Chinnappan, A., & Ramakrishna, S., (2020). Pseudocapacitive efficiency of covalently Cr-complex with L-histidine-methyl ester as a ligand graphene oxide blended with conducting polymer (POAP) as electrode material in supercapacitor. *Journal of Molecular Liquids, 315*, 113697.
70. Zhu, J., Zhang, Q., Chen, H., Zhang, R., Liu, L., & Yu, J., (2020). *Setaria viridis*-inspired electrode with polyaniline decorated on porous heteroatom-doped carbon nanofibers for flexible supercapacitors. *ACS Applied Materials & Interfaces, 12*, 43634–43645.
71. Islam, M. M., Mollah, M. Y. A., Susan, M. A. B. H., & Islam, M. M., (2020). Frontier performance of in situ formed α-MnO_2 dispersed over functionalized multi-walled carbon nanotubes covalently anchored to a graphene oxide nanosheet framework as supercapacitor materials. *RSC Advances, 10*, 44884–44891.
72. Wen, J., Ding, Y., Zhong, J., Chen, R., Gao, F., Qiao, Y., & He, H., (2020). Ice-interface assisted large-scale preparation of polypyrrole/graphene oxide films for all-solid-state supercapacitors. *RSC Advances, 10*, 41503–41510.
73. Zhu, L., Hao, C., Wang, X., & Guo, Y., (2020). Fluffy cotton-like GO/Zn–Co–Ni layered double hydroxides form from a sacrificed template GO/ZIF-8 for high performance asymmetric supercapacitors. *ACS Sustainable Chemistry & Engineering, 8*, 11618–11629.
74. Das, P., Shi, X., Fu, Q., & Wu, Z. S., (2020). Substrate-free and shapeless planar micro-supercapacitors. *Advanced Functional Materials, 30*, 1908758.
75. Kumari, S., Maity, S., Vannathan, A. A., Shee, D., Das, P. P., & Mal, S. S., (2020). Improved electrochemical performance of graphene oxide supported vanadomanganate (IV) nanohybrid electrode material for supercapacitors. *Ceramics International, 46*, 3028–3035.
76. Wang, M., Yang, J., Liu, S., Li, M., Hu, C., & Qiu, J., (2020). Nitrogen-doped hierarchically porous carbon nanosheets derived from polymer/graphene oxide hydrogels for high-performance supercapacitors. *Journal of Colloid and Interface Science, 560*, 69–76.
77. Xu, M., Guo, H., Xue, R., Wang, M., Wu, N., Wang, X., & Yang, W., (2021). Sandwich-like GO@ Co $(OH)_2$/PANI derived from MOFs as high-performance electrode for supercapacitors. *Journal of Alloys and Compounds, 863*, 157699.
78. Al-Zohbi, F., Ghamouss, F., Schmaltz, B., Abarbri, M., Zaghrioui, M., & Tran-Van, F., (2021). Enhanced storage performance of PANI and PANI/graphene composites synthesized in protic ionic liquids. *Materials, 14*, 4275.
79. Azizighannad, S., Wang, Z., Siddiqui, Z., Kumar, V., & Mitra, S., (2021). Nano carbon doped polyacrylamide gel electrolytes for high performance supercapacitors. *Molecules, 26*, 2631.
80. Elmessiry, K. S., El-Aassar, M. R., Nassr, A. B. A. A., Kenawy, E. R., Moharam, B. E., & Ali, N., (2021). Free-standing working electrodes for supercapacitors based on composite polymer nanofibers and functionalized with graphene oxide. *Journal of Electronic Materials, 50*, 5599–5611.

81. Haghshenas, M., Mazloum-Ardakani, M., Amiri-Zirtol, L., & Sabaghian, F., (2021). Arginine-functionalized graphene oxide for green and high-performance symmetric supercapacitors. *International Journal of Hydrogen Energy, 46*, 30219–30229.
82. Zhao, Z., Wang, H., Huang, H., Li, L., & Yu, X., (2021). Graphene oxide/polypyrrole/polyaniline composite hydrogel synthesized by vapor-liquid interfacial method for supercapacitors. *Colloids and Surfaces A: Physicochemical and Engineering Aspects, 626*, 127125.
83. Konwar, G., Deka, J., Raidongia, K., & Mahanta, D., (2021). Fabrication of polyaniline–graphene oxide hybrid nanocomposites by green interfacial polymerization for all-solid-state supercapacitors and enzymatic glucose sensors. *New Journal of Chemistry, 45*, 17909–17917.
84. Chen, X., Xu, L., & Liu, G., (2021). Graphene oxide-mesoporous silica-polyaniline composite with intercalated structures as supercapacitor electrode materials. *Russian Journal of Physical Chemistry A, 95*, 623–629.
85. Xu, J., Liang, J., Zou, Y., Xu, F., Chen, Q., Xiang, C., & Sun, L., (2021). Layer-by-layer self-assembled GO-$MoS_2Co_3O_4$ three-dimensional conducting network for high-performance supercapacitors. *Journal of Energy Storage, 43*, 103195.
86. Zeplin, G., & Neiva, E. G., (2021). One-pot green synthesis of graphene oxide/MnO_2/polyaniline nanocomposites applied in aqueous and neutral supercapacitors and sensors. *Journal of Electroanalytical Chemistry*, 115776.
87. Lu, M., Cao, Y., Xue, Y., & Qiu, W., (2021). Preparation of graphene Oxide/$La_2Ti_2O_7$ composites with enhanced electrochemical performances for supercapacitors. *ACS Omega*.
88. Tang, Z., Tang, S., Li, Q., Wei, Z., & Zhou, T., (2021). Synergistic effect of microwave heating and hydrothermal methods on synthesized Ni_2CoS_4/GO for ultrahigh capacity supercapacitors. *Journal of Colloid and Interface Science, 582*, 312–321.
89. Zhao, X., Li, W., Li, F., Hou, Y., Lu, T., Pan, Y., & He, J., (2021). Wearable yarn supercapacitors coated with twisted PPy@ GO nanosheets and PPy@ PAN-GO nanofibres. *Journal of Materials Science, 56*, 18147–18161.

CHAPTER 7

APPLICATIONS OF RECYCLED SOLID SUBSTANCES OF SOLAR PHOTOVOLTAIC MODULES AND THEIR ENVIRONMENTAL IMPACT ANALYSIS

MANISHA SHEORAN[1], PANCHAM KUMAR[2], SUSHEELA SHARMA[1], and PRABHAT RANJAN[3]

[1]*Department of Basic Science, Bhartiya Skill Development University, Jaipur, Rajasthan, India*

[2]*School of Electrical Skills, Bhartiya Skill Development University, Jaipur, Rajasthan, India*

[3]*Department of Mechatronics Engineering, Manipal University, Jaipur, Rajasthan, India*

ABSTRACT

With the increasing population density, energy requirements have heightened drastically, and it was fulfilled by renewable energy technology. In this, solar photovoltaic technology proved as the main contributing factor. Solar photovoltaic installations in India started from late 1993. It further hiked exponentially from early 2010 to 2018. Till December 2019, total solar installations were reported to 33.730 GW. This chapter pertains to the state-wise installations of solar photovoltaic systems in the country. At the end life cycle, solar photovoltaic modules will bring damaging effects on the ecosystem; however, these effects can be curtailed by persuading

Nanotechnology-Enhanced Solid Materials: Design, Synthesis, Properties, Applications, and Perspectives.
Lionello Pogliani, Ann Rose Abraham, A. K. Haghi, & Prabhat Ranjan (Eds.)

towards recycling methods. A feasibility analysis is carried out on the recycled products so as to bring out their utilization in various aspects. The recovery product can be further utilized for applications like making window frames of Aluminum parts, using copper wires for new connections, making decorative items of silicon panels, also utilizing the solar panels for making the buildings, tables, cupboards, and toys working under low voltage requirements. The damaged solar panels can be used in making monuments and dustbins for public use. The panels can be used for making vehicle's body, and travel bags and also can be utilized for making roads after the removal of harmful elements from the used and damaged solar panels. Recovery product of recycling is analyzed quantitatively so as to know the payback of recycling procedures. Existing recycling methods of solar photovoltaic technology are assessed and their environmental impacts are studied. Landfilling option is also taken under consideration, but due to lower economic value, it cannot be taken forward on a commercial scale. The chapter lays emphasis upon the environmental impact of the solar photovoltaic technology along with their recycling economic analysis. Recycling came out to be less burdensome than landfilling procedure. Value can be made from the damaged and decommissioned solar panels. Further recycling brought out the environmental benefits in terms of human toxicity potential, global warming potential, abiotic resource depletion potential, acidification potential, etc. Solar waste disposal, it's handling, and coordination of the economy make recycling a real phenomenon.

7.1 INTRODUCTION

The accelerated growth of solar photovoltaic installations in India has been attributed to Jawaharlal Nehru National Solar Mission launched in 2010 by the Indian government [1–3]. The main objective of this mission was to establish India as a leader in the field of solar energy. Solar installations in India started from early 1993. Overconsumption of the fossil fuels lead to their exploitation and paved the way for alternative renewable sources of energy. Among these, solar energy is considered to be a clean and prominent source of energy as its freely available from the Sun. Being free from harmful gaseous emissions during the operational phase solar photovoltaic technology is considered as an important clean source of energy among the other renewable technologies.

The solar power installations have gone up to 3 MW from 2010 to 26,869 MW in 2018 in India and are expected to proliferate to 479 GW by 2047 [4].

Along with this, the waste accumulation in the country will reach 2,00,000 tons in 2030 and 1.8 million tons by 2050 [5]. Due to their restricted lifespan of 25 to 30 years, the amount of waste accumulation has been left unnoticed. The amount of solar photovoltaic waste accumulated is directly proportional to their installation [6–8]. Due to further negligence, solar photovoltaic waste will become a disaster in the approaching times. Solar photovoltaic waste is included in the Waste Electrical and Electronic Equipment directive by the European Union. Further, it was given the responsibility to the manufacturers to look after the waste so generated from the solar photovoltaic panels [9].

State-wise installations of solar photovoltaic technology are shown in Figure 7.1. From the figure data, it is analyzed that Andhra Pradesh, Karnataka, Madhya Pradesh, Maharashtra, Rajasthan, Tamil Nadu, Telangana, Uttar Pradesh, and Gujarat are leading states in solar power installations.

While Meghalaya, Sikkim, Mizoram, Lakshadweep, and Nagaland have lesser installations in comparison to other states. With the surplus installations in the country, the land requirement is also a major issue that needs emphasis as the amount of waste generated at the end life cycle of solar photovoltaic panels is to be dumped on land before sending it to any recycling firm [10–12].

7.2 ENVIRONMENTAL IMPACT OF SOLAR PHOTOVOLTAIC TECHNOLOGY

Elements like selenium, cadmium, lead, and tellurium present in first and second-generation solar photovoltaic modules are categorized as hazardous as these affect the aquatic and terrestrial inhabitants as it affects the kidney, bones, and lungs due to their higher absorption capacity [13–16]. Cadmium and lead pose a danger due to further leaching into soil from decommissioned and damaged solar photovoltaic panels. During the manufacturing phase, emissions from the firm also affect the health of working people [16, 17]. Safe disposal of solar photovoltaic waste is an impeding factor for the environment. Analysis of environmental impact provides the viable impulse to the manufacturer for completely ensuring the safe disposal of hazardous components of solar panels.

7.3 RECYCLING OF SOLAR PHOTOVOLTAIC PANELS

Analysis of recycling methods requires the determination of the economic feasibility of crystalline silicon, amorphous silicon, copper indium gallium

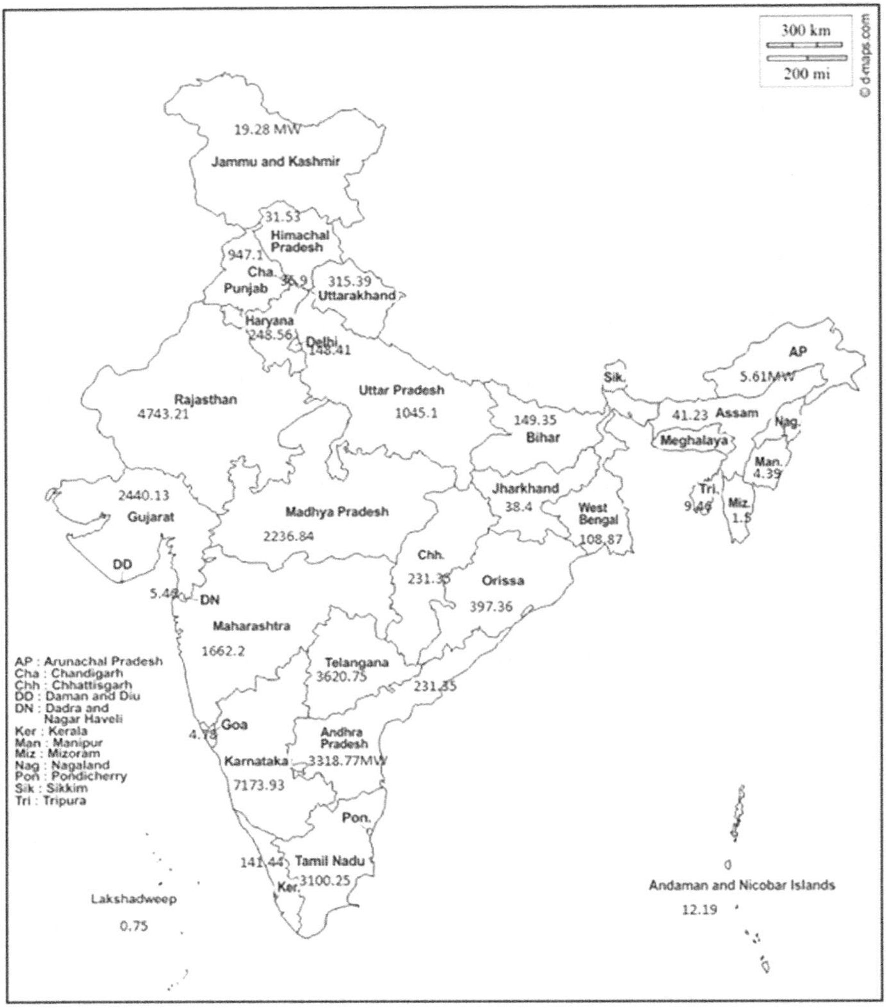

FIGURE 7.1 State-wise installations of solar photovoltaic technology (MNRE).

diselenide, cadmium telluride, and polycrystalline silicon photovoltaic panels. The recycling process of thin film photovoltaic panels involves the sweltering, acid bath for the recovery of elements like Ga, Se, and In. Glass can also be removed by decomposition through high temperature, and acid dissolution can also be used for the removal of EVA polymer layer [18–20]. Stripping with the help of chemicals is processed in the case of cadmium telluride panels for the removal of metals. Further separation of these metals can be done by precipitation, electrodeposition, and evaporation [21, 22]. Due to the maximum installations of crystalline silicon photovoltaic modules in India, emphasis is laid on its economic recycling analysis and also on the solar panels containing elements of economic benefit like Gallium, Indium, Silicon, Cadmium, Tellurium, Aluminum, etc. [23]. Frame made of glass and aluminum separated by the physical process is not taken into consideration for economic recycling analysis. For the determination of economic feasibility of recycling on various generation of the solar cells like a-Si, c-Si, CIGS, CdTe, GaAs an analysis is carried out.

The analysis is carried out on a 1 m^2 area of solar panel [8]. The recovered product mass is depicted as:

$$m_r = At_s\rho_s Z_s \text{ [g/module]} \tag{1}$$

where; m_r is the recovered product mass; A is the area in cm^2; t_s is the thickness in cm; ρ_s is the semiconductor material density in g/cm^3; and Z_s is the material recovered in percentage.

The profit made from selling of the recovered semiconductor is estimated by:

$$P_s = m_r v_s \text{ [\$/module]} \tag{2}$$

where; v_s is the money earned by reselling.

Other than semiconductor materials, profit can also be earned by recovering the glass:

$$m_{g.r} = At_g\rho_g Z_g \text{ [g/module]} \tag{3}$$

where; $m_{g \cdot r}$ is the recovered product mass; A is the area; t_g is the thickness of glass in cm; ρ_g is the glass density in g/cm^3; Z_g is the glass recovered, it is taken as 100%.

$$P_g = m_g v_g \text{ [\$/module]} \tag{4}$$

where; P_g is the profit from recycling glass; v_g is the money earned by selling recovered glass.

Along with recycling the disposal cost for landfill can be calculated depending on the waste generated by a module.

$$W = AEw/N_p \text{ [kg/module]} \tag{5}$$

where; E is the power of each module in per unit area in W/m^2; w is the weight of the module in Kg; Np is the nominal power in Watt.

The overall disposal cost (D) is calculated as:

$$D = WT \text{ [\$/module]} \tag{6}$$

where; W is the solar module waste mass; T is the tipping fees paid for the disposal site.

The cumulative profit (P_c) earned by the recycling:

$$P_c = (P_s + P_g) + D - C \text{ [\$/module]} \quad (7)$$

where; C is the recycling cost of the material prevailing in the market.

The recovered mass from different PV modules like c-Si (279.6 g), a-Si (1.17 g), CIGS (5.23 g of Ga and 8.62 g of In), CdTe (8.98 g of Cd and 9.15 g of Te) [24]. The recovered product mass from each of the above solar panels is 16.64 kg from c-Si, 17.680 kg from CIGS and 16.64 kg from CdTe panels. Ps of c-Si panels is 7.54, in CIGS panels 25.85 from Ga and 15.70 from In, in CdTe panels 0.23 from Cd and 2.02 from Te. Pg from c-Si, CIGS, CdTe are 0.06, 0.07, and 0.06, respectively [9]. Weight (kg) of different panels were 12.32 (c-Si), 11.64 (mc-Si), 13.43 (a-Si), 17.5 (CIGS), 16.72 (CdTe) [25]. Pc of c-Si, mc-Si, a-Si, CIGS, CdTe were –23.96, –23.99, 0.73-C, 22.25, and –0.24, respectively [9]. The overall disposal cost of c-Si, mc-Si, a-Si, CIGS, and CdTe panels came out to be 0.61, 0.58, 0.67, 0.87, 6.45 for each in $ and 43.31, 41.18, 47.57, 61.77, 457.95 in rupees, respectively.

From the above, it is concluded that CIGS panels are economically feasible for recycling. However, disposal costs for the CdTe panels are the highest and they also possess environmental hazards due to the presence of Cd. Recycling of a-Si panels isn't practiced on a commercial level due to less recyclable semiconductor material. About 16–17 kg/m^2 of glass can be recovered. Environmental friendly recycling needs to be practiced along with aiming for economic value. The recycled products so obtained and their reuse is depicted in Figure 7.2.

7.4 CONCLUSION

With the increasing installations of solar photovoltaic plants in the country amount of waste gathered will reach to 2,00,000 million tons in 2030 and 1.8 million tons by 2050. Negligence towards solar photovoltaic waste will be the main culprit for adverse effects on the environment. Value can be generated from solar photovoltaic waste by their recycling at the end of life. Recycling techniques need to be practiced at a commercial scale so as to recover valuable materials as it will boon the employment sector by establishing a recycling industry, and a further new industry will be set up to utilize recycled materials. The construction industry can utilize the recovered aluminum in making doors, window panes, and floor tiles. The automotive industry can utilize Al, Cu, glass, etc. The packaging industry can also utilize recycled solar panel parts. Interior design can be furnished by

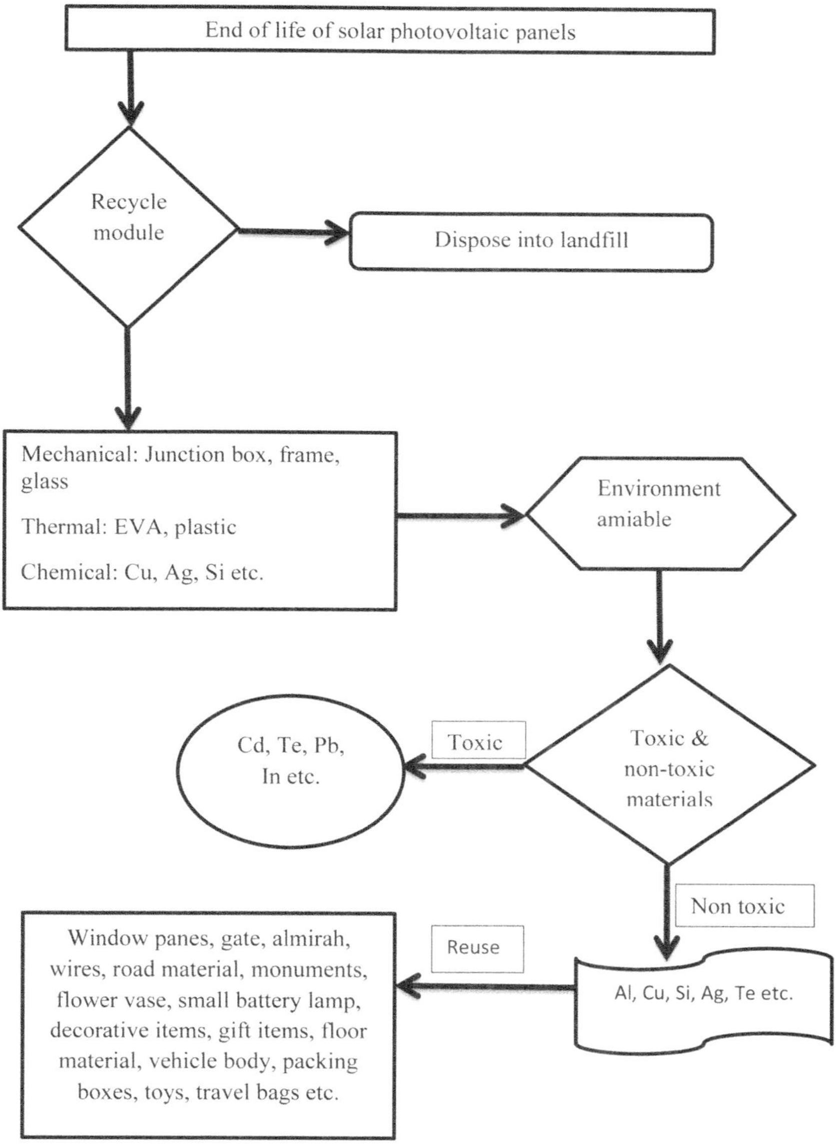

FIGURE 7.2 Solar PV recycled materials and their application.

recycled products. Recycling of the CIGS solar panels is most economical due to the presence of the CdS buffer layer; however, CdTe panel recycling is not economically beneficial, but it's a must for the reduction of toxicity occurring due to disposal of CdTe panels in the environment. Crystalline silicon panels, though they contain fewer amounts of valuable materials other than silver, are non-toxic to the environment. Recycling should be taken ahead with focusing on both environmental and economic benefits to attain sustainability in terms of development.

KEYWORDS

- **amorphous silicon**
- **crystalline silicon**
- **environmental impact**
- **solar panel**
- **solar photovoltaic**
- **solar PV module recycling**

REFERENCES

1. Manju, S., & Sagar, N., (2017). Progressing towards the development of sustainable energy: A critical review on the current status, applications, developmental barriers and prospects of solar photovoltaic systems in India. *Renewable and Sustainable Energy Reviews, 70*, 298–313. https://doi.org/10.1016/j.rser.2016.11.226.
2. Sahoo, S. K., (2016). Renewable and sustainable energy reviews solar photovoltaic energy progress in India: A review. *Renewable and Sustainable Energy Reviews, 59*, 927–939. https://doi.org/10.1016/j.rser.2016.01.049.
3. Sharma, A., (2011). A comprehensive study of solar power in India and world. *Renewable and Sustainable Energy Reviews, 15*(4), 1767–1776. https://doi.org/10.1016/j.rser.2010.12.017.
4. Arora, N., Bhattacharjya, S., Bakshi, S. K., Anand, M., Dasgupta, D., Gupta, A., Prasad, N. S., et al., (2018). *Greening the Solar PV Value Chain*. www.resourceefficiencyindia.com (accessed on 19 September 2022).
5. Suresh, S., Singhvi, S., & Rustagi, V., (2019). *Managing India's PV Module Waste* [Analytical]. Bridge To India.
6. Blanco, C. F., Cucurachi, S., Peijnenburg, W. J. G. M., Beames, A., & Vijver, M. G., (2019). Are technological developments improving the environmental sustainability of photovoltaic electricity? *Energy Technology*, ente.201901064. https://doi.org/10.1002/ente.201901064.
7. Chowdhury, Md. S., Rahman, K. S., Chowdhury, T., Nuthammachot, N., Techato, K., Akhtaruzzaman, Md., Tiong, S. K., Sopian, K., & Amin, N., (2020). An overview of

solar photovoltaic panels' end-of-life material recycling. *Energy Strategy Reviews, 27*, 100431. https://doi.org/10.1016/j.esr.2019.100431.

8. Deng, R., Chang, N. L., Ouyang, Z., & Chong, C. M., (2019). A techno-economic review of silicon photovoltaic module recycling. *Renewable and Sustainable Energy Reviews, 109*, 532–550. https://doi.org/10.1016/j.rser.2019.04.020.
9. McDonald, N. C., & Pearce, J. M., (2010). Producer responsibility and recycling solar photovoltaic modules. *Energy Policy, 38*(11), 7041–7047. https://doi.org/10.1016/j.enpol.2010.07.023.
10. Held, M., & Ilg, R., (2011). Update of environmental indicators and energy payback time of CdTe PV systems in Europe. *Progress in Photovoltaics: Research and Applications, 19*(5), 614–626. https://doi.org/10.1002/pip.1068.
11. Pagnanelli, F., Moscardini, E., Granata, G., Abo, A. T., Altimari, P., Havlik, T., & Toro, L., (2017). Physical and chemical treatment of end of life panels: An integrated automatic approach viable for different photovoltaic technologies. *Waste Management, 59*, 422–431. https://doi.org/10.1016/j.wasman.2016.11.011.
12. Sica, D., Malandrino, O., Supino, S., Testa, M., & Lucchetti, M. C., (2018). Management of end-of-life photovoltaic panels as a step towards a circular economy. *Renewable and Sustainable Energy Reviews, 82*, 2934–2945. https://doi.org/10.1016/j.rser.2017.10.039.
13. Faircloth, C. C., Wagner, K. H., Woodward, K. E., Rakkwamsuk, P., & Gheewala, S. H., (2019). The environmental and economic impacts of photovoltaic waste management in Thailand. *Resources, Conservation and Recycling, 143*, 260–272. https://doi.org/10.1016/j.resconrec.2019.01.008.
14. Fthenakis, V. M., Kim, H. C., & Alsema, E., (2008). Emissions from photovoltaic life cycles. *Environmental Science & Technology, 42*(6), 2168–2174. https://doi.org/10.1021/es071763q.
15. Goe, M., & Gaustad, G., (2016). *Estimating Direct Human Health Impacts of End- of -life Solar Recovery*. Golisano Institute for Sustainability.
16. Prakash, A., & Rayabagi, S., (2015). Challenges in manufacturing and end-of-life recycling or disposal of solar PV panels. *IOSR-JEEE, 10*(4), 7.
17. Contreras-Lisperguer, R., Muñoz-Cerón, E., Aguilera, J., & Casa De, J. La., (2017). Cradle-to-cradle approach in the life cycle of silicon solar photovoltaic panels. *Journal of Cleaner Production, 168*, 51–59. https://doi.org/10.1016/j.jclepro.2017.08.206.
18. Granata, G., Pagnanelli, F., Moscardini, E., Havlik, T., & Toro, L., (2014). Recycling of photovoltaic panels by physical operations. *Solar Energy Materials and Solar Cells, 123*, 239–248. https://doi.org/10.1016/j.solmat.2014.01.012.
19. Klugmann-Radziemska, E., (2012). Current trends in recycling of photovoltaic solar cells and modules waste/recykling zużytych ogniw i modułów fotowoltaicznych—stan obecny. *Chemistry-Didactics-Ecology-Metrology, 17*(1, 2), 89–95. https://doi.org/10.2478/cdem-2013-0008.
20. Mahmoudi, S., Huda, N., Alavi, Z., Islam, M. T., & Behnia, M., (2019). End-of-life photovoltaic modules: A systematic quantitative literature review. *Resources, Conservation and Recycling, 146*, 1–16. https://doi.org/10.1016/j.resconrec.2019.03.018.
21. Marwede, M., Berger, W., Schlummer, M., Mäurer, A., & Reller, A., (2013). Recycling paths for thin-film chalcogenide photovoltaic waste – Current feasible processes. *Renewable Energy, 55*, 220–229. https://doi.org/10.1016/j.renene.2012.12.038.
22. Paiano, A., (2015). Photovoltaic waste assessment in Italy. *Renewable and Sustainable Energy Reviews, 41*, 99–112. https://doi.org/10.1016/j.rser.2014.07.208.

23. Kim, H., Cha, K., Fthenakis, V. M., Sinha, P., & Hur, T., (2014). Life cycle assessment of cadmium telluride photovoltaic (CdTe PV) systems. *Solar Energy, 103*, 78–88. https://doi.org/10.1016/j.solener.2014.02.008.
24. Fthenakis, V. M., (2003). *CdTe PV: Facts and Handy Comparisons.* Brookhaven National Laboratory.
25. Xsunx, (2009). *Preliminary Specification Sheet CIGS Thin Film Solar Modules.* Available at: http://www.xsunx.com/pdf/CIGS Brochure-draft.pdf (accessed on 19 September 2022).
26. *Physical Progress (Achievements).* Ministry of New & Renewable Energy. https://mnre.gov.in/physical-progress-achievements (accessed on 19 September 2022).

CHAPTER 8

CORE-SHELL STRUCTURE, SUPER PARA MAGNETISM, AND FUNCTIONALIZATION OF MAGNETIC SOLID NANOPARTICLES AND THEIR APPLICATION IN TREATMENT OF WASTEWATER

ADITI SINGH[1], RAJPREET KAUR[1,2], ANITA GUPTA[1], CHRISTINE JEYASEELAN[1], and POONAM KHULLAR[2]

[1]*Department of Chemistry, Amity University, Noida, Uttar Pradesh, India*

[2]*Department of Chemistry, B.B.K. D.A.V. College for Women, Amritsar, Punjab, India*

ABSTRACT

Nanoparticles, with their sub-microscopic size, have unique magnetic properties which enable their application in a wide area, including engineering, medicine, and environmental remediation. The core-shell structure of iron oxide (IO) magnetic nanoparticles (MNPs) and their superparamagnetic behavior is a platform for their implementation *in vivo* and *in vitro* applications. Control over the shape and size of IO MNPs has proved their importance in various fields. Application in wastewater treatment, removal of organic contaminants from water, and how surface functionalization of MNPs provides them stabilization have been briefly discussed.

Nanotechnology-Enhanced Solid Materials: Design, Synthesis, Properties, Applications, and Perspectives. Lionello Pogliani, Ann Rose Abraham, A. K. Haghi, & Prabhat Ranjan (Eds.)

8.1 INTRODUCTION

Nanoscience captures an important area of modern science's research with its reach at the cellular and molecular level leading to its advancement in medical and life sciences. Materials made out of nanoparticles have been providing merits because of their uniqueness in size and their physiochemical properties [1]. Among these various sorts of nanomaterials, magnetic nanoparticles (MNPs) have always caught the eyes of scientists, as exhibited by an unusual increase in papers of research that focuses on these kinds of materials [2]. A magnetic field is used to manipulate this class of nanomaterials. A magnetic material such as cobalt, iron, and nickel with a chemical entity having such functionality together form such particles. Magnetic nanobeads are clusters of MNPs 50–200 nm in diameter comprising individual MNPs.

NPs can be synthesized using two methods: The first method is the 'top-down' method, in which large materials are broken down for the generation of NPs. This method of synthesis is normally used for ceramic materials where shape and size are not essential elements for their application as the particles obtained through this synthesis have assorted shapes, and there is a wide distribution of particle size. The other method of synthesis is called the 'bottom-up' method. Larger NPs are formed by the fusion of atoms and molecules. Using this method NPs can be obtained that have fewer defects in them with more homogeneous chemical composition and uniformity in the distribution of shape and size [3].

MNPs possess both the features of nanoparticles (NPs) and the properties of a magnet. Metal alloy NPs, metal oxide NPs, and metal NPs are the main class of MNPs. FePt, FeCo, etc., come under metal alloy NPs and Iron oxides (γ-Fe_2O_3 and Fe_3O_4), and Ferrites ($CoFe_2O_4$ and $MnO.6ZnO.4Fe_2O_4$) are included mainly under Metal oxide NPs. The shape and particle size of Fe_2O_3 and Fe_3O_4 are easily controllable and can be prepared manually and also these are the MNPs that are used extensively. For the preparation of a diverse number of Ferrite NPs ($MnO.6ZnO.4Fe_2O_4$, $Mn_3Zn_7Fe_2O_4$, etc.), some of the metallic elements such as Zinc (Zn) and Manganese (Mn) can be added to nanosized structures of iron oxide (IO).

MNPs with their behavior being non-poisonous and non-immunogenic can show surface effects with great specific surface area and prove to be a good carrier for drugs, fragments of DNA, and modified compounds. MNPs also prove to be biocompatible and some of them possess superparamagnetism [4]. Various methods and techniques for the advancement in the production of controlled size and monodispersed MNPs (e.g., Fe_3O_4, FePt, and γ-Fe_2O_3) have to lead to an increase in MNP's application in various fields [5].

NPs of IO with superparamagnetic property can be used for various *in vivo* applications such as drug delivery, tissue repairment, cell separation, etc. In such applications, the value of magnetism for NPs should be high, particle size less than 100 nm, and distribution of the size of the particle should be narrow. Also, for such applications, MNPs have to coat the surface peculiarly; being biocompatible and non-toxic, and localization of particles must be such that it allows a targeted delivery [6]. Particle structure, the functionality of the surface, and the resultant magnetic properties are needed to be understood for all those applications to measure and understand by implicating the properties on the challenges being faced.

There has been a lack of brief review of characterization methods and simplified demonstration of various fields in support of MNPs which focus on characterization, synthesis, and MNPs application [7].

In this book chapter, we discuss the core-shell structure of MNPs, magnetic properties and the structure of IO NPs, their superparamagnetic behavior, their functionalization, their removal from organic contaminants, and their application in wastewater treatment.

8.2 CORE-SHELL STRUCTURE OF MNPs

The structure of the core-shell of MNPs plays an important role for its characteristic properties of magnetism, bio-recognition, and biocatalytic features with its core part being made of inorganic magnetic material and composition of the shell of biological or biomolecular species which are being chemically bound to the core with organic linkers [8]. The magnetic structure of a single domain is usually acquired by MNPs. Regarding the study of magnetic properties that are size-dependent and without the involvement of complex domain wall movement; MNPs have been providing great opportunities to understand magnetic properties at the atomic level. Understanding fundamentally the magnetic properties of NPs can have a great impact on modern technology and certainly, interpretation of some fundamental issues, such as quantum tunneling of magnetization could be done by the study of MNPs. They have potential towards modern science, such as applications as a drug carrier for delivery of site-specific drugs [9]. Study on the architecture of core-shell has been done extensively for various nanomaterials. Besides varying shape, size, and composition, this methodology turns into a predominant technique by enabling us to freely address entering technical demands on properties of magnetism like magnetization, thermal stability, and coercivity. Understanding magnetic interactions and achieving the desirable characteristics of magnetism would be certainly easy on exploring different combinations of core-shell [10]. Categorizing spinels based on the distribution of cation, they are normal, inverse, or partially inverse. In a normal spinel structure, the tetrahedral sites are occupied by 8 bivalent cations and the octahedral sites are occupied by 16 trivalent cations, whereas, in an inverse spinel structure, 8 octahedral sites are occupied by 8 bivalent cations, and distribution of 16 trivalent cations is among 8 octahedral and 8 tetrahedral sites. i=0, for a normal spinel, and i=0, for an inverted spinel. It is said to be partially inverted if both the octahedral and tetrahedral sites are occupied by the divalent cations and $0 < i < 1$ [11].

In the spinel structure of ferrites, 24 cations (Fe^{2+}, Co^{2+}, Ni^{2+}, Zn^{2+}, Mn^{2+}, Mg^{2+}, Gd^{3+}, Fe^{3+}) and 32 anions of O^{2-} form a unit cell in which the cations can acquire positions in 96 possible ways (32 octahedral and 64 tetrahedral positions) [12]. Nanoferrites, the chemical formula being MFe_2O_4 (M = Zn, Mn, Co, Ni, etc.), have shown noteworthy properties converse to its bulk behavior. Magnetite (Fe_3O_4) NPs being chemically inert towards oxidation and in-built magnetic properties have been descended by Ferrite systems of Cobalt Ferrite ($CoFe_2O_4$; CFO) and Manganese Ferrite ($MnFe_2O_4$) [13]. $NiFeO_4$ and $CoFe_2O_4$ have inverse spinel structure with Nickel Ferrite being soft having high electrical resistance used in applications related to high frequency and the surface properties influence their NPs and thus their fabrication becomes important for their catalytic applications. For controlling the shape and size surfactants, organic solvents, organic molecules, and organometallic compounds are used and also for the synthesis of metal ferrite NPs. The representation for spinel structures for Nickel and Cobalt Ferrite is $(Fe^{3+})[Ni^{2+}Fe^{3+}]O_4$ and $Fe^{3+})[Co^{2+}Fe^{3+}]O_4$, respectively [14].

Strong magnetism is possessed by some of the elements such as Fe, Co, Ni, etc., and a few of them show poor magnetism, e.g., Mg, Mo, and Li.

8.3 MAGNETIC PROPERTIES AND STRUCTURE OF IRON OXIDE (IO) NANOPARTICLES

Among the eight IOs being known, magnetite, hematite, and maghemite are the most favorable ones due to their uniqueness in catalytic, biochemical, magnetic, and other properties that are appropriate for biological and technical applications [15].

Materials possess five types of magnetism categorized as: paramagnetism, diamagnetism, ferrimagnetism, ferromagnetism, and anti-ferromagnetism. A property of a substance in which it tends to oppose an applied magnetic field is referred to as diamagnetism in which the magnetic susceptibility is negative, small (10^{-6}), and temperature independent. Also, such materials have no such electrons that are unpaired, and the direction of spins of the paired electrons is opposite, which offsets the magnetic moment. Whereas, paramagnetic substances have electrons that are unpaired with spins having random directions of magnetic moments. On application of magnetic field, the alignment of the spin is in the direction of the applied magnetic field with a value of magnetic susceptibility small (0–0.01) and positive. Curie-Weis law describes its behavior and variation with temperature. Ferromagnetic materials have similar behavior as paramagnetic substances having unpaired electrons and their alignment in the direction of the applied magnetic field. The difference lies in the magnetic moment of ferromagnetic substances, which remain in the lower-energy state and parallel to each other and this state remains even after the removal of the applied magnetic field. The characteristics of a ferrimagnetic magnet are that it is capable of keeping magnetic moment in lack of magnetic field and the direction of spin of neighboring electron pairs are opposite. The net magnetic moment in an antiferromagnet is zero, which is due to the neighbor valence electrons having an intrinsic magnetic moment in the opposite direction [11].

In 3D orbitals of iron, four electrons are unpaired providing a strong magnetic moment to it. Various magnetic states can arise from the formation of crystals from iron atoms. In the paramagnetic state, the net magnetic moment possessed by the crystal is zero with the random alignment of the individual magnetic moments concerning each other. A small net magnetic moment is achieved by the crystal when being subjected to an external magnetic moment resulting in the alignment of some of these moments. There is no need for an external magnetic field in the case of ferromagnetic crystals as all the individual moments are aligned even without it [10]. Superparamagnetism is exhibited by NPs comprising ferromagnetic materials size being less than 10–20 nm and these kinds of materials include elemental metals, oxides, alloys, and other chemical compounds [9]. In a ferrimagnetic crystal, the arrangement of moments is antiparallel with net magnetic moment from two different atoms. The crystal is referred to as an antiferromagnetic when the magnitude of the antiparallel magnetic moments is the same and they possess no net magnetic moment.

Paramagnetism is exhibited by hematite above Curie temperature of 956 K. It possesses weak ferrimagnetism at room temperature and goes through a phase transition at 260 K (the Morrin temperature) to a state of antiferromagnetism. Size of particles, crystallinity, and extent of cation substitution are the factors of the dependence of the behavior of magnetism in hematite. With the decrease in the size of the particle of hematite, there is also a decrease in Morrin temperature. Curie temperature and Morrin temperature tend to be lowered by factors such as cation substitution and poor crystallinity at all temperatures.

Ferrimagnetism is possessed by bulk magnetite and maghemite. Magnetite possesses ferrimagnetism at room temperature with a Curie temperature of 850 K. Superparamagnetic behavior is seen in particle size less than 6 nm in particles of magnetite at room temperature. However, the synthesis methods determine the properties of magnetism possessed by the particles. Maghemite possesses ferrimagnetism at room temperature but is unstable at higher temperatures and its magnetic susceptibility is lost with time. When doped with ions of other metals stabilization can be achieved. The particle size of maghemite less than 10 nm exhibits superparamagnetism at room temperature [10].

In the crystal structure, tetrahedral and octahedral sites are occupied by closely packed planes of iron cations and oxygen anions. In hematite, Fe (III) ions occupy the octahedral sites with oxygen ions in a hexagonal close-packed arrangement. In magnetite and maghemite, the arrangement of oxygen ions is cubic close-packed. The structure of magnetite is an inverse spinel with face-centered cubic (Fcc) unit cell and Fe (II) ions occupying the octahedral sites, and Fe (III), the tetrahedral and octahedral sites being randomly distributed. $Fe^{3+}(A)[Fe^{2+}Fe^{3+}](B)O_4$ is the formula deduced from the arrangement. In this, Fe^{3+} occupies the tetrahedral site denoted by A with four O atoms surrounding it, a mixture of Fe^{2+}/Fe^{3+} occupies the octahedral site being denoted as B with 6 O atoms surrounding it [10, 11]. The occupancy of iron ions in the tetrahedral and octahedral sites of the spinel structure of magnetite (Fe_3O_4) affects its magnetic properties. In the octahedral sites, the magnetic spins of the ions are ferromagnetically coupled and tetrahedral sites are antiferromagnetically coupled. Due to the same number of Fe^{3+} ions in octahedral and tetrahedral sites, the magnetic spins cancel each other. Therefore, contribution towards net magnetic moment is only by the magnetic spins of Fe^{2+} ions in the octahedral sites. Replacement of Fe^{2+} ions with other divalent transition metal ions, Co^{2+}, Mn^{2+}, Zn^{2+}, and Ni^{2+}, is one of the ways to control the magnetic properties of NPs with spinel structure (Figure 8.1(a)). Except for Zn^{2+} in which the net magnetic moments are proportional to the magnetic spins. Among the divalent ions, Mn^{2+} ions exhibit the strongest magnetic moment which has five unpaired electrons in its d-orbital.

$ZnFe_2O_4$ has a normal spinel structure in which Zn^{2+} and Fe^{2+} ions occupy the tetrahedral and octahedral sites, respectively. Due to the nonmagnetic property of Zn^{2+} ions and antiferromagnetic coupling between Fe^{3+} ions, $ZnFe_2O_4$ is not suited to be used as an MRI contrast agent. However, the net magnetization increases due to the partial substitution of Fe^{2+} and Zn^{2+} ions as the magnetic spins of Fe^{3+} are not perfectly canceled (Figure 8.1(b)) [16].

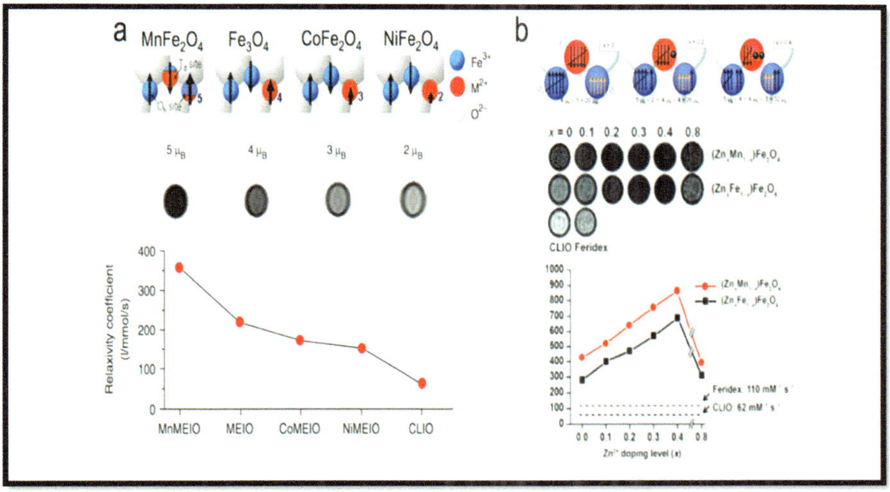

FIGURE 8.1 (a) Schematic representation of the spin alignments of magnetic ions in various ferrite structures. $MnFe_2O_4$ has the highest magnetization with a magnetic spin magnitude of 5 μB. Therefore, $MnFe_2O_4$ nanoparticles exhibit the strongest T_2 contrast effect. (b) Schematic representation of the spin alignments of magnetic ions in Zn-doped ferrite structures. Zn doping increases the number of noncancelled Fe^{3+} ions progressively, resulting in higher magnetization. The T_2 contrast effect is enhanced up to the Zn doping level of x = 0.4.

Source: Reprinted with permission from Ref. [16]; ©2015, American Chemical Society.

Maghemite has a similar spinel structure to magnetite but the cation sublattice has some vacancies in it. All or most of the Fe atoms in maghemite are in Fe (III) state. There is regularity in Fe (III) ions arrangement with occupancy in two-thirds of the sites but one vacant site is being followed with two filled sites. It was found by Haneda and Morry that the degree of vacancy ordering decreases with decreasing particle size, with no vacancy ordering in maghemite smaller than about 20 nm [10, 11].

8.4 SUPERPARAMAGNETIC MNPS

Small ferrimagnetic or ferromagnetic NPs exhibit a kind of magnetism known as superparamagnetism in which a random flip of the direction of magnetization is possessed under the temperature influence. With the decrease in NPs size below a critical value (around 20–30 nm), the behavior of separate NPs is as such a single magnetic domain, and the behavior of superparamagnetism is exhibited. They tend to respond rapidly to the magnetic field being applied but negligible coercivity and residual magnetism are exhibited by them [17]. There is an increase in the ratio of surface-to-volume and also the number of atoms on the surface increases than present in the bulk [18]. IO NPs possess superparamagnetism due to a reduction in the size of NPs than in the bulk state. On applying an external magnetic field, there is an alignment of magnetic moment in the direction of the field applied, and on its removal; the magnetic moment immediately becomes zero confirming the property of superparamagnetism in IO NPs. No coercivity is exhibited by superparamagnetic particles therefore there is no observation of hysteresis in the magnetization curve. The temperature below which the stable magnetization is observed is known as blocking temperature which is used to outline the starting behavior of superparamagnetism [17].

Hematite (α-Fe_2O_3), maghemite (β-Fe_2O_3), and magnetite (Fe_3O_4) with different critical properties, like molecular size, and various magnetic specifications are the oxides of iron that shape the core of superparamagnetic iron oxide nanoparticles (SPIONs) [19]. Modifications of the active surface of SPIONs lead to an increase in their colloidal stability in aqueous media. These modifications may involve their coating with a biocompatible, hydrophilic polymer or capping SPIONs with organic acids. Figure 8.2 illustrates the conjugation of coated SPIONs with various biomolecules [20].

Ferrimagnetism is exhibited by magnetite in its natural form possessing the strongest properties of magnetism of all oxides of transition metals. After subjection of a ferrimagnetic material to an external magnetic field, its magnetization tends to remain up to some level even after the removal of the field. This is due to the alignment of magnetic moments even if the magnetic field is absent. Therefore, a hysteresis loop is shown by its magnetization moment in the function of an external magnetic field. SPIONs show their properties of magnetism when being subjected to an external magnetic field, but in the case of magnetite, it preserves the properties even in the absence of a magnetic field. Therefore, to be superparamagnetic the diameter of the crystals of Fe_3O_4 should be 20 nm or less [19]. Unique size-dependent magnetic properties are exhibited by Fe_3O_4 as shown in Figure 8.3. Bulk Fe_3O_4 is ferrimagnetic with a multidomain structure and Curie temperature of 858 K. With the decrease in particle size below ~100 nm, Fe_3O_4 particles become a single domain structure with maximization in coercivity. When the size of the particles decreases below ~20 nm, the magnetization of Fe_3O_4 NPs is randomized by thermal energy so that they become superparamagnetic [21].

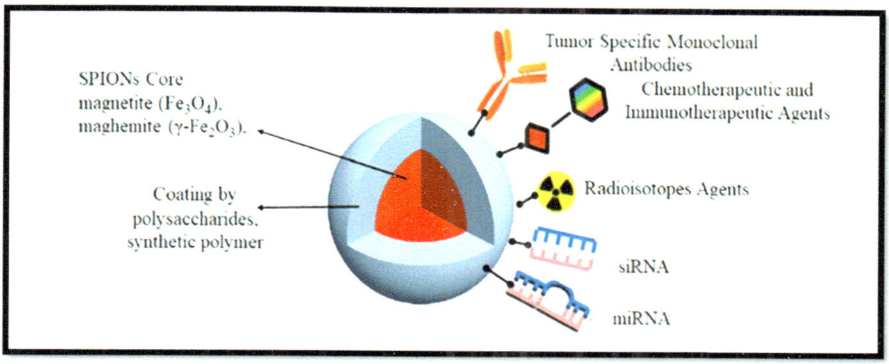

FIGURE 8.2 Schematic of SPION structure for therapeutic applications. SPIONs core (magnetite, maghemite, hematite) can be coated with various synthetic polymers (such as PEG, PVP, dextran, chitosan, and polysaccharide). They can be used as chemotherapeutic, immunotherapeutic, and radiotherapeutic agents and be conjugated with tumor-specific monoclonal antibodies/siRNA/miRNA in targeted drug delivery.

Source: Reprinted with permission from Ref. [20]; ©2020, American Chemical Society.

Likewise, Fe^{3+} is present in both hematite and maghemite possessing different properties in the bulk form and after the formation of tiny crystals show superparamagnetism [19].

8.5 FUNCTIONALIZED MNPS

The ideality of superparamagnetic NPs in biomedicine prepared via organic synthesis is because of monodispersity in shape, size, and composition. The stabilization acquired by these superparamagnetic NPs is by surfactants and their hydrophobic nature. Surface functionalization is required to make the NPs hydrophilic which is done via the addition of surfactant or exchange of surfactants. An amphiphilic molecule comprises of both hydrophobic component that forms a double-layered structure with the initial chain of hydrocarbon and hydrophilic segment which is in the outer periphery of NPs making them water-soluble, and this process of adsorption results in the addition of surfactants [19, 22]. In surfactant exchange, a bifunctional surfactant directly replaces the initial surfactant [22]. To modify the surface of superparamagnetic particles of FeCo having a high magnetic moment, a modification in the coating of phospholipid micelle with PEG has been recently developed by Dai's group [23]. In the same way for the transfer of hydrophobic NPs to aqueous systems from the organic solvents, different amphiphilic polymers, like poly(styrene)-block-poly(acrylic acid) (PS-PAA), polyethylene glycol-2-tetradecyl ether, and tetradecylphosphonate are used.

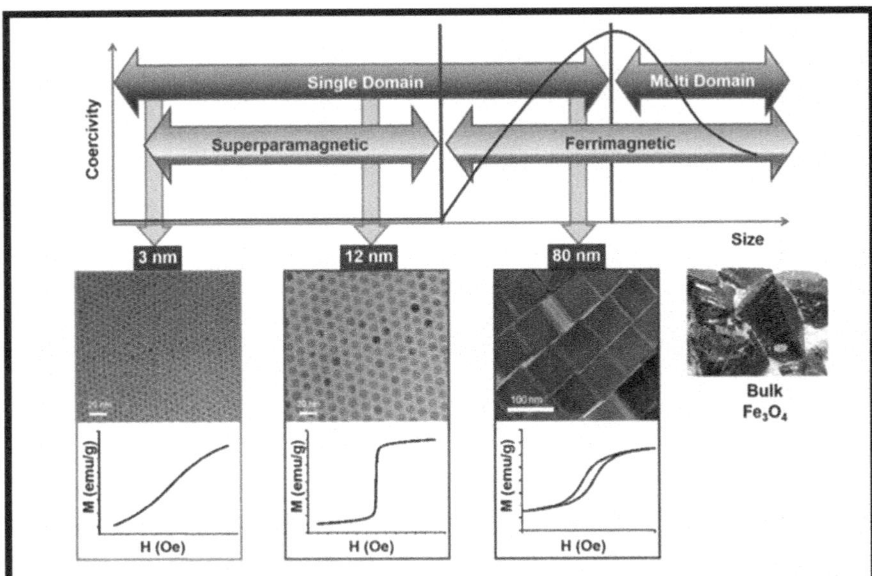

FIGURE 8.3 Surfactant-based size dependent properties of iron oxide nanoparticles.

Source: Inset TEM images and magnetization curves of magnetic nanoparticles were reprinted with permission from Ref. [21]; ©2015, American Chemical Society.

To steer clear of agglomeration organic materials are often chosen to passivate the IO NPs surface during or after the preparation procedure. Also, the surface being hydrophobic with a wide ratio of surface-area-to-volume, cause the reason for their agglomeration if not properly coated and also the hydrophobic interaction between the NPs. IO NPs coated with organic materials provide higher potentials to be applied in many fields. The structure of this type of magnetic IO NPs includes two main parts: magnetic IOs with preserved magnetic properties and organic materials whose properties are also preserved. Also, these types of functionalized NPs not only provide the basic characteristics of magnetism but also hold good biodegradability and biocompatibility of the functional organic compounds.

For functionalization based upon small molecules and surfactants, it can be simply categorized based on characteristics of the surface into three kinds: water-soluble, oil-soluble, and amphiphilic. Oil-soluble ones are the ones having weak attraction towards the solvent domain and are mostly hydrophobic groups such as alkylphenol, fatty acids, etc. This type of functionalization work upon preventing or decreasing agglomeration of NPs and monodispersity arises due to an increase in their stability and are relatively easily preparable and controllable. Whereas water-soluble ones are those having strong attraction among the chemical compounds in the solvent domain and are mostly hydrophilic groups like polyol, ammonium salt, lycine, etc. These can be broadly used in biodetection and bioseparation. Those comprising both hydrophobic and hydrophilic chemical compounds are referred to as amphiphilic [24].

8.6 APPLICATIONS IN WASTEWATER TREATMENT

Due to lower cost and higher efficiencies, technology-based upon nanomaterials has proven to be one of the favorable techniques for the treatment of wastewater and also to meet the rising stringent water quality standards [15]. Recently, several methods have shown their efficiency towards the removal of heavy metals from water, such as chemical precipitation, photocatalytic oxidation, membrane filtration, adsorption, ion exchange, and electrochemical techniques. Adsorption has emerged as a crucial technique for the removal of heavy metal from wastewater, leading to high-standard treatment of effluents due to its flexible operation and design. Also, regeneration of adsorbents is possible by satisfactory processes of desorption that are highly efficient, easily operational, and the cost of maintenance being low [25, 26]. Wastewater persists in contaminants that include organic and inorganic compounds, heavy metals, including pollutants of complex compounds. The release of these contaminants in the environment harms humans and our ecological surroundings. Therefore, it is a must to remove these contaminants to tackle the problems related to water pollution [26].

Selection of materials that are harmless with hydrophilic coating on them should be done carefully so that they can bind or react to substances that are toxic in wastewater. Magnetic nanomaterials have been used for the removal of various pollutants from wastewater through catalysis and adsorption methods. For the removal of heavy metals from aqueous solutions, nanosized metal oxides (NMOs) are potent towards their removal which include nanosized magnesium oxides, ferric oxides, titanium oxides, aluminum oxides, and cerium oxides are categorized as emerging ones. NMOs show very favoring sorption to heavy metals in respect of selectivity and high capacity, as suggested in recent research. From these studies, when size is reduced to nanometer from micrometer, stability decreases due to increased surface area and due to interaction, such as Vanderwaal forces, they become prone to agglomeration.

Leaving behind earlier NMOs, the easy separation of magnetic NMOs from aqueous systems under a magnetic field has grabbed attraction. NMOs that have been studied widely for the adsorption process include manganese oxides, IOs, titanium oxides, and aluminum oxides [27]. Iron oxide magnetic nanoparticles (IOMNPs) with their strong capacity of adsorption, high reactivity, high surface area-volume ratio, and also an additional magnetic property serve as a simple and less costly process of separation for their extraction from aqueous systems because they tend to aggregate on applying an external magnetic field. With superparamagnetic behavior of NPs redispersion can be easily done as removal of the magnetic field leads to loss of magnetic field. Superparamagnetic behavior is exhibited by magnetite and maghemite being small in size and they can function as nanoadsorbents or as the core element of structures of core-shell where the NPs act as magnetic separation and for the adsorption of pollutants, desired functionality is provided by the core-shell [15]. Treatment of water using SPIONs relies on surface functional groups, side effects, and also on applying external magnetic field easy removal from water is possible. SPIONs have also been potent towards metal ions removal from water by surface modification with chelating ligands to heavy metal ions and further removed by an external magnetic field. The capacity of adsorption can be enhanced by decreasing the diameter of the particle and specific surface modification [28]. Through surface modification, functional groups on the surfaces of NPs can be altered providing specific functionalization for selective uptake of ions and enhancing their capability for uptake of heavy metals in treatment of wastewater treatment in addition to the improvements in their stability against agglomeration. Various coating materials such as organic, inorganic molecules, polymeric, and nonpolymeric materials are used for coating NPs based on Van der Waal's interaction, covalent attachment, and electrostatic interactions for their stabilization and also to prevent oxidation [2, 26].

8.7 MNPS FOR REMOVAL OF ORGANIC CONTAMINANTS

Numbers of dyes and dye-related stuff have been causing significant concern due to its harmful effects on humans by polluting the aqueous systems [29]. For this purpose, MNPs are chosen as with the help of magnetic separation their recollection from water proves to be easy. Dyes are mostly stable towards biodegradation, photodegradation, and oxidizing agents. Therefore, for removal, the process of adsorption is efficient and the method costs less [30]. Although, adsorption has problems related to the disposal of adsorbents, it is simply an effective method without the concern of generation of unwanted by-products, and a good percentage of dye can be removed. Figure 8.4 shows the differences between physical and chemical adsorption. Physical adsorption (physisorption) is caused by weak Van der Waal's force and on the surface of adsorption contains multilayers of the adsorbate. The adsorption is known as chemical adsorption (chemisorption) when the attractive force existing between the particles of adsorbent and adsorbate are almost of the same strength as the covalent bonding, which is single-layer adsorption [31].

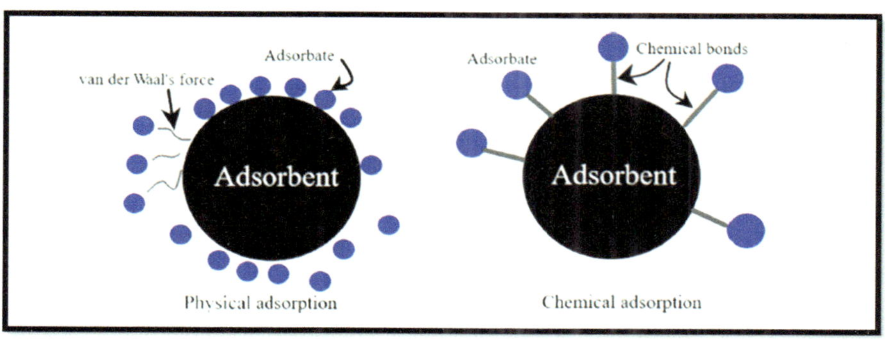

FIGURE 8.4 Physical and chemical adsorption.

Source: Reprinted with permission from Ref. [31]; ©2020, American Chemical Society.

Dyes can be categorized into two groups: cationic and anionic which can be removed from the aqueous solutions by using MNPs as an adsorbent [29].

Cationic dyes are harmful as they can interact easily with the surface of cell membranes that carry a negative charge on them and ought to be toxic. At pH 5–12 their removal from the aqueous system is quick and highly efficient [30].

8.8 CONCLUSIONS AND FUTURE PROSPECTS

MNPs provide a high surface-area-to-volume ratio offering a large surface for physical and chemical adsorption. Due to their number of physiochemical properties, MNPs have been a particular area of interest for the purification of water. Being cost-efficient and easy availability of functionalization, the water purification process using MNPs is safe and clean. Surface functionalization improves chemical stability and biocompatibility and also tailor's solubility in water and dispersibility. These properties have been proven effective towards applications such as the extraction of metal nanoparticles using MNPs, and the extraction of dyes using MNPs. In these applications, surface, active, and aqueous bulk solubilized MNPs can be used for the extraction process. The extraction process is facilitated when there is simultaneous operation of both hydrophobic and hydrophilic interactions.

For minimizing the toxic effects of magnetic nanomaterials, safety measures should be taken during their fabrication, or their use, as any kind of negligence, can lead to threats towards humans and the environment. The potential towards their reusability and regeneration should be looked over for designing easy steps of regeneration. Progress in functional material is needed to be done, which could be environmentally acceptable and cost-effective.

KEYWORDS

- **core-shell**
- **functionalization**
- **iron oxide nanoparticles**
- **magnetic nanoparticles**
- **superparamagnetic**
- **water-treatment**

REFERENCES

1. Akbarzadeh, A., Samiei, M., & Davaran, S., (2012a). Magnetic nanoparticles: Preparation, physical properties, and applications in biomedicine. *Nanoscale Research Letters, 7*(1). https://doi.org/10.1186/1556-276X-7-144.
2. Gul, S., Khan, S. B., Rehman, I. U., Khan, M. A., & Khan, M. I., (2019). A comprehensive review of magnetic nanomaterials modern day theranostics. *Frontiers in Materials, 6*, 1–15. https://doi.org/10.3389/fmats.2019.00179.
3. Hernández-Hernández, A. A., Aguirre-Álvarez, G., Cariño-Cortés, R., Mendoza-Huizar, L. H., & Jiménez-Alvarado, R., (2020). Iron oxide nanoparticles: Synthesis, functionalization, and applications in diagnosis and treatment of cancer. *Chemical Papers, 74*(11), 3809–3824. https://doi.org/10.1007/s11696-020-01229-8.
4. Guo, T., Lin, M., Huang, J., Zhou, C., Tian, W., Yu, H., Jiang, X., et al., (2018). The recent advances of magnetic nanoparticles in medicine. *Journal of Nanomaterials, 2018*. https://doi.org/10.1155/2018/7805147.
5. Clemons, T. D., Kerr, R. H., & Joos, A., (2019). Multifunctional magnetic nanoparticles: Design, synthesis, and biomedical applications. *Comprehensive Nanoscience and Nanotechnology, 1–5*(8), 193–210. https://doi.org/10.1016/B978-0-12–803581-8.10462-X.
6. Laurent, S., Forge, D., Port, M., Roch, A., Robic, C., Vander, E. L., & Muller, R. N., (2010). Erratum: Magnetic iron oxide nanoparticles: Synthesis, stabilization, vectorization, physicochemical characterizations, and biological applications, *Chemical Reviews*, 2008, *108*, p. 2064; *Chemical Reviews, 110*(4), p. 2574. https://doi.org/10.1021/cr900197g.
7. Sandler, S. E., Fellows, B., & Thompson, M. O., (2019). Best practices for characterization of magnetic nanoparticles for biomedical applications. *Analytical Chemistry, 91*(22), 14159–14169. https://doi.org/10.1021/acs.analchem.9b03518.
8. Evgeny, K. Synthesis, properties and applications of magnetic nanoparticles and nanowires—A brief introduction. *Magnetochemistry, 5*(4), 61. https://doi.org/10.3390/magnetochemistry5040061.
9. Liu, C., & Zhang, Z. J., (2001). Size-dependent superparamagnetic properties of Mn spinel ferrite nanoparticles synthesized from reverse micelles. *Chemistry of Materials, 13*(6), 2092–2096. https://doi.org/10.1021/cm0009470.
10. Song, Q., & Zhang, Z. J., (2012). Controlled synthesis and magnetic properties of bimagnetic spinel. *J. Am. Chem. Soc. 2012, 134*(24), 10182–10190.
11. Carta, D., Casula, M. F., Falqui, A., Loche, D., Mountjoy, G., Sangregorio, C., & Corrias, A., (2009). A Structural and magnetic investigation of the inversion degree in ferrite nanocrystals mfe2o4 (M = Mn, Co, Ni). *Journal of Physical Chemistry C, 113*(20), 8606–8615. https://doi.org/10.1021/jp901077c.
12. Issa, B., Obaidat, I. M., Albiss, B. A., & Haik, Y., (2013). Magnetic nanoparticles: Surface effects and properties related to biomedicine applications. *International Journal of Molecular Sciences, 14*(11), 21266–21305. https://doi.org/10.3390/ijms141121266.
13. Ansari, S. M., Ghosh, K. C., Devan, R. S., Sen, D., Sastry, P. U., Kolekar, Y. D., & Ramana, C. V., (2020). Eco-friendly synthesis, crystal chemistry, and magnetic properties of manganese-substituted $CoFe_2O_4$ nanoparticles. *ACS Omega, 5*(31), 19315–19330. https://doi.org/10.1021/acsomega.9b02492.
14. Ortiz-Quiñonez, J. L., Pal, U., & Villanueva, M. S., (2018). Structural, magnetic, and catalytic evaluation of spinel Co, Ni, and Co-Ni ferrite nanoparticles fabricated

by low-temperature solution combustion process. *ACS Omega, 3*(11), 14986–15001. https://doi.org/10.1021/acsomega.8b02229.

15. Wu, W., Wu, Z., Yu, T., Jiang, C., & Kim, W. S., (2015). Recent progress on magnetic iron oxide nanoparticles: Synthesis, surface functional strategies and biomedical applications. *Science and Technology of Advanced Materials, 16*(2). https://doi.org/10.1088/1468-6996/16/2/023501.
16. Lee, N., Yoo, D., Ling, D., Cho, M. H., Hyeon, T., & Cheon, J., (2015). Iron oxide based nanoparticles for multimodal imaging and magnetoresponsive therapy. *Chemical Reviews, 115*(19), 10637–10689. https://doi.org/10.1021/acs.chemrev.5b00112.
17. Bao, N., & Gupta, A., (2011). Self-assembly of superparamagnetic nanoparticles. *Journal of Materials Research, 26*(2), 111–121. https://doi.org/10.1557/jmr.2010.25.
18. Seabra, A. B., Pelegrino, M. T., & Haddad, P. S., (2017). Antimicrobial applications of superparamagnetic iron oxide nanoparticles: Perspectives and challenges. In: *Nanostructures for Antimicrobial Therapy*. Elsevier Inc. https://doi.org/10.1016/B978-0-323-46152-8.00024-X.
19. Dulińska-Litewka, J., Łazarczyk, A., Hałubiec, P., Szafrański, O., Karnas, K., & Karewicz, A., (2019). Superparamagnetic iron oxide nanoparticles-current and prospective medical applications. *Materials, 12*(4). https://doi.org/10.3390/ma12040617.
20. Nabavinia, M., & Beltran-Huarac, J., (2020). Recent progress in iron oxide nanoparticles as therapeutic magnetic agents for cancer treatment and tissue engineering. *ACS Applied Bio Materials, 3*(12), 8172–8187. https://doi.org/10.1021/acsabm.0c00947.
21. Ling, D., Lee, N., & Hyeon, T., (2015). Chemical synthesis and assembly of uniformly sized iron oxide nanoparticles for medical applications. *Accounts of Chemical Research, 48*(5), 1276–1285. https://doi.org/10.1021/acs.accounts.5b00038.
22. Frey, N. A., Peng, S., Cheng, K., & Sun, S., (2009). Magnetic nanoparticles: Synthesis, functionalization, and applications in bioimaging and magnetic energy storage. *Chemical Society Reviews, 38*(9), 2532–2542. https://doi.org/10.1039/b815548h.
23. Chen, Z., Wu, C., Zhang, Z., Wu, W., Wang, X., & Yu, Z., (2018). Synthesis, functionalization, and nanomedical applications of functional magnetic nanoparticles. *Chinese Chemical Letters, 29*(11), 1601–1608. https://doi.org/10.1016/j.cclet.2018.08.007.
24. Wu, W., He, Æ. Q., & Jiang, Æ. C., (2008). *Magnetic Iron Oxide Nanoparticles: Synthesis and Surface Functionalization Strategies*, 397–415. https://doi.org/10.1007/s11671-008-9174-9.
25. Gutierrez, A. M., Dziubla, T. D., & Hilt, J. Z., (2018). *Recent Advances on Iron Oxide Magnetic Nanoparticles as Sorbents of Organic Pollutants in Water and Wastewater Treatmen*t, *32*, 111–117. https://doi.org/10.1515/reveh-2016-0063.
26. Xu, P., Ming, G., Lian, D., Ling, C., Hu, S., & Hua, M., (2012). Science of the total environment use of iron oxide nanomaterials in wastewater treatment: A review. *Science of the Total Environment, 424*, 1–10. https://doi.org/10.1016/j.scitotenv.2012.02.023.
27. Hua, M., Zhang, S., Pan, B., Zhang, W., Lv, L., & Zhang, Q., (2012). Heavy metal removal from water/wastewater by nanosized metal oxides: A review. *Journal of Hazardous Materials, 211, 212*, 317–331. https://doi.org/10.1016/j.jhazmat.2011.10.016.
28. Li, X., Xu, G., Liu, Y., & He, T. (2011). Magnetic Fe_3O_4 Nanoparticles: Synthesis and Application in Water Treatment. *Nanosci. Nanotechnol. Asia, 1*(1), 14–24.
29. Sharma, M., Kalita, P., Senapati, K. K., & Garg, A., (2018). Study on magnetic materials for removal of water pollutants. *Emerging Pollutants - Some Strategies for the Quality Preservation of Our Environment*. https://doi.org/10.5772/intechopen.75700.

30. Abkenar, S. D., Khoobi, M., Tarasi, R., Hosseini, M., Shafiee, A., & Ganjali, M. R., (2015). Fast removal of methylene blue from aqueous solution using magnetic-modified Fe_3O_4 nanoparticles. *Journal of Environmental Engineering, 141*(1), 04014049. https://doi.org/10.1061/(asce)ee.1943-7870.0000878.
31. Moosavi, S., Lai, C. W., Gan, S., Zamiri, G., Akbarzadeh, P. O., & Johan, M. R., (2020). Application of efficient magnetic particles and activated carbon for dye removal from wastewater. *ACS Omega, 5*(33), 20684–20697. https://doi.org/10.1021/acsomega.0c01905.

CHAPTER 9

SOLID-STATE CARBON NANOTUBES

MIHAI CIOCOIU

Textiles-Leather and Industrial Management at Gheorghe Asachi Technical University of Iasi, Romania

ABSTRACT

Bionanocomposites and carbon nanotubes (CNTs) by virtue of their uniqueness make them better placed against other conventional materials for specialty applications. There are numerous examples of the prevailing commercial uses of CNTs and their potential applications in the near future. Nanotubes are synthesized in different types ranging from multiwalled nanotubes possessing various degrees of perfection to the close-to-perfect single-walled nanotubes (SWNTs) molecular in nature. The past decade of research has brought to the fore many feasible applications of these materials, such as electronic, field emission, electrochemical devices, interconnects, namely, supercapacitors (SCs) and batteries, nano-sensors, electromechanical actuators, targeted drug delivery systems, etc. The present status of this frontier area and the value-added applications in existence as against the revolutionary and state-of-the-art applications of different types of CNTs within different timelines are reviewed in this chapter. The open opportunities challenges are thrown, and the main bottlenecks, covering manufacturing of large amounts of nanotubes, will be introduced along with the applications space for CNTs. The summary of this chapter captures the broad picture of CNT applications. Moreover, a few of the concerns pertaining to health, environment, and safety protocols for CNTs too are part of the discussion.

Nanotechnology-Enhanced Solid Materials: Design, Synthesis, Properties, Applications, and Perspectives.
Lionello Pogliani, Ann Rose Abraham, A. K. Haghi, & Prabhat Ranjan (Eds.)

9.1 INTRODUCTION

The cumulative effect of structure, topology, and dimensions generates various physical properties in CNTs which are quite unique than most of the other known materials. After research of more than a decade which showed that such tiny quasi one-dimensional (1-D) structures exhibit lot much of potential in a host of applications areas, like nano-probes, molecular reinforcing materials in composites, display, molecular, electronic devices, sensors, energy-harvesting [1–3]. There have been huge modifications in synthesis and refinement protocols, which can yield qualitative nanotubes on a mass scale. There is a huge and exciting database derived from theoretical calculations and experimentations which opens up numerous unique possibilities in the applications of carbon depending upon their property profile. On looking at the applications landscape for CNTs and CNT-based materials, it can be break into several categories depending upon different conditions: incumbent ones that are present in the marketplace, short-term ones that will come up in the marketplace in the coming decade, while long-term ones refer to those post a decade down the line. Beyond these clearly foreseeable practical applications, there have been significant advancements in the knowledge of the chemistry science and physics underlying carbon-based materials that have worked as a front runner for the area of nanotechnology, dragging a lot much of attention right from the academicians to industries alike.

Carbon nanotubes (CNTs) came up as a unique research outcome in carbon science, apart from the Buckminsterfullerene molecules [4]. Fullerenes have so far stimulated and motivated the huge scientific community, and their area of applications are very extensive up to now. While, graphite and carbon fibers [5], the analogs of nanotubes, have been in use for commercial uses over decades. Their application profile ranges from conductive fillers and mechanical structural fillers in composites (e.g., aerospace industry) to electrode materials (e.g., batteries). Although primarily the bonding in graphite, carbon fibers and nanotubes remains same (i.e., the sp^2-hybridized, hexagonal layer-like honeycomb lattice pattern); nanotubes are distinct from the other allotropic forms of carbon by virtue of their unique morphology, scale, and deviations in crystal lattices. CNTs resemble with the theoretical in-plane characteristics possessed by graphite, which involve the carbon-carbon bond that is free from defects, which make the bond quite strong in nature. The substantially large modulus, electrical, and thermal conductivity as well as elasticity of graphite are all found as the intrinsic properties of nanotubes [6].

There are essentially two types of nanotubes that are used currently [7]. The multiwalled carbon nanotubes (MWNTs), which have shown the lot much of promise. These are produced catalytically by using chemical vapor deposition (CVD) technique. The so-formed multi-layered graphitic tubular structures are subjected to annealing to reduce defect density. They may be many tens of nanometers in diameter, but sometimes structures such as single-layered tubular structures smaller in dimensions are also obtained. In the 1990s, Iijima et al. [8, 9] obtained smaller, but largely graphitized multi-walled structures by employing high-temperature arc-discharge technique. It was followed by the discovery of SWNTs (single-walled nanotubes) [10, 11] in 1993 by the NEC and IBM teams using catalytic vapor deposition method to obtain single-layer tubular structures with controlled growth of graphene having dimensions as little as 1 nm. To grow such defect-free graphene tubes smaller in dimensions made researchers to prescribe numerous applications for such structures, characterized by chirality, defect-free, and small dimension structures which showed future prospects of using them as quantum wires.

Carbon-based fibrous structures, on the other hand unlike the conventional graphite-like materials possess molecular structures that find applications in non-conventional fields such as nanoelectronics. Today, nanotubes are prepared with a precise tuning of a number of layers, like SWNTs and double-walled nanotubes (DWNTs), and there is a lot much headway in terms of scalability of such materials. It has also become possible to manufacture CNTs on a mass scale these days that are doped with elements like nitrogen and find use in selective applications. As far as applications of nanotubes are concerned, they are divided into conventional (where nanotubes are value-added as against carbon fibers and other allotropes of carbon) and unconventional (whereas nanotubes with unique applications) uses, regardless of the timelines expected for applications.

9.2 APPLICATIONS OF CARBON NANOTUBES (CNTS)

Over the last two decades of research in fullerenes and nanotubes has brought to the fore plethora of applications for such materials. However, the applications of carbon nanomaterials have been limited in reaching to the marketplace, due to problems in processing, scale-up, and the inability to assess the timeline that is required to reach new products to market, particularly when such products lead to disruptive technologies. Various applications of fullerenes and nanotubes belong to this category, like drug delivery and nanoelectronics using fullerenes and nanotubes, respectively. Moreover, there are innate challenges in the nano-regime that have reduced the rate at which the applications of such new materials grow.

Likewise, some applications of CNT's are commercialized in reality today. The well-known examples of applications of CNTs in bulk are the application of CVD-grown much bigger MWNTs (multiwalled nanotubes) especially in lithium-ion batteries (LIBs) and plastics in electrostatic discharge (ESD) applications. The excellent mechanical or electronic properties of perfect nanotubes are not made use of in both of these applications, but as additives to the products (e.g., mechanical stability to batteries and static charge dissipation in plastics). These applications are in accord with bulk volumes of nanotubes, and it has led to manufacturing of such material on a mass scale over the last few years.

Presently, the large volumes of MWNTs turn out to be hundreds of tons per year. The condition of SWNTs however is different with the volume of material manufactured/year is less than a few kilograms. Few indicators predict that there may be a quantum jump in the manufacture of SWNTs in following years will be driven by the demand for this material in selective application areas. One of the potential applications of SWNTs that requires material on large scale is the field emission display, which has been extensively considered as the major application of SWNTs; and prototypes have already been showcased by big companies namely, Samsung and Motorola. However, the product is not yet commercialized for unknown reasons.

Some of the key application areas of both MWNTs and SWNTs are dealt with. The important applications that arise by virtue of the intrinsic properties of nanotubes will be put to use in the range of areas such as electronics, energy, composites, sensors, field emission, biology, etc.

9.3 ELECTRONICS APPLICATIONS

The integration of nanotubes with electronic devices takes a long application time and therefore commercialization of products may take a time span of 10 years or so. However, nanotubes in particular, SWNTs have come up as a potential class of electronic materials by virtue of their dimensions in the nanoscale and unique properties, like ballistic electronic conduction [12, 13] and insensitivity to electro-migration [14, 15]. SWNTs (metallic and semiconducting) reveal electrical properties that are much better placed than the incumbent best of the electronic materials. The latest advancements in the separation [16, 17] of SWNTs having various electronic properties have made it possible for the nanoscientists to design the field effect transistors (FET) and interconnects on a mass scale. In the quest of Moore's law, giants in the semiconductor industry such as Intel, IBM, AMD, UMC, and TSMC are commercializing silicon-based transistors to 65 nm scale node (Q1, 2007). Intel is aimed at commercializing transistors to 45 nm node till 2007 [18, 19]. The smallest silicon transistor that can be conceived today is 16 nm and is expected to reach markets not before 2018 as a part of an evolutionary process. To bring about a revolution in speeding up this process to envision realistically only nanotechnology can come to rescue. Single-walled carbon nanotubes, of 1 nm diameter, have been found to work as field-effect transistors almost 10 years back, and thereafter, the FETs have grown in a steady manner in terms of versatility and promise and will meet the futuristic needs of nanoelectronics.

The earliest room-temperature research in CNT-based transistors was undertaken by the Delft [20] and IBM teams [21]. They showed that semiconducting SWNTs possess a channel conductance that can be enhanced by many folds (10^5) on application of gate voltage externally, similar to MOSFET. Such CNT-based FETs are p-type, and the charge transport was regulated with the changes in the Schottky barrier height at the electrical contact in between the nanotube and the metal probe on subjecting it to a gate voltage. Subsequently, continuous research is underway to grow n-type and ambipolar CNT-FETs. Till 2001, an important milestone was achieved in building nanotubes from transistors to logic circuits. Transistor circuits containing up to three transistors which exhibit a plethora of digital logic operations, were showed namely:

- an inverter;
- a logic NOR;
- a static random-access memory cell; and
- an ac-ring oscillator [22].

IBM devised a technique to prepare an array of CNT-based transistors, without the use of distinct metallic and semiconducting nanotubes [23]. Eventually many technological progresses attracted the use of CNTFETs (carbon nanotube field-effect transistors) in diverse applications. CNT-powered transistors were found to have lower switching resistance and more current-handling capacity 20 and 200 times respectively as compared to traditional power MOSFETs. A University of Maryland team has shown that CNTs have significant carrier mobility of the order of 10^5 cm^2/Vs. Other investigators could further the feasibility of employing nanotubes-based

FETs ranging from GHz to THz frequency. More innovations, such as nanotube Y-junctions to develop switches [24], and functionalizing chemically to alter nanotubes conductance have brought better flexibility in design terms in the ultimate chip-forming process.

Apart from transistors, nanotubes can also work as interconnects because of the metallic characteristics possessed by them. As interconnect feature downsizes, copper resistivity exacerbates which is attributed to their scattered grain-boundary and surface heterogeneities [25] unlike CNTs that show a ballistic flow of electrons having electron mean-free paths of many micrometers, and have an ability to conduct massive current densities [26]. CNTs withstand current densities of the order of 1,009 A/cm^2, superseding copper by 1,000 times. A bunch of densely packed aligned metallic CNTs, or a MWNT large enough and free from defects are to be employed for the next generation interconnects. A bunch of 400 SWNTs can be as thin as 20 nm in diameter and using in bundles and connecting them serve two major purposes. Firstly, connecting in parallel overcomes the issue of the more contact resistance of a nanotube and secondly bundles lower capacitance as against that possessed by a nanotube [27]. Nanotubes are characterized by high aspect ratio structures such as vias, having maximum current densities. Apart from on-chip (short and via) interconnects, because of the mechanical flexibility and compliance of nanotubes in bundles, they too are used for back-end (die to package) flexible interconnects and these can be obtained in a timeline rapidly as compared to the integrated on-chip nanotubes-enabled structures. The innate electromechanical properties of nanotubes can be used to advantage in the preparation of electrode arrays that find a host of applications in the microelectronics industry.

FETs containing semiconducting CNTs are used in the form of channels in transistors [28, 29]. Such carbon-nanotubes FETs have successfully demonstrated a conductance, G_{ON}/G_{OFF} ratio [30]. In addition, thin films made up of high dielectric materials, like ZrO_2 and HfO_2, are favorable for gate dielectric integration in FETs as they lead to higher ON-state current densities (speed) and lower working power consumption. The integration of high-κ dielectric materials, however, is quite challenging in planar MOSFETs because of the intrinsic mobility decrease of the Si channels. CNTs have a unique advantage in the form of compatibility with high-κ dielectric materials. The absence of dangling bonds at the interface between nanotubes and high-κ dielectrics, accompanied by the poor non-covalent interactions prevent any hindrance to the electron scavenging in CNTs [30].

Another important electronic application of CNT's is the production of an electromechanical memory device by Nanterro Inc. By taking the advantage of mechanical resilience and electrical conductivity, CNTs can work as sensitive nano-electromechanical devices namely, non-volatile memory. With a crossbar layered configuration, Nanterro is fabricating a bi-stable memory device which is actuated electrostatically [31]. This device, is co-fabricated by LSI Logic, will reach the market in the near future. Invisible circuits driven by transparent transistors hold a lot of promise and potential in consumer, military, and industrial electronic devices.

The flexible and transparent transistors prepared using SWNTs have led to numerous exciting achievements. SWNTs have been in use to prepare the conducting as well as semiconducting films, of thin-film transistors (TFT) [32]. Such devices are designed on plastic substrate surfaces by piling of layers and transfer printing of SWNT's networks developed by CVD method on optimization. The unique properties possessed by SWNTs networks have imparted electrical (better efficiency on plastic), optical (clarity in the visible region), and mechanical (can be bent readily) properties to such TFTs which are impossible to attain in traditional materials. A work of similar kind [33] has demonstrated that SWNTs electrodes can be integrated with organic semiconducting materials to produce TTFTs (transparent thin film transistors) fabricated on plastic substrate surfaces.

Such devices have exhibited excellent electrical properties like abysmally low contact resistance, mobilities greater than 0.5 $cm^2V^{-1}s^{-1}$, flexibility (curvature radius < 5 mm) and optical transparency (69%). Moreover, further efforts aimed at producing ultrathin, transparent, uniform, conducting layers of pure SWNTs and DWNTs and fabricating such films on different substrates were quite successful [34]. The performance of such films in respect of conductivity and transparency was in semblance with the commercially employed indium tin oxide (ITO) films although supersede in transparency in the vital IR bandwidth. Therefore, the transparent CNT films would replace ITO, in the application areas like displays, touch screens and (Light-Emitting Diodes) LEDs in coming years. Getting reasonably higher transparency (>90%) accompanied by high conductivity is challenging.

The necessary electrical applications of CNTs are on those fronts wherein high aspect ratio, and electrical conductivity of CNTs with better dispersion in polymer matrices make them useful in applications such as ESD and electrical shielding [35]. For ESD applications, the nanotubes should have the ability to prevent the buildup of charge on the plastic substrate surfaces. Currently, huge amounts of multiwalled nanotubes and nanofibrils are prepared by the Hyperion Catalysis International and procured in the form of premixed material with plastics to the general electric (GE) for using it in engineering plastics which are further used in making automobile components and is one of the largest application of nanotubes presently. The MWNTs (Multi-Walled Nano Tubes) are commercially used for the first time in making use of their properties to solve the problem of ESD. Further applications cover microelectronics products that are high-end wherein static buildup of electric discharge from packaging materials can cause destruction of electronic chip parts. Apart from the ESD, electromagnetic interference (EMI) shielding applications too have been explored for nanotubes with more loadings, taking the advantage significantly higher conductivity of nanotubes. In fact, carbon fibers are used on this front however, nanotubes can have a comparative advantage as against the conventional fillers, it is yet unknown whether nanotube-enabled plastics would supersede traditional materials in the area of EMI shielding or otherwise.

Another front where nanotubes can bring about a sea of change may be in the future is in the area of on-chip heat management. Large consumption of power and consequent issue of heat dissipation is major problem in microprocessors presently. Many scientists reported that efficient cooling can be accomplished in microprocessors with the help of multiwalled carbon nanotube (MWCNT) arrays that are aligned (MWNTs) [36]. Small cooling elements can be manufactured and grown on silicon chips to ensure substantially high-power dissipation from the chips heated up and hot spots generated. The cooling efficiency showed by nanotube-based elements are the useful candidates for on-chip heat dissipation. The problem in this case is that of thermal interfaces at which scattering responsible for dissipation of heat. To improve performance on this count, the thermal contact resistance has to be reduced. The high thermal anisotropy possessed by nanotubes-based structures (high longitudinal conductivity vis-a-vis weak transverse conductivity) may either pose a challenge or a comparative benefit while manufacturing cooling elements out of nanotubes.

9.4 ENERGY APPLICATIONS

Conventional carbon-based materials have been in use either in the form of electrode materials or the filler, which is conducive to the active materials in numerous electrochemical energy-harvesting applications because of their properties such as better chemical stability and electrical conductivity. Thus, as a natural choice CNTs have been used as the preferred substitute for electrode materials attributed to their unique electrical and electronic properties, in addition to broad electrochemical stability window, and more availability of surface area. With reference to energy production and storage, nanotubes exhibit huge potential in supercapacitors (SCs), Li-ion batteries, solar cells, and fuel cells. In general energy applications would be the largest application domain wherein nanotubes can be used in bulk. Hydrogen-storage was considered as one of the major energy applications which generated a great promise for nanotubes however now it is not found to be a viable option as the thorough investigations showed that the hydrogen storage capacity of nanotubes at ambient temperature is quite low and therefore it is not much of practical interest as far as storage of hydrogen in nanotubes is concerned.

The CNTs have a huge potential as the electrode materials in LIBs [37–40] as they showed an increase in reversibility up to 1,000 mAh/g [41]. However, the larger reversible capacity, the hysteresis, and the lack of a voltage plateau, found in a hard carbon [42], do not find nanotubes a right candidate for the electrode materials in LIBs. Alternatively, CNTs have been commercially used as the conductive fillers of choice in the materials for anodes in LIBs [43]. CNTs on dispersion in graphite (3% by wt.) develop a homogeneous conductive network along with a mechanically stubborn electrode, leading to the acquisition of a two-fold energy efficiency in LIBs [44]. Nitrogen-doped CNTs (N-CNTs) and nanofibers too have exhibited higher efficiency in reversible energy storage in LIBs (480 mAh/g), larger than commercially used carbon-based materials used in LIBs (330 mAh/g).

Energy storage systems apart from batteries, supercapacitors have been explored extensively as they have the capacity to store and deliver energy on demand fast and with efficiency over a longer life cycle by adopting a simple charge separation technique [45]. Moreover, their broad spectrum of power capacity makes it suitable to couple with other energy-storage systems namely, batteries, and fuel cells (FCs). Although they find use these days in a host of electronic devices (e.g., memory backup), from computers to vehicles, their low energy density has limited the adoption in the market on a larger scale. In such situation, supercapacitors containing carbon-nanotube electrodes have been researched for increasing their energy density [46, 47]. E.g., a SWNT electrode showed the highest capacitance of the order of 180 F/g and a power density varying between 6.5 to 7 Wh/kg. Similarly, chemically modified MWNTs are in the infancy stage of commercialization to employ them as a conductive filler particularly in porous carbon-based materials in SCs [48].

Silicon-based photovoltaics have attained solar-energy conversion efficiencies of 25%; however, since it is an environmentally hazardous technology, and expensive and therefore not suitable as a substitute to fossil fuels for extensive applications. Another option is to employ organic thin films of conjugated polymers, which have low conversion efficiencies (5%), but have lower cost in comparison with silicon-based technology. Conjugated polymers an ability to absorb a light and create an exciton which can split into an electron-hole pair on subjecting it to the high electric field observed at polymer/metal interface. One of the soundest designs for an organic solar cell makes use of a bulk hetero-junction composite made out of a conjugated polymer containing electron acceptors a nanoscale producing a percolating network via the thin film. This enhances the area at the interfaces between the metal as a filler and the polymer, which exacerbates charge separation and provides for scavenging of the electrons via the network to collect at the electrode. An organic solar cell can be commercially viable and comparable with silicon-based technologies it needs to attain an energy conversion efficiency of the order of 5 to 10%, and possess a minimum lifetime (five years) and cost (< 1/peak watt).

CNTs can be an excellent candidate for designing bulk hetero-junctions in organic solar cells that is attributed to the high aspect ratio possessed by them which provides for the development of a percolating network of CNTs at lower dopant concentration quantity in a polymer composite [49, 50]. As 1-D nanostructures, nanotubes are suitable for electron transport too as they show quasiballistic properties [51, 52]. The bulk heterojunction (BHJ) organic solar cell is in general designed as a sandwich comprising of a transparent conducting electrode, normally ITO or fluoride tin oxide (FTO), deposited on a glass or PET substrate, a film of poly(3,4-ethylene dioxythiophene) (PEDOT) or poly(styrenesulfonate) (PSS), a film of the composite material, and a thermally evaporated electrode in the top, commonly aluminum. Ultrasonication is employed for the dispersion of nanotubes in the conjugated polymer, and then it is subjected to centrifugation to get rid of larger aggregates. The dispersion is then coated on the transparent electrode by using either drop or spin casting. Kymakis et al. [53, 54] proved for the first time that SWNTs on dispersing in poly(3-octylthiophene) (P_3OT) demonstrated a rise in short circuit current (ISC), by two folds as against P_3OT in isolation in an ITO/composite/aluminum sandwich-type device. An increase in the intensity of light brings an increase in the photocurrent of the device, which is attributed to the generation of excitons in the composite. The device performance was improved further in 2005 by depositing the ITO electrode with the conductive co-polymer (PEDOT:PSS) that brings about the promotion of injection of hole into the ITO [55]. This device revealed a rise in power conversion efficiency from 2.5×10^{-5} to 0.1% as against P_3OT alone.

By adjusting the concentration levels of SWNTs in the composite matrix, the highest efficiency attained was with 1% SWNTs, as further loading caused the limitation in the photocurrent due to the lower rate of photogeneration from the polymer [56, 57].

CNTs find use in various conjugated polymer matrices leading to enhancement in charge transport and separation. Numerous research teams have attained a 2–6-fold rise in the photocurrent of the composite materials in differentiation with the pure polymer [58–65]. Devices fabricated from composites containing polymers that are water-soluble and conducting enabled to sustain the better environmentally friendly processing and cut down in costs of production. The performance of such devices; however, is not good enough to be of commercial interest currently, but in the long run, the enhancement will pay off. Bulk hetero-junctions of nanotubes in organic conjugated polymers have brought about improvement in the exciton splitting and charge transport; however, such devices are yet to be optimized so as make them absorb light over the solar spectrum in entirety. Although the solar spectrum shows a light intensity to the maximum between the wavelength range 500 to 800 nm, however, many of the semiconducting polymers won't absorb significantly within this region. One of the approaches adopted to better the photoresponse within 500–800 nm is to introduce organic dyes with higher absorption coefficients at the nanotube/conducting polymer interface that can absorb to the maximum [66, 67].

Bhattacharyya et al. reported the enhancement in performance of SWNT or MWNT-based solar cells by introducing a dye containing aromatic pyrenyl groups N-(1-pyrenyl) maleimide (PM) into a SWNT/P_3OT device. Kymakis et al. used Naphthalocyanine (NaPc) as the sensitizing agent in a SWNT/P_3OT system. The enhancement of the order of 20 of the photocurrents was obtained in this device with the dye in connection with no dye use condition. Naphthalocyanine finds use in MWNT-poly(3-hexylthiophene) (P3HT) composites to significantly enhance the absorption of light in the visible area of the spectrum, furthering a greater production rate of excitons. Nanotube-based hybrids are also prepared to promote exciton formation and charge separation with the help of quantum dots (QDs) and nanoparticles. QD-SWNT hybrid systems were prepared by the Landi et al. research team to use QD electron affinity to the advantage and the lower dopant concentration levels of SWNTs. Photovoltaic efficiency shown by this composite resemble with the SWNT/P_3OT but with a restricted efficiency, due to interfacial defects at the junctions in the device. Titania (TiO_2) nanoparticles too find use in enhancing the performance of the device by using electrostatic interactions. Feng et al. [68] used MWNTs, which are functionalized with acidic groups to bind titania nanoparticles and found a doubly increased photocurrent in a device with no nanotubes.

Semiconducting CNTs can produce excitons on absorption of light in the near-infrared region, which in turn can further effective exciton formation in the device and enlarge the bandwidth of light that is absorbed [69, 70]. Spataru et al.; Ando et al.; Kazaoui et al. [71] have showed with a SWNT/poly[2-methoxy, 5-(2-ethylhexoxy)-1,4-phenylene vinylene] (MEHPPV) or SWNT/P_3OT composite having Al and ITO electrodes in a sandwich structure, that the photo-response bandwidth can be broadened beyond the visible region to 300–1,600 nm. The photo-response shown by such devices in a region 900–1,500 nm is in tandem with that shown by semiconducting SWNTs. Below 900 nm the photocurrent is driven by exciton formation from MEHPPV of P_3OT.

CNTs, therefore, have been promising candidates for hetero-junctions in organic thin-film solar cells in bulk. Such devices have exhibited marked improvement in charge separation and scavenging, however, presently work with comparatively low performances. Even with such low efficiencies, there is an increase in interest, however, to go ahead with the commercialization, particularly in low-power consumption applications in the time to come. Better enhancement and optimization of the nanotube/polymer interfaces in such materials can help in reaching energy conversion efficiency of the order of 15 to 20% in the coming decade. The emphasis in the current development is laid upon improvement in device performance in addition to the longevity of the cell such that a competitive technology using silicon cells is developed. At this juncture the potential pathways are dye sensitization and nanoparticle hybrids use to better exciton formation and charge separation. In fuel-cells, CNTs hold a lot of promise as complementary materials for Pt catalyst [72, 73]. CNTs [74] and nanohorns piled up [75] with a greater part of reactive edge sites were found to be quite effective in anchoring uniformly distributed tiny particles of Pt catalysts, which has led to a doubling of fuel-cell efficiency [76]. The National Institute of Standards and Technology (NIST), carbon nanotechnologies, Inc., Motorola, Inc. and Johnson Matthey FCs, Inc. in collaboration are working to design a free-standing carbon-nanotube electrodes that may be applied in micro-FCs so as to fulfill the ever-increasing demand for power and longer run times in customized microelectronic devices. Proton-exchange membrane fuel cells (PEMFC) with greater efficiencies have been designed with Pt catalysts grown on MWNTs [77]. This application seems to be very promising and has tremendous market potential of five years.

9.5 MEDICAL APPLICATIONS

The applications of CNTs in mechanical and structural field hold a lot much promise as the largest mass-scale applications for the material. Yamamoto et al. described the mechanical, and thermal properties of CNTs in detail. Nanotubes are ideal candidates for fibers with outstanding mechanical properties as against the excellent carbon fibers, which already has the specific strength (strength/density) 50 times that of steel and have excellent load-bearing capacity of fillers in composites. Carbon fibers have been in use as fillers in high strength, performance, and light weight composites. The products wherein they find use vary from costlier tennis rackets to spacecraft as well as aircraft components. In reality, nanotubes are much better in their performance for more than carbon fibers in mechanical applications. The tensile strength of nanotubes may be in the vicinity of a few 100 GPa while their modulus of elasticity lies in the tera-pascal range [78–80], which is far more than conventional carbon fibers. Nanotubes have higher strain sustenance too when subjected to compression than conventional carbon fibers.

The deformation reversibility like buckling, has been researched [81]. Flexibility and compliance of nanotubes are governed by the geometric factors of the CNT's, which could be coordinated. The flexibility of nanotubes when subjected to mechanical force is vital as far as their potential application as nanoprobes and in electromechanical fields are concerned [82]. The properties of nanotubes have induced many research teams, such as NASA, to invest huge amounts of resources in designing CNT-based composites for a host of applications in the aerospace industry. Although nanotubes have close to ideal structures and higher aspect ratios, there are challenges to in adopting such materials in structural applications. In spite of that, there has not been much substantial change, in fabricating nanotube-based composites that supersede the most conventional carbon fiber-based composites.

Nanotube-reinforced polymer composites have advantages of their own however, the major problem is in forming a better interface between nanotubes and the polymer matrix and provide much better force-transfer from the matrix to the nanotubes during loading [83–86]. It is so because firstly nanotubes are smooth at the atomic level and if engineered well, the interface becomes strong. Secondly, nanotubes are at times bundled into aggregates (concentric alignment within MWNTs and SWNTs as bundles) that show variation in load response, in comparison with the free nanotubes. Previous studies of load transfer with Raman spectroscopy hinted at just mixing nanotubes into a polymer that however doesn't assure sound load transfer. Therefore, latest approaches are all dependent upon chemical modification of the surface functional groups of nanotubes so as to create stronger interfaces [87].

Apart from these intrinsic issues, processing of nanotube-reinforced polymers (e.g., epoxy) is not easy as even a low wt.% of the nanotubes (< 5%) improves viscosity significantly, making processing quite uneasy. Lastly, even when aspect ratios are very high, nanotubes essentially behave like short fibers, that are not perfect structural fillers in comparison with the continuous fiber fillers in composites. The function of nanotubes in composites may vary as against active load-carrying fibers in the composites, nanotubes however an important role as a matrix additive, offering multifunctionality to the composite. There are many companies in the USA (e.g., Zyvex corporation) and Japan (e.g., Mistui Corporation, Toray) which are producing nanotube-dispersed plastics for getting specialty sporting goods and the products having far better performance are in the market already. These use the chemical vapor deposition (CVD-driven) multi-wall carbon nanotubes (MWNTs) and Nano-fibrils as additives.

Nanotubes in such products work more as enhancers of matrix over the primary load-sustaining structures. In an approach, it has been observed that nanotubes can better the properties of 3D composites. Here nanotubes are not the major load-carrying fibers, but add value to the composites in the transverse direction to the fibers plane. Conventional fiber-filled composites consisting of fiber stacks aligned and embedded in a polymer matrix possess outstanding in-plane characteristics but are poor in through-thickness (transverse to the fiber stacks direction) characteristics. There have been numerous strategies, particularly through innovative fiber-packing architectures, have been however successful to a limited extent. Oriented inter-laminar arrays of carbon-nanotubes (CVD coated onto SiC fiber stacks) offer improved multifunctional characteristics along the direction of thickness. The arrays of carbon nanotubes provide for the fastening of adjoining stacks in the 3D-composites, offering the highly sought after interlaminar strength and toughness when subjected to varying loading situations. The fabricated 3D-composites wherein nanotubes are embedded, marked enhancements in the properties such as inter-laminar fracture toughness, de-lamination resistance, damping, thickness transverse to the fiber stacks direction, and other mechanical properties without compromising the in-plane characteristics of the composites simultaneously.

Such 3D composites using nanotubes add value by incorporating distinct geometries to conventional composites of nanotubes find applications as structural polymer composites. The interfacial tension between nanotubes and polymer matrices can be used to advantage by applying it on the front of vibration damping. Damping eradication is a main problem in aerospace and automobile industry, and the systematic application of nanotubes for vibration damping [88, 89]. Research has demonstrated that the nanotube-polymer matrix sliding energy-dissipation mechanism can increase the loss modulus of a polymer matrix by nearly three folds with a comparatively little (1–2%) weight fraction of properly dispersed SWNT as a reinforcing agent. An increase in damping in epoxy and other high-temperature resins takes place with SWNT fillers. Mass scale production and use of nanotube-based composites, wherein nanotubes are used as fillers and as damping elements which would be an application in the bulk. Such applications of nanotubes in composites are under development still it may take several years to assess the efficacy of nanotube composites as compared to conventional fiber-reinforced composites. Nanotubes in composites would offer multifunctionality (matrix stiffness, damping, conductivity, ESD, etc.), instead of merely working as mechanical fillers.

Pristine nanotubes offer weak interfaces with polymers, therefore there have been efforts to dope nanotubes for the acquisition of more interfacial strength, which is very vital in composites. Preliminary investigations on the synthesis of epoxy composites by using N-doped MWNTs demonstrated a significant rise in the glass transition temperature with the introduction of little amount (2.5 wt.%) of nitrogen. It has been observed that it is possible to decorate polystyrene (PS) on N-doped MWNTs [90] and nitroxide-driven free radical polymerization mechanism [91] with no acid treatment that is required otherwise to create defects in CNTs. It shows that in-plane doping is of importance in the formation of covalent linkages at the nanotube-polymer interfaces. Latest mechanical and electrical characterization have shown that PS-grafted doped nanotubes demonstrate improved characteristics as against blends of PS and doped nanotubes. These explorations have given an impetus to the design of strong, and conducting composites. Terrones et al. and Kavan et al. described the chemical and electrochemical doping of CNTs for various applications.

The potential applications of nanotube-reinforced polymers in the field of medicine, like microcatheter, have been studied [92]. Identically dispersed CNTs in nylon polymer matrix make micro-catheters to acquire external diameters of about 0.52 mm size, with two-fold increase in mechanical strength, color intensity, and decrease in thrombogenicity and blood coagulation ability, in comparison with a pure polymer containing microcatheter. Moreover, stable biological responses of nanotube-reinforced microcatheters were used to measure the systematic T-cells and a histopathological examination [93]. Therefore, it is found that the carbon-nanotube-based composite driven catheter demonstrated excellent properties in comparison with a pure polymer-based catheter, and it is predicted that such systems will have extensive uses in numerous medical devices.

9.6 SENSING APPLICATIONS

CNTs have been quite effective as sensors making use of their electrical, electrochemical, and optical properties. Very low concentration levels of hazardous gases are detected, which is of importance from environmental and chemical safety point of view. SWNTs have been considered as promising candidates as gas sensors because of their 1-D electronic structure. SWNTs have a comparative advantage over traditional metal-oxide-dependent sensors in respect of consumption of power, detection limits, miniaturization, and mass production reliability [94]. Kong et al. [95] reported a change in conductance by three folds than individual semiconducting SWNTs on exposing to NH_3 and NO_2 gas within few seconds. This prototype SWNT gas sensing element [96], consisted of a system dependent upon a field-effect transistor (FET). The variation in conductance of a back-gated SWNT channel on the adsorption of an analyte is followed through source/drain electrodes. The SWNT-FET that is untreated selectively demonstrates a p-type behavior, with limits of voltage being altered on exposure to gas [97, 98]. Electron charge transfer takes place between a SWNT and an analyte that is adsorbed which can be proved both theoretically [99] and experimentally [100]. Nevertheless, it is yet to be clear up as how the signal transduction takes place in a SWNT-FET.

On exposing toward electron-lending gases namely, [NH_3] shifts the Fermi level to a higher energy level from the valence band (VB), resulting into hole depletion and conductance reduction. Electron-accepting gasses (i.e., NO_2), on the other hand, the shift the Fermi level nearer to the VB generated more hole carriers and conductance increase. Heinze et al. [101] demonstrated that a SWNT-FET works as a Schottky barrier transistor, numerous works have been presented on the function of the SWNT-metal interface. Bradley et al. [102] for the first time proved experimentally that the signal generated was not from the interface, but on exposing interface-passivated sensing elements to NH_3 and polyethyleneimine (PEI). Results of similar kind were derived for thionyl chloride. Anyway, a latest work by Zhang and others [103] proved that NO_2 sensing takes place at the interface in preference between the nanotubes and electrodes. Thorough investigations on numerous analyzes need to be done to get a better understanding about the mechanism underlying sensing.

Most of the initiatives till today have been aimed at bettering the sensitivity and selectivity. A commonly employed approach is the functionalization of the nanotubes with selectively attaching ligands like polymers. NO_2 at the PPT level was detected in preference over NH_3 using a PEI-functionalized SWNT array device, on the other hand NH_3 selectivity was accomplished by nafion-functionalization. SWNT array device becomes a n-type on functionalization using electron-lending PEI, which allows for better binding affinity with the electron-accepting NO_2. The array was grown with nanoparticles of transition metal for the analytes under examination possessing a bit of electron-donating or accepting behavior (i.e., CH_4, CO) [104]. It is suggested that the transition metal-analyte complex formed promotes the charge transfer between the nanotubes and metal nanoparticles. Staii et al. [105] utilized DNA-grown SWNTs for many analytes with the tailored sensitivity by changing the DNA sequences. Robustness in the sensor is a vital parameter for applications on long term. It is observed for various analytes that the signal derived from the sensor won't get restored to its starting value, which is due to adsorption of an analyte irreversibly onto the nanotube. Therefore, the nanotube array has to be reactivated and freed from such analytes. Kong et al. could bring about the removal of such analytes by thermal desorption. A photodesorption at the molecular level by UV irradiation [106] and a gate bias [107] were used to reactivate the nanotube surface. Hydrolysis is a readily used method for the remediation of moisture sensitive analytes.

Snow et al. [108] showed the use of an innovative SWNT chemi-capacitor for gas sensing. An optically transparent and electrically continuous single-walled nanotubes (SWNT) network was decorated on silicon oxide substrate by CVD, and electrodes were patterned on the SWNT film. Analytes are subjected to polarization by application of an electric field from the SWNT network, resulting into rise in capacitance. This chemi-capacitor is characterized by high sensitivity, rapid, and reversible. Functionalizing with chemo-selective polymers made it possible to detect dimethyl methylphosphonate (DMMP) at ppb level, which is used as a stimulant for the nerve agent sarin. Further this configuration has an added advantage of the possibility of simultaneous monitoring of both the capacitance and conductance. More study showed the ratio of the conductance to capacitance response is a characteristic of a given analyte and is independent of concentration [109] which confirms an unknown gas can be made-out by considering the ratio of conductance to capacitance.

Carbon-nanotube gas sensors are available commercially in the market. A sensor, by nanomix, is meant for the detection of hydrogen at industrial level wherein nanotube networks were decorated on a silicon substrate and in electrical contact. A top layer fabricated for preferential binding of hydrogen was then coated to the array. A substantial enhancement in cross-sensitivity, in comparison with other commercialized products in the form of catalytic beads or metal oxides, has been showing with interferants namely, CO, CO_2, H_2S, NH_3, and CH_4. Tailoring the top layer for recognition purpose led to the development of a CO_2 sensing device for breath analysis. Preparation of strong preferential binding ligands, coupled with a nanotube separation achieved recently using electronic structure, is very vital for further advancements on this front.

CNTs are outstanding materials for electrodes in electrochemical sensors attributed to rapid electron transfer kinetics by a host of electroactive moieties. A plethora of sensor architectures have been designed varying from carbon nanotube-based paste electrodes (CNTPE), glassy carbon electrodes (GCE) altered using CNTs, metal nanoparticle-altered CNT electrodes, to CNTs reinforced in a conducting polymer matrix. Many sensing technologies consist of enzymes immobilized on the surface of CNT electrodes, either by means of electrostatic interactions or through covalent bonding. Sensing of Glucose is an example of the electrochemical biosensor. Most of the electrochemical glucose sensors involve immobilization of the enzyme glucose oxidase (GOx), and catalyzes the oxidation of β-D-glucose to D-glucono-1,5-lactone using hydrogen peroxide (H_2O_2) as a by-product of the reaction. The liberated H_2O_2 is further subjected to oxidation at the electrode surface, sensed, and measured as current generated. Such a sensor configuration was designed for the first time by Sotiropoulou and Chaniotakis with the help of MWCNTs decorated directly on the surface of a Pt electrode. The device was developed by binding GOx to carboxylic acid functional groups attached to the MWNTs. Wang et al. [110] encapsulated MWNTs using a Teflon and GOx grown on a GCE, and showed a substantial rise in detection of glucose in comparison with a similar device prepared from graphite. Further, Gao et al. [111] reinforced MWNTs and GOx in a conducting polypyrrole (PPy) matrix and demonstrated a nearly linear response relationship to glucose concentration level in the range of 0–20 mM. Tang et al. applied another sensing architecture by introducing Pt nanoparticles in a Nafion/GOx/MWNT matrix probe [112]. The Pt nanoparticles were used further in SWNT by Hrapovic et al. for Nafion/GOx/SWNT on the surface of a GCE and obtained a sensitivity of 0.5 μ [113]. The sensitivity was improved further by using Cu nanoparticles to the extent of 250 μM [114].

The major limitation of these devices is the narrow linear response range of glucose concentration. Moreover, numerous devices have been designed for the electrochemical sensing of DNA. The "ssDNA" or single-stranded DNA is made to immobilize on the surface of the electrode and the resulting current changes are measured on hybridization of DNA with a complementary strand. Normallt, a redox indicator having more affinity for double stranded DNA (dsDNA) is made use of to alter the electrochemical response. Cai et al. [115] fabricated a sensor first of its kind using a MWNT/GCE and daunomycin (Dmc) as an intercalate. Methylene blue (MB) too has been in use as a redox indicator to detect DNA [116]. Methods using no reagent for the detection of DNA have also been monitored by following the oxidation of guanine [117, 118]. A host of other analyzes have been analyzed by using electrochemical responses. Ajayan et al. have used CNTs for the first time as electrochemical sensors [119], who found reversible oxidation of dopamine on the surface an electrode made from multiwalled CNT. Many researchers have suggested that directly electrons are transferred to nanotube electrodes from redox enzymes namely, GOx [120, 121] and cytochrome C [122]. Analytes like NAD^+ [123], cholesterol [124], and indole acetic acid [125] too have been detected. Since electrochemical sensors use nanotubes, they are considered as a mature technology that is CNT-based for bio-sensing. However, the use of CNTs for commercial purposes in biosensors has more challenges and a difficult task as well as it will take its own time as far as their availability is concerned in the near future.

Semiconducting SWNTs have been developed as field-effect transistor (FET) devices. These devices are quite sensitive and are adsorption-based on the surface of the nanotube. Besteman et al. fabricated the first SWNT-FET-based device for biosensing [126] by binding GOx, through a linker, to solitary SWNTs, exhibiting a bit of change in conductance when glucose is added. Chen et al. [127] showed protein detection selectively by first functionalization of the nanotubes by a suitable ligand, such as biotin or staphylococcal protein A or U1A antigen, and demonstrated a quantifiable change in conductance on attaching protein to the target ligand. However, further research by the same team hinted at the adsorption of protein on the surface of the gold electrodes as the underlying mechanism of sensing, since on passivating the electrodes, there is no measurable conductance change on binding the protein [128]. Star et al. developed a SWNT-FET-based DNA hybridization sensor and showed detection of a solitary nucleotide polymorphism, responsible for a host of diseases [129]. Tang et al., through a similar kind of work, demonstrated that the mechanism underlying deoxyribonucleic acid (DNA) hybridization sensing on the surface of single-walled nanotube-filled effect transistor (SWNT-FETs) was hybridization at the surface of gold electrodes, thereby questioning the role of SWNT-FETs in detecting DNA hybridization [130].

The latest nanotube biosensor developed takes the advantage of photoluminescence (PL) exhibited by the semiconducting SWNT. Lefebvre et al. described the science of PL shown by CNTs. SWNTs have a competitive edge over traditional fluorophores; which fluoresce within a given region of the electromagnetic spectrum wherein blood and tissue won't fluoresce, and will not photobleach thereby are sensitive to variations in the localized dielectric environment. Strano et al. brought about the adsorption of GOx and work as a reaction facilitator, and quenches single-walled nanotube (SWNT) PL, to dissolve single-walled nanotube (SWNT) individually. H_2O_2 liberated on the oxidation of glucose followed by its reaction with GOx and restoration of SWNT PL. A DNA hybridization sensor was also developed, by allowing adsorption of a target DNA strand in the first place onto the SWNT. Hybridization with the complement led to a hypsochromic shift (blue shift) of the order of 2 meV at the max emission wavelength. Ultimately, Bivalent metal ions are added to DNA-dispersed SWNT which resulted into a bathochromic shift (red shift) of the SWNT PL. This shift was detectable by encapsulating the single-walled nanotube in living cells as well. The application of SWNT PL opens up fresh opportunities of detecting, however, commercial availability of such sensor may take another 10 years or so.

Apart from the FET-based sensors, there have also been research initiatives to design sensors made up of nanotubes that function as electrodes, to make good use of the field-ionization properties possessed by nanotubes towards gases. E.g., the gas-ionization sensor, wherein the arrays of nanotubes replace transition metal-based electrodes [131]. The sharp tips of nanotubes produce large amounts of electric fields at comparatively lower voltages, reducing break-down voltages by many folds as against the conventional electrodes, and consequently making compact, battery-powered, and safe operability of these sensors. These sensors exhibit better sensitivity and outstanding selectivity, and are not influenced by parameters like temperature, humidity, and gas flow.

Such simple, cost effective, sensor electrodes depending upon oriented MWNTs vertically could be used for numerous gas-sensor applications, like environmental monitoring and in the chemical industry. Nanotubes aligned vertically, by virtue of their marked elastic compressibility, provide for fabricating electromechanical pressure sensors. The resistivity in phase determined in accordance with the cyclic compressive strain on arrays of nanotubes aligned vertically has been sensitive to little variations in the strain levels [132]. These strain sensors are quite useful in MEMS devices. The response in the form of conductivity of the nanotube to pressure too has been made use of inflow sensing, but the challenge of factoring out the effects of only pressure is posed in comparison with the temperature, adsorption, etc., in influencing the nanotube conductance.

Electromechanical actuators have been suggested by using sheets of SWNTs and MWNTs [133, 134]. Nanotube sheets infiltrated by using polymer binders have been found to work as an excellent electrochemical actuator, resembling the actuator mechanism observed in natural muscles. The nanotube-based actuators are better placed over conducting polymer-based ones, as the former don't require an ion intercalation (that curtails actuator life). Several applications have been suggested ranging from nanotube-based micro-cantilevers to artificial muscles which are stable at elevated temperatures. The electromechanical and electrochemical applications of nanotubes are being researched and have stiff challenges because of intrinsic limitations as against novel ferroelectric and organic materials.

9.7 FIELD EMISSION AND LIGHTING APPLICATIONS

CNTs have been prescribed as best field emitters attributed to their lower threshold voltage, better emission stability and longer emitter lifetime [135–138]. Ma et al. contributed to the discussion of electron relaxation processes shown by CNTs. These properties make them advantageous in designing cathode-ray lighting systems and flat panel displays. The cathode ray tube (CRT) as a prototype lighting device is devised by using arc-based MWNTs as cold electron sources showed stable electron emission, sufficient luminance, and longer life of the emitters [139]. Samsung fabricated a 4.5-inch diode type flat panel display with the help of arc-derived SWNTs [140]. Although single-walled nanotubes were found to possess best emitting performance, they were vulnerable to degradation at elevated emission current [141]. Double-walled nanotubes and thin multiwalled nanotubes have been considered as the excellent field emitting materials as they possess a low threshold voltage in tune with SWNTs and a comparatively better structural stability than SWNTs, resembling with that of MWNTs [142–144]. Hiraoka et al. [145] successfully devised a DWNT forest with the help of Ni-based alloys using Fe or Cr as catalysts, followed by measurement of the field emission from a CNT/buckypaper cathode to an anode. They proposed that the uniform emission from the DWNT electrode is due to the better electrical contact of DWNTs with the grid substrate. Printable DWNT or thin

MWNT-based field emitters have a competitive edge for designing larger-area field emission displays (FEDs) at comparatively lower cost. Although some technical issues need to be sorted out, still nanotube-based larger-area color FEDs will be a reality soon in the market.

Pfeiffer et al. described the science underlying DWNTs. Doping of nanotubes would also be of great help in the field emission. Charlier et al. [146] have proved both theoretical and experimental terms that B-doped MWNTs show improved field emission (turn-on voltage of 1.4 V/μm) as compared to pure MWNTs (turn on voltage of 3 V/μm). It is attributed to B atoms at the tip of nanotubes, which leads to improved density of states nearing the Fermi level. Theoretical tight-binding and ab-initio-based computations show that the work function of B-doped SWNT is too low (1.7 eV) as compared to pure MWNTs. Moreover, it has been observed that N-doped MWNTs in bundles have the ability to generate electrons at comparatively lower turn-on voltage (2 V/μm) and higher current density (0.2–0.4 A/cm^2) [147]. Further, isolated N-doped MWNTs have also exhibited outstanding field emission property profile at 800°K; work functions of the order of 5 eV and emission currents of magnitude 100 nA experimentally were derived at ±10 V [148].

9.8 BIOLOGICAL APPLICATIONS

The optical properties of nanotubes make them useful in bio-imaging applications in live cells and tissues. The distinct photostability of SWNT PL provides for lengthening of excitation time at more laser fluence as against organic fluorophores and QDs. Opaque tissue shows largely attenuated absorption, auto-fluorescence, and scattering properties in the wavelength range of 700–1,400 nm. This range overlaps with the fluorescence profiles of a host of semiconducting NTs (nanotubes), providing for their detection in blood and dense tissue [149]. Cherukuri et al. [150] captured images of nanotubes on introduction by macrophages, witnessing a lack of photobleaching – a phenomenon that is detrimental to traditional bio-imaging. Nanotube fluorescence was utilized to capture images of SWNT in tissue sections and to determine their strength in blood [151]. The tremendous resonance-driven Raman scattering effect shown by nanotubes is applied to tissues as resonance boost takes place at near-IR wavelengths [152]. Saito et al. studied in detail about resonance Raman scattering science. Moreover, Heinz described Rayleigh scattering in CNTs, while Hartschuh explained supporting techniques, like near-field optics.

The effect of drug therapeutics is consistently on the rise via the fabrication of novel targeted drug delivery systems. Earlier, such systems covered viral vectors, liposomes, cationic lipids, polymers, and nanoparticles. Viral vectors consist of an inbuilt transfection capacity, however. There were safety issues regarding their applications, resulting in new avenues in the form of other vehicles [153]. Although nonviral carriers are versatile in terms of shape, size, and materials, a major concern is the weak penetrability of few therapeutics into cells. CNTs are easily internalized by cells; and on surface functionalization, they show lower cytotoxicity for a few days [154–161]. Moreover, they offer a more surface area to volume ratio, offering nanotubes the ability to conjugate with more functional groups than spheres and to have more loadings of therapeutics. Therefore, nanotubes have dragged a lot much of attention as promising candidates for carriers used in targeted drug delivery.

A right protocol for conjugating the drug with the nanotube is asked for to translate CNTs into viable delivery carriers. The nanotubes are characterized by hydrophobicity and therefore not soluble in biological media in liquid form, so numerous methods have been employed for practical application of the NTs (nanotubes), both by using covalent and noncovalent interactions to bring about their dissolution. Noncovalent methods, involve molecules possessing hydrophobic functional groups, covering RNA, DNA [162], Triton-X-100 [163], and 1-pyrenebutanoic acid [164], made to adsorb reversibly onto the nanotube surfaces, while the hydrophilic functional groups make the conjugates soluble. Covalent functionalization is a very common method with different underlying chemistries depending upon the generation of carboxyl [165, 166] and amino functional groups at the ends and walls of the nanotubes. The carboxyl sites, are bonded with amino [167–169] and thiol functional groups, to attach numerous biomolecules to the nanotubes. An important step for nanotube drug delivery is to ensure conjugation of the nanotubes such that the ligancy of the biomolecules is preserved. The covalently functionalized NTs (nanotubes) tend to work as sites of condensation for plasmid DNA that has been considered as a starting material for nanotube targeted gene-delivery carriers. Enzymes like Biliverdin/Xbeta reductase has also been vulnerable to conjugate with functionalized nanotubes. Salvador-Morales et al. [170] also presented that nanotubes can regenerate a response in the complementary portion of the human immune system, as shown from the adsorption of selective fibrinogen and apolipoproteins.

Researchers recently have demonstrated that the conjugates of nanotubes can penetrate into cell and release the therapeutics to ensure delivery of drugs successfully. Wu et al. presented the application wherein passive cellular uptake is used to deliver multiwalled carbon nanotubes in conjugation with amphotericin B (AmB), an antibiotic in connection with fungal infections effectively, however the same conjugate is toxic to mammalian cells when it is isolated in solution. The conjugate decreased the toxic potential at Jurkat cells, while retaining or increasing in some instances, and thereby inhibit the growth of *C. parapsilosis*, *C. albicans*, and *C. neoformans*.

The incorporation of foreign DNA into cells is one more key area for the delivery of therapeutic agents by using CNTs. Cai et al. applied nanotube spearing, using the magnetic characteristics of nickel-dispersed single-walled carbon nanotubes (SWNTs), for the delivery of DNA plasmid vectors having the sequence for EGFP (a fluorescent protein) to Bal17 and Mouse splenic B cells, which are not divisible cells, and consequently unable to transfect. They found expression of EGFP in the cells, while a vector with no EGFP didn't exhibit any measurable fluorescence. Moreover, the commercial product, Lipofectamine 2000, too didn't show any detectable fluorescence from EGFP, demonstrating the tremendous efficiency of the nanotube-spearing technique. Cai et al. too showed that the nanotubes tiny in size provide for easy penetration with very little hindrances to the cell, and the magnetic spearing may be better placed over the ballistic method.

Single-walled nanotubes (SWNTs) show strong absorption at Near-IR wavelengths and exhibit localized heating. This study demonstrated that NIR radiation is responsible for the nanotubes to release the DNA, providing for its internalization into the nuclei of the cell. This technique was also employed to bring about cell death. Nanotubes on functionalization with a folate moiety made them to be internalized specifically by cells having folate receptor tumor markers. NIR irradiation caused the death of cells because of localized heating due to the nanotubes. Kam et al. applied a technique that made use of a disulfide linker for the conjugation of oligonucleotide-SWNT and siRNA-SWNT. These conjugated nanotubes were passively endocytosed inside the cells, followed by the cleavage of the disulfide bond due to the presence of thiol-reducing enzymes and the acidic medium present in the liposome. NTs (nanotubes) have also been employed for the delivery of proteins, namely, cyt-c, to cells to cause apoptosis, or cell death that is programmed. In this study, cyt-c-SWNT were endocytosed into NIH-3T3 cells. The cyt-c-SWNTs were then released on addition of chloroquine to the cell environment, leading to an increase in the pace of apoptosis [171].

Substantial progress has been made on the front delivery of therapeutics via carbon nanotubes. However, there are many problems areas too that are yet to be sorted out before the potential of such materials is fully tapped in clinical applications. The field would see furtherance in the growth with the advancements of a clear ways for the release of nanotubes from endosomes into cells. Passive methods provide for constant up taking of nanotubes by cells, a regulated, but nondestructive, technique of nanotube mobility with respect to selecting the target area would enhance the therapeutic potential of such materials. Better efficient methods of release would enhance the efficacy too. More research is required to estimate the effects in the long-term on conjugation of nanotubes *in-vivo*, given problems namely, toxicity, and interaction with biological entities such as blood and tissue. The metabolism of such materials, and the probability of *in-vivo* NTs (nanotube) release on delivering the therapeutics are of importance. While such problem areas may require years together to address, still the field is quite promising, as NTs (nanotubes) have already been showing large and versatile loading power for therapeutics, specific unloading, and more cell penetration as compared to a host of other materials used in the targeted delivery vehicles.

Apart from biomedical applications, the diffusion of nanotubes into cells is of pivotal importance. The mainstay mechanism of the cellular mobility of CNTs is the transportation. Comprehending this mechanism will help give an impetus to drug fabrication and gene delivery, other than to enable specific cellular targeting in bio-imaging applications. There is a huge controversy and confusion about the underlying mechanism by which nanotubes transport into cells. Dai et al. have presented that SWNTs that are functionalized with acidic groups can transport into the cells by following the endocytosis route; however, Bianco et al. prescribes a passive, endocytosis-free mechanism for SWNTs and multiwalled nanotubes (MWNTs) functionalized with ammonium group. Multiwalled nanotubes (MWNTs), have a size limit beyond the endocytosis line, therefore the suggested mechanism is one wherein the flipping of lipid molecules of the membrane takes place, providing for an access to the cells [172].

The coating on the nanotube surface is a parameter that is instrumental in the NT (nano-tube) and cell interaction. Serum in cell environment has numerous proteins, like albumin, fibronectin, and transferrin, which will be adsorbed on the surface of nanotubes. In the systems with no serum, the cell generates adequate protein fast to coat the nanotube. Such proteins are available *in vivo* too and rapidly coat any foreign material incorporated into the body. In both of these cases, the actual surface made available to the cell is the protein-altered surface [173], and it has been confirmed in the gold nanoparticles [174]. The morphology of a particle too influences the way particles interact with the cells. Phage cells present in the immune system consume foreign particles (phagocytosis) with size larger than many hundreds of nm, and is suggested one of the pathways for cellular uptake by CNTs. Next potential mechanism is receptor-driven endocytosis. In case of nanotube cellular mobility, clathrin-driven endocytosis is also another largely exploited pathway mediated by receptors [175]. The clathrin naturally takes a molecular size limiting to about 150 nm for the cages developed during the course of endocytosis on the membrane, which prescribes a cap on the size of particles. Moreover, physical forces, covering van der Waals, electrostatic, and hydrophobic interactions are responsible for the attraction of towards cells and induce the uptake efficiency, e.g., the cations on a nanoparticle boosts the optimal wrapping of the clathrin to nanoparticles with the size greater than 25 nm [176]. If the free energy of adsorption between the nanoparticles and the cell membrane is greater than a given receptor-ligand linkage, the optimal wrapping size may reduce too [177]. However, MWNTs, having a 10 nm radius and 200 nm length, are not amenable to follow an endocytotic route, but possibly instant switching of lipid molecules in the membrane. One more unaddressed issue is the fate of nanotubes on getting into the cell. The majority of biological applications demand the subsequent expulsion of nanotubes from the cell; although there is no clarity in the literature about it. Extensive research work needed on nanotube cellular diffusion before such problems are addressed, allowing exact knowledge and control of nanotube transport into the cells. As such, the use of nanotubes for the biological applications has a lot much of promise and potential in the long run, although not in the foreseeable future.

9.9 MISCELLANEOUS APPLICATIONS

One of the key applications of nanotubes is nanotube-based electrode tips. Given very small sizes, better conductivity, mechanical strength and elasticity, nanotubes have demonstrated huge potential to be used as nano-electrodes [178–180]. The nanotube-based tips allow for higher resolution for imaging and longevity due to their elasticity, as against traditional silicon tips that crash frequently. Currently, MWNT-based AFM probe tips are purchased from manufacturers of scanning-probe microscopes. These are however are very costly, e.g., many 100 dollars/tip and can be looked upon as a specialty application with low demand. There have been many offers for the use of nanotube tips in futuristic technologies, say in imaging, nanolithography [181] and nanoelectrodes. The nanotube tips have been employed for the cause of imaging of surfaces in addition to biological sample matrices, namely, DNA at higher resolution. The nanotube tips are conducting and can be modified by chemical processing to use it in chemical imaging and in maneuvering the molecular moieties on surfaces. It is expected that the future of nanotube probes will rely upon how do we fabricate these tips (either decorated directly or transferred onto traditional tips) on a mass scale.

One more exciting application that has been suggested for nanotubes is their application as membrane filters. Scientists suggested that membranes prepared from oriented arrays of nanotube (both SWNTs and MWNTs) could be made use of for selective mobility of species [182–184]. Activated and porous carbons have been conventionally employed as filters (e.g., water treatment) but the nanotubes are advantageous because of their uniform and tiny pore sizes, which facilitate the separation of quite small molecules becomes. Although the pore size is small, the flow rate is not decreased because of the distinct nature of the solvents flow via the cavities of the nanotubes. One more advantage is that the nanotube surfaces can be chemically functionalized so as to ensure the selective separation. Although research is underway to design nanotube-derived membranes for filtration, molecular, and gas separation, these are still the areas yet to be explored more and the products commercialized.

Research work has been undertaken in obtaining stable dispersions of nanotubes in different solvents (e.g., nanotube inks) [185], that can either be spun or printed on numerous substrates. Readily printable (e.g., inkjet printing) circuits and films are quite useful for a host of applications varying from flexible electronics to antennas. Nanotubes can be spun into fibers too by employing different techniques by extradition of slurries of nanotubes or by drawing from oriented arrays. Dispersions of SWNTs with more weight percentage have either been spun or polymer binders for drawing fibers of macro-scale lengths [186, 187]. The fibers can be woven into sheets, membranes, etc. Many applications have been found for these nanotube-based fibers and other derived products, namely, incandescent bulbs with high brightness [188], fillers in polymers, shielding elements as well as ballistic protection. Aligned nanotubes can be decorated directly and employed to prepare unique products, like brushes [189]. The small brushes with bristles made out of nanotubes have been suggested for use in a wide range of application areas covering cleaning micro-spaces, conducting contact brushes, and probe arrays. The next application of oriented nanotube arrays is dry adhesive tapes if nanotube arrays are engineered well then, they have excellent adhesive strengths superseding gecko feet even [190]. The range of assembled and dispersed products derived from nanotubes would meet the needs of many industrial applications in the time to come and would provide for consistent and roll-to-roll manufacturing introducing nanotubes. A host of such products (preferentially the spun fibers) could reach the marketplace within the coming decade [191].

Research in CNTs has offered opportunities in designing novel carbon nanostructures, which would have specialty applications. Nanotubes reinforced with different materials have been prepared by adopting strategies like capillarity infiltration [192, 193]. Hybrid structures consisting of nanowires (NWs) and nanotubes have been developed and even branched nanostructures have been prepared [194, 195]. All such structures allow for the variety of building blocks that could be applied in nanotechnology-based products in the coming decade or so. However, few structures, like carbon nanohorns (fullerene-alike structures with lower density and higher surface area) as explained in detail by Yudasaka et al. [196] which can find use in applications such as electrodes and SCs and a range of other applications. These materials can be easily synthesized on a mass production using the pyrolysis of carbon precursors and further commercialized within the 10-year time span.

9.10 ENVIRONMENTAL AND HEALTH APPLICATIONS

CNTs hold a lot of promise in the fields such as electronics, environment, energy, and biomedicine. One of the problems yet to be solved is the safety and biocompatibility of the NT-based materials [197–205]. It has been proposed recently that a simple and effective way to assess the toxicity and biocompatibility of CNTs is by determination of CD^{4+} and CD^{8+} T-cells in peripheral blood and their histopathological examination on carbon-nanotube implanted tissues. Most importantly, in the procedure proposed, the mortality of animals was quite low, and no chronological variations of body weight were noticed amongst nanotube-administered rats against control over a period three months after implantation. Time-based difference in peripheral T-cells, which consisted of CNTs *in situ*, was related to a phase of granuloma development. Although the potential toxicity was examined and found quite low for CVD-decorated MWNTs, further thorough and systematic (long-term) research pertaining to the toxic nature of numerous kinds of CNTs (covering direct inhalation of the nanotubes in lungs) have to be undertaken to solve the safety issues of CNTs.

Nanotube cytotoxicity appears as situational, based largely on the extent and kind of functionalization, state of aggregation, and the availability of metal catalyst particles left over after synthesis. Pristine nanotubes are presented to affect oxidative stress and reduce cell viability [206, 207], although there is an indication that remaining catalyst particles are responsible for this effect [208]. The cytotoxicity can be largely reduced by functionalization. Sayes et al. [209] showed that the more cytotoxic potential of pristine SWNTs can be decreased to zero by enhancing coverage with a covalently bonded polar functional group. The selection of functional groups can modify a spectrum of cell responses covering the activation of primary immune cells by PEG-coated SWNTs [210]. The toxic potential of modified CNTs using noncovalent interactions depends on the kind of the adsorbate. The DNA-coated SWNTs and encapsulated leave cells free on penetration of the complex [211]. Toxicity studies are also carried out by Terrones et al. on N-doped nanotubes.

9.11 CONCLUSION

The applications of nanotubes are summarized based on the timeline that depend on mass-scale manufacturing from those that are dependent upon regulated fabrication. There are challenges, too, in the successful transfer of understanding of commercialization of applications. The promise with which nanotubes has enriched the research community should pay off in the future. This demands the tapping of the full potential of nanotubes as a commercial entity. The applications of nanotubes, can be classified in two categories; one, based on the commercialization timeline and second, based on the scale of manufacturing (i.e., bulk vs. limited volume).

There are a plethora of the issues and challenges in exploring the carbon-nanotube applications since its inception. These challenges have reduced the rate at which nanotube commercialization takes place. The topmost issue is in regard to the manufacturing. The large-scale applications of nanotubes essentially will depend upon the capacity of manufacturing nanotubes in large volume, at the industrial scale. In such a case CVD-decorated MWNTs (Endo fibers) mass-scale manufacturing has been fairly successful. Company giants like Hyperion in the USA and Mitsui Corporation and Showa-Denko in Japan have erected manufacturing concerns that can manufacture many tons of product/year so as to meet the demand for nanotubes in particular from the ESD and battery-based applications.

The volume of CNTs is small as compared to carbon fibers and other carbons that are commercialized. As far as SWNTs are concerned the situation is different. After pursuing the research for more than a decade on this front, still no method is able to manufacture this material in a given volume.

The quantities are yet confined to some grams, restricting any really an application in the bulk for this material. There have been exciting initiatives by many manufacturing concerns in the USA, Canada, and China promising the procurement of SWNT in required volumes to the availability of MWNT in the coming decade or so, but such claims need to be substantiated. Apart from the availability there are there are other bottlenecks like cost (SWNTs still cost around 100 USD/gram), in comparison with the few dollars/kg for the CVD-decorated MWNTs) in addition to basic materials science related issues and compatibility, e.g., the aspect of interfaces pertaining to polymer composites and furtherance of research in engineering to address all concerned issues.

The applications which will hold lot much of promise in the near future will surface from value added applications, like the use of nanotubes as additives in battery electrodes to impart mechanical stability. The applications that demand organized nanotube structures should emphasize upon integration pathways. The revolutionary applications of nanotubes, involve nanotube electronics, which lay more focus on regulated assembly and integration. There are bottlenecks in this route, e.g., the tailored growth of nanotubes with selective chirality or diameter. There are still challenges in synthesizing nanotubes with regulated length, oriented alignment, location control, etc. Processing too is a problem, e.g., the temperatures that are maintained to process nanotubes currently are too high than for silicon semiconductor processing and therefore integration of nanotubes in the present processes in the industry poses challenges. These high value-added applications using nanotubes will take its own time to realize. In the near future, there would be nanotube-based devices in frontier applications, namely, field emission displays, sensors, membranes, coatings, etc., for example, electromechanical memories and transparent conductive films which are dependent upon random organization of nanotubes. To make them commercially viable, consistency in the manufacturing cost-effectiveness is required, such that they compete with other material options present in the market today. Scientists carried out a survey on the synthesis and processing means of CNTs.

There are a host of other issues which need to be looked into seriously as we carry on with applications of nanotubes. The issues of pivotal importance are the environmental and health-safety ones regarding nanomaterials. Experiments in isolation have demonstrated that the nanotubes are non-toxic. However, we need to have standard protocols as regards how to handle such materials, both on the laboratory and an industrial scale. Well-defined regulations have to be in vogue when huge amounts of the material are required to be dealt with routinely. The impact of these on health in the long run need to be examined. Standard procedures have to be laid down on the way nanotube-based materials need to be packed and transported as well as the ways to handle them.

The ensuing decade will be the nanotechnology decade which ask for further research in nanometrology as suggested by many scientists. CNTs and CNTs-derived products will have a deciding role to play. Huge amounts of resources are invested and will be invested in the time to come for understanding and production of the CNT-based materials. Several years till now have been quite exciting. The new scientific discoveries and understanding developed with nanotubes have been significant and fulfilling. We are at the crossroads now where the laboratory research findings need to be translated into products of societal importance. Some of such initiatives will emerge as successful and some may fail miserably. The crucial test for the nanotube materials will be whether it will live up to the expectations and promises that have been raised depending upon its interesting and exciting properties.

KEYWORDS

- **carbon nanotubes**
- **chemical vapor deposition**
- **double-walled nanotubes**
- **field effect transistors**
- **multiwalled carbon nanotubes**
- **thin-film transistors**

REFERENCES

1. Baughman, R. H., Zakhidov, A. A., & Zakhidov, W. A., (2002). Carbon nanotubes–the route toward applications. *Science, 297*, 787–792.
2. Yakobson, B. I., & Smalley, R. E., (1997). Fullerene nanotubes: C1,000,000 and beyond. *Am. Sci., 85*, 324.
3. Ajayan, P. M., (1999). Nanotubes from carbon. *Chem. Rev., 99*, 1787–1800.
4. Kroto, H. W., Heath, J. R., O'Brien, S. C., Curl, S. C., & Smalley, R. E., (1985). C60: Buckminsterfullerene. *Nature, 318*, 162, 163.
5. Dresselhaus, M. S., Dresselhaus, G., Sugihara, K., Spain, I. L., & Goldberg, H. A., (1988). *Graphite Fibers and Filaments.* Springer, Berlin, Heidelberg.

6. Oberlin, A., Endo, M., & Koyama, T., (1976). Filamentous growth of carbon through benzene decomposition. *J. Cryst. Growth, 32*, 335–349.
7. Saito, R., Dresselhaus, G., & Dresselhaus, M. S., (1998). *Physical Properties of Carbon Nanotubes*. Imperial College Press, London.
8. Iijima, S., (1991). Helical microtubules of graphitic carbon. *Nature, 354*, 56–58.
9. Ebbesen, T. W., & Ajayan, P. M., (1992). Large scale synthesis of carbon nanotubes. *Nature, 358*, 220.
10. Iijima, S., & Ichihashi, T., (1993). Single shell carbon nanotubes of 1-nm diameter. *Nature, 363*, 603–605.
11. Bethune, D. S., Klang, C. H., deVries, M. S., Gorman, G., Savoy, R., Vazquez, J., & Beyers, R., (1993). Cobalt catalysed growth of carbon nanotubes with single atomic layer wells. *Nature, 363*, 605–607.
12. Dekker, C., (1999). Carbon nanotubes as molecular quantum wires. *Phys. Today, 52*, 22–28.
13. Javey, A., (2003). Ballistic carbon nanotube field-effect transistors. *Nature, 424*, 654–657.
14. McEuen, P. L., Fuhrer, M. S., & Park, H. K., (2002). Single-walled carbon nanotube electronics. *IEEE Transact. Nanotechnol., 1*, 78–85.
15. Kreup, F., Graham, A. P., Liebau, M., Duesberg, G. S., Seidel, R., & Unger, E., (2004). *Carbon Nanotubes for Interconnect Applications* (pp. 683–686). IEDM.
16. Arnold, M. S., Green, A. A., Hulvat, J. F., Stupp, S. I., & Hersam, M. C., (2006). Sorting carbon nanotubes by electronic structure using density differentiation. *Nature Nanotechnol., 1*, 60–65.
17. Zhang, G. Y., (2006). Selective etching of metallic carbon nanotubes by gas-phase reaction. *Science, 314*, 974–977.
18. Wei, B. Q., Vajtai, R., & Ajayan, P. M., (2001). Reliability and current carrying capacity of carbon nanotubes. *Appl. Phys. Lett., 79*, 1172.
19. URL http://en.wikipedia.org/wiki/45nm URL: http://en.wikipedia.org/wiki/65nm17 (accessed on 19 September 2022).
20. Tans, S. J., Verschueren, A. R. M., & Dekker, C., (1998). Room-temperature transistor based on a single carbon nanotube. *Nature, 393*, 49–52.
21. Martel, R., Schmidt, T., Shea, H. R., Hertel, T., & Avouris, P., (1998). Single- and multiwall carbon nanotube field-effect transistors. *Appl. Phys. Lett., 73*, 2447.
22. Bachtold, A., Hadley, P., Nakanishi, T., & Dekker, C., (2001). Reports, logic circuits with carbon nanotube transistors. *Science, 294*, 1317–1320.
23. Collins, P. G., Arnold, M. S., & Avouris, P., (2001). Engineering carbon nanotubes and nanotube circuits using electrical breakdown. *Science, 292*, 706–709.
24. Klinke, C., Hannon, J. B., Afzali, A., & Avouris, P., (2006). Field effect transistors assembled from functionalized nanotubes. *Nano Lett., 6*, 906–910.
25. Rossnagel, S. M., & Kaun, T. S., (2004). Alteration of Cu conductivity in the size effect regime. *J. Vac. Sci. Technol. B, 22*, 240–247.
26. White, C. T., & Todorov, T. N., (2001). Quantum electronics – nanotubes go ballistic. *Nature, 411*, 649–651.
27. Naeemi, A. R., Sarvari, R., & Meindl, J. D., (2005). Performance comparison between carbon nanotube and copper interconnects for gigascale integration (GSI). *IEEE Electron. Device Lett., 26*, 84–86.
28. Lin, Y. M., Appenzeller, J., Chen, Z., Chen, Z. G., Cheng, H. M., & Avouris, P., (2005). High-performance dual-gate carbon nanotube FETs with 40-nm gate length. *IEEE Electron. Device L., 26*, 823–825.

29. Javey, A., Guo, J., Farmer, D. B., Wang, Q., Yenilmez, E., Gordon, R. G., Lundstrom, M., & Dai, H., (2004). Self-aligned ballistic molecular transistors and electrically parallel nanotube arrays. *Nano Lett. 4,* 1319–1322.
30. Javey, A., Kim, H. S., Brink, M., Wang, Q., Ural, A., Guo, J., McIntyre, P., et al., (2002). High-kappa dielectrics for advanced carbon-nanotube transistors and logic gates. *Nature Mater., 1,* 241–246.
31. Nantero. Website http://www.nantero.com/ (accessed on 19 September 2022).
32. Cao, Q., Hur, S. H., Zhu, Z. T., Sun, Y. G., Wang, C. J., Meitl, M. A., Shim, M., & Rogers, J. A., (2006). Highly bendable, transparent thin-film transistors that use carbon-nanotube-based conductors and semiconductors with elastomeric dielectrics. *Adv. Mater., 18,* 304–309.
33. Cao, Q., Zhu, Z. T., Lemaitre, M. G., Xia, M. G., & Rogers, M. S. J. A., (2006). Transparent flexible organic thin-film transistors that use printed single-walled carbon nanotube electrodes. *Appl. Phys. Lett., 88,* 113511.
34. Wu, Z., Chen, Z., Du, X., Logan, J. M., Sippel, J., Nikolou, M., Kamaras, K., et al., (2004). Transparent, conductive carbon nanotube films. *Science, 305,* 1273–1276.
35. URL: http://www.fibrils.com/ (accessed on 19 September 2022).
36. Kord´as, K., Tth, G., Moilanen, P., Kumpumki, M., Vhkangas, J., Uusimki, A., Vajtai, R., & Ajayan, P. M., (2007). Chip cooling with integrated carbon nanotube microfin architectures. *Appl. Phys. Lett., 90,* 123105.
37. Leroux, F., Metenier, K., Gautier, S., Frackowiak, E., Bonnamy, S., & Beguin, F., (1999). Electrochemical insertion of lithium in catalytic multi-walled carbon nanotubes. *J. Power Sources, 81,* 317–322.
38. Claye, A. S., Fischer, J. E., Huffman, C. B., Rinzler, A. G., & Smalley, R. E., (2000). Solid-state electrochemistry of the Li single wall carbon nanotube system. *J. Electrochem. Soc., 147,* 2845–2852.
39. Shimoda, H., et al., (2002). Lithium intercalation into opened single-wall carbon nanotubes: Storage capacity and electronic properties. *Phys. Rev. Lett., 88,* 015502.
40. Lu, W., & Chung, D. D. L., (2001). Anodic performance of vapor-derived carbon filaments in lithium-ion secondary battery. *Carbon, 39,* 493–496.
41. Gao, B., Kleinhammes, A., Tang, X. P., Bower, C., Fleming, L., Wu, Y., & Zhou, O., (1999). Electrochemical intercalation of single-walled carbon nanotubes with lithium. *Chem. Phys. Lett., 307,* 153–157.
42. Sato, M., Noguchi, A., Demachi, N., Oki, N., & Endo, M., (1994). A mechanism of lithium storage in disordered carbons. *Science, 264,* 556–558.
43. Endo, M., Kim, C., Nishimura, K., Fujino, T., & Miyashita, K., (2000). Recent development of carbon materials for Li ion batteries. *Carbon, 38,* 183–197.
44. Endo, M., Kim, Y. A., Hayashi, T., Nishimura, K., Matsushita, T., Miyashita, K., & Dresselhaus, M. S., (2001). Vapor-grown carbon fibers (VGCFs) basic properties and battery application. *Carbon, 39,* 1287–1297.
45. Conway, B. E., (1999). *Electrochemical Supercapacitors-Scientific Fundamentals and Technological Applications.* Kluwer, New York.
46. Niu, C., Sickel, E. K., Hoch, R., Moy, D., & Tennent, H., (1997). High power electrochemical capacitors based on carbon nanotube electrodes. *Appl. Phys. Lett., 70,* 1480–1482.
47. An, K. H., Kim, W. S., Park, Y. S., Moon, J. M., Bae, D. J., Lim, S. C., Lee, Y. S., & Lee, Y. H., (2001). Electrochemical properties of high-power supercapacitors using single-walled carbon nanotube electrodes. *Adv. Funct. Mater., 11,* 387–392.

48. Kim, Y. J., Kim, Y. A., Chino, T., Suezaki, H., Endo, M., & Dresselhaus, M. S., (2006). Chemically modified multi-walled carbon nanotubes as an additive for supercapacitors. *Small, 2*, 339–345.
49. Landi, B. J., Castro, S. L., Ruf, H. J., Evans, C. M., Bailey, S. G., & Raffaelle, R. P., (2005). CdSe quantum dot-single wall carbon nanotube complexes for polymeric solar cells. *Sol. Energ. Mater. Sol. C., 87*, 733–746.
50. Kymakis, E., & Amaratunga, G. A. J., (2006). Electrical properties of single-wall carbon nanotube-polymer composite films. *J. Appl. Phys., 99*, 084302.
51. Xu, Z. H., Wu, Y., Hu, B., Ivanov, I. N., & Geohegan, D. B., (2005). Carbon nanotube effects on electroluminescence and photovoltaic response in conjugated polymers. *Appl. Phys. Lett., 87*, 263118.
52. Bachtold, A., Fuhrer, M. S., Plyasunov, S., Forero, M., Anderson, E. H., Zettl, A., & McEuen, P. L., (2000). Scanned probe microscopy of electronic transport in carbon nanotubes. *Phys. Rev. Lett., 84*, 6082–6085.
53. Kymakis, E., Alexandrou, I., & Amaratunga, G. A. J., (2003). High open-circuit voltage photovoltaic devices from carbon-nanotube-polymer composites. *J. Appl. Phys., 93*, 1764–1768.
54. Kymakis, E., & Amaratunga, G. A. J., (2002). Single-wall carbon nanotube/conjugated polymer photovoltaic devices. *Appl. Phys. Lett., 80*, 112–114.
55. Kymakis, E., & Amaratunga, G. A. J., (2005). Carbon nanotubes as electron acceptors in polymeric photovoltaics. *Rev. Adv. Mater. Sci., 10*, 300–305.
56. Bhattacharyya, S., Kymakis, E., & Amaratunga, G. A. J., (2004). Photovoltaic properties of dye functionalized single-wall carbon nanotube/conjugated polymer devices. *Chem. Mater., 16*, 4819–4823.
57. Kymakis, E., & Amaratunga, G. A. J., (2003). Photovoltaic cells based on dye sensitization of single-wall carbon nanotubes in a polymer matrix. *Sol. Energ. Mater. Sol. C., 80*, 465–472.
58. Rahman, G. M. A., Guldi, D. M., Cagnoli, R., Mucci, A., Schenetti, L., Vaccari, L., & Prato, M., (2005). Combining single wall carbon nanotubes and photoactive polymers for photoconversion. *J. Am. Chem. Soc., 127*, 10051–10057.
59. Raffaelle, R. P., Landi, B. J., Harris, J. D., Bailey, S. G., & Hepp, A., (2005). Carbon nanotubes for power applications. *Mater. Sci. Eng. B-Solid, 116*, 233–243.
60. Miller, A. J., Hatton, R. A., & Silva, S. R. P., (2006). Water-soluble multiwall-carbonnanotube-polythiophene composite for bilayer photovoltaics. *Appl. Phys. Lett., 89*, 123115.
61. Miller, A. J., Hatton, R. A., & Silva, S. R. P., (2006). Interpenetrating multiwall carbon nanotube electrodes for organic solar cells. *Appl. Phys. Lett., 89*, 133117.
62. Lee, S. B., Katayama, T., Kajii, H., Araki, H., & Yoshino, K., (2001). Electrical and optical properties of conducting polymer-C-60-carbon nanotube system. *Synth. Met., 121*, 1591, 1592.
63. Landi, B. J., Raffaelle, R. P., Castro, S. L., & Bailey, S. G., (2005). Single-wall carbon nanotube-polymer solar cells. *Prog. Photovoltaics, 13*, 165–172.
64. Rud, J. A., Lovell, L. S., Senn, J. W., Qiao, Q., & Mcleskey, J. T., (2005). Water soluble polymer/carbon nanotube bulk heterojunction solar cells. *J. Mater. Sci., 40*, 1455–1458.
65. Valentini, L., & Kenny, J. M., (2005). Novel approaches to developing carbon nanotube based polymer composites: Fundamental studies and nanotech applications. *Polymer, 46*, 6715–6718.

66. Spiekermann, S., Smestad, G., Kowalik, J., Tolbert, L. M., & Grtzel, M., (2001). Poly(4-undecyl-2,2'-bithiophene) as a hole conductor in solid state dye sensitized titanium dioxide solar cells. *Synth. Met., 121*, 1603, 1604.
67. Gratzel, M., (2001). Photoelectrochemical cells. *Nature, 414*, 338–344.
68. Feng, W., Feng, Y., Wu, Z., Fujii, A., Ozaki, M., & Yoshino, K., (2005). Optical and electrical characterizations of nanocomposite film of titania adsorbed onto oxidized multiwalled carbon nanotubes. *J. Phys.-Condens. Matter., 17*, 4361–4368.
69. Spataru, C. D., Ismail-Beigi, S., Benedict, L. X., & Louie, S. G., (2004). Quasiparticle energies, excitonic effects and optical absorption spectra of small-diameter single-walled carbon nanotubes. *Appl. Phys. A-Mater., 78*, 1129–1136.
70. Ichida, M., Mizuno, S., Tani, Y., Saito, Y., & Nakamura, A., (1999). Exciton effects of optical transitions in single-wall carbon nanotubes. *J. Phys. Soc. Jpn., 68*, 3131–3133.
71. Kazaoui, S., Minami, N., Nalini, B., Kim, Y., & Hara, K., (2005). Near-infrared photo-conductive and photovoltaic devices using single-wall carbon nanotubes in conductive polymer films. *J. Appl. Phys., 98*, 084314.
72. Britto, P. J., Santhanam, K. S. V., Rubio, A., Alonso, A., & Ajayan, P. M., (1999). Improved charge transfer at carbon nanotube electrodes. *Adv. Mater., 11*, 154–157.
73. Che, G., Lakshmi, B. B., Fisher, E. R., & Martin, C. R., (1999). Carbon nanotube membranes for electrochemical energy storage and production. *Nature, 393*, 346–349.
74. Endo, M., Kim, Y. A., Ezaka, M., Osada, K., Yanagisawa, T., Hayashi, T., Terrones, M., & Dresselhaus, M. S., (2003). Selective and efficient impregnation of metal nanoparticles on cup-stacked-type nanofibers. *Nano Lett., 3*, 723–726.
75. Yoshitake, T., Shimakawa, Y., Kuroshima, S., Kimura, H., Ichihashi, T., Kubo, Y., Kasuya, D., et al., (2002). Preparation of fine platinum catalyst supported on single-wall carbon nanohorns for fuel cell application. *Physica B, 323*, 124–126.
76. Kim, C., Kim, Y. J., Kim, Y. A., Yanagisawa, T., Endo, M., & Dresselhaus, M. S., (2004). Pt/Ru supported on cup-stack typed carbon nanotubes as an electrode for fuel cell applications. *J. Appl. Phys. 96*, 5903–5905.
77. Wang, C., Waje, M., Wang, X., Tang, J. M., Haddon, R. C., & Yan, Y. S., (2004). Proton exchange membrane fuel cells with carbon nanotube based electrodes. *Nano Lett., 4*, 345–348.
78. Treacy, M. M. J., Ebbesen, T. W., & Gibson, J. M., (1996). Exceptionally high young's modulus observed for individual carbon nanotubes. *Nature, 381*, 678–681.
79. Yu, M., Lourie, O., Dyer, M. J., Moloni, K., Kelly, T. F., & Ruoff, R. S., (2000). Strength and breaking mechanism of multiwalled carbon nanotubes under tensile load. *Science, 287*, 637–640.
80. Wagner, H. D., Lourie, O., Feldman, Y., & Tenne, R., (1998). Stress–induced fragmentation of multiwall carbon nanoutbes in a polymer matrix. *Appl. Phys. Lett., 72*, 188–190.
81. Yakobson, B. I., (1998). Mechanical relaxation and intramolecular plasticity in carbon nanotubes. *Appl. Phys. Lett., 72*, 918.
82. Cao, A., Dickrell, D. L., Sawyer, W. G., Ghasemi-Nejhad, M. N., & Ajayan, P. M., (2005). Super-compressible foamlike carbon nanotube films. *Science, 310*, 1307–1310.
83. Ajayan, P. M., Schadler, L. S., Giannaris, C., & Rubio, A., (2000). Single-walled carbon nanotube-polymer composites: Strength and weakness. *Adv. Mater., 12*, 750.
84. Veedu, V. P., Cao, A., Li, X., Ma, K., Soldano, C., Kar, S., Ajayan, P. M., & Ghasemi-Nejhad, M. N., (2006). Multifunctional composites using reinforced laminae with carbon-nanotube forests. *Nature Mater., 5*, 457–462.

85. Suhr, J., Zhang, W., Ajayan, P. M., & Koratkar, N. A., (2006). Temperature-activated interfacial friction damping in carbon nanotube polymer composites. *Nano Lett., 6*, 219–223.
86. Schadler, L. S., Giannaris, S. C., & Ajayan, P. M., (1998). Load transfer in carbon nanotube epoxy composites. *Appl. Phys. Lett., 73*, 3842.
87. Eitan, A., Jian, K. Y., Dukes, D., Andrews, R., & Schadler, L. S., (2003). Surface modification of multiwalled carbon nanotubes: Toward the tailoring of the interface in polymer composites. *Chem. Mater., 15*, 3198–3201.
88. Suhr, J., Koratkar, N., Keblinski, P., & Ajayan, P. M., (2005). Viscoelasticity in carbon nanotube composites. *Nature Mater., 4*, 134.
89. Gibson, R. F., Ayorinde, E. O., & Wen, Y. F., (2007). Vibrations of carbon nanotubes and their composites: A review. *Compos. Sci. Technol., 67*, 1–28.
90. Fragneaud, B., Masenelli-Varlot, K., González-Montiel, A., Terrones, M., & Cavaillé, J. Y., (2005). Efficient coating of N-doped carbon nanotubes with polystyrene using atomic transfer radical polymerization. *Chem. Phys. Lett., 419*, 567.
91. Dehonor, M., Varlot, K. M., Montiel, A. G., Gauthier, C., Cavaillé, J. Y., Terrones, H., & Terrones, M., (2005). Nanotube brushes: Polystyrene grafted covalently on CNx nanotubes by nitroxide-mediated radical polymerization. *Chem. Commun., 42*, 5349–5351.
92. Endo, M., Koyama, S., Matsuda, Y., Hayashi, T., & Kim, Y. A., (2005). Thrombogenicity and blood coagulation of a micro-catheter prepared from carbon nanotubenylon based composite. *Nano Lett., 5*, 101–106.
93. Koyama, S., Haniu, H., Osaka, K., Koyama, H., Kuroiwa, N., Endo, M., Kim, Y. A., & Hayashi, T., (2006). Medical application of carbon nanotube-filled nanocomposites: Microcatheter. *Small, 2*, 1406–1411.
94. Qi, P., Vermesh, O., Grecu, M., Javey, A., Wang, Q., Dai, H., Peng, S., & Cho, K. J., (2003). Toward large arrays of multiplex functionalized carbon nanotube sensors for highly sensitive and selective molecular detection. *Nano Lett., 3*, 347–351.
95. Kong, J., Franklin, N. R., Zhou, C., Chapline, M. G., Peng, S., Cho, K., & Dai, H., (2000). Nanotube molecular wires as chemical sensors. *Science, 287*, 622–625.
96. Barone, P. W., Baik, S., Heller, D. A., & Strano, M. S., (2005). Near-infrared optical sensors based on single-walled carbon nanotubes. *Nature Mater., 4*, 86–92.
97. Jeng, E. S., Moll, A. E., Roy, A. C., Gastala, J. B., & Strano, M. S., (2006). Detection of DNA hybridization using the near-infrared band-gap fluorescence of single walled carbon nanotubes. *Nano Lett., 6*, 371–375.
98. Heller, D. A., Jeng, E. S., Yeung, T. K., Martinez, B. M., Moll, A. E., Gastala, J. B., & Strano, M. S., (2006). Optical detection of DNA conformational polymorphism on single-walled carbon nanotubes. *Science, 311*, 508–511.
99. Peng, S., & Cho, K. J., (2000). Chemical control of nanotube electronics. *Nanotechnol., 11*, 57–60.
100. Lee, C. Y., Baik, S., Zhang, J., Masel, R. I., & Strano, M. S., (2006). Charge transfer from metallic single-walled carbon nanotube sensor arrays. *J. Phys. Chem. B, 110*, 11055–11061.
101. Heinze, S., Tersoff, J., Martel, R., Derycke, V., Appenzeller, J., & Avouris, P., (2002). Carbon nanotubes as Schottky barrier transistors. *Phys. Rev. Lett., 89*, 106801.
102. Bradley, K., Gabriel, J. C. P., Star, A., & Grüner, G., (2003). Short-channel effects in contact-passivated nanotube chemical sensors. *Appl. Phys. Lett., 83*, 3821–3823.

103. Zhang, J., Boyd, A., Tselev, A., Paranjape, M., & Barbara, P., (2006). Mechanism of NO_2 detection in carbon nanotube field effect transistor chemical sensors. *Appl. Phys. Lett., 88*, 123112.
104. Lu, Y., Li, J., Han, J., Ng, H. T., Binder, C., Partridge, C., & Meyyappan, M., (2004). Room temperature methane detection using palladium loaded single-walled carbon nanotube sensors. *Chem. Phys. Lett., 391*, 344–348.
105. Staii, C., & Johnson, A. T., (2005). DNA-decorated carbon nanotubes for chemical sensing. *Nano Lett., 5*, 1774–1778.
106. Chen, R. J., Franklin, N. R., Kong, J., Cao, J., Tombler, T. W., Zhang, Y., & Dai, H., (2001). Molecular photodesorption from single-walled carbon nanotubes. *Appl. Phys. Lett., 79*, 2258–2260.
107. Novak, J. P., Snow, E. S., Houser, E. J., Park, D., Stepnowski, J. L., & McGill, R. A., (2003). Nerve agent detection using networks of single-walled carbon nanotubes. *Appl. Phys. Lett., 83*, 4026–4028.
108. Snow, E. S., Perkins, F. K., Houser, E. J., Badescu, S. C., & Reinecke, T. L., (2005). Chemical detection with a single-walled carbon nanotube capacitor. *Science, 307*, 1942–1945.
109. Snow, E. S., & Perkins, F. K., (2005). Capacitance and conductance of single-walled carbon nanotubes in the presence of chemical vapors. *Nano Lett., 5*, 2414–2417.
110. Wang, J., & Musameh, M., (2003). Carbon nanotube/teflon composite electrochemical sensors and biosensors. *Anal. Chem., 75*, 2075–2079.
111. Gao, M., Dai, L. M., & Wallace, G. G., (2003). Biosensors based on aligned carbon nanotubes coated with inherently conducting polymers. *Electroanal., 15*, 1089–1094.
112. Tang, H., Chen, J., Yao, S., Nie, L., Deng, G., & Kuang, Y., (2004). Amperometric glucose biosensor based on adsorption of glucose oxidase at platinum nanoparticle modified carbon nanotube electrode. *Anal. Biochem., 331*, 89–97.
113. Hrapovic, S., Liu, Y., Male, K. B., Luong, & T, J. H., (2004). Electrochemical biosensing platforms using platinum nanoparticles and carbon nanotubes. *Anal. Chem., 76*, 1083–1088.
114. Male, K. B., Hrapovic, S., Liu, Y., Dashan, W., Luong, & T, J. H., (2004). Electrochemical detection of carbohydrates using copper nanoparticles and carbon nanotubes. *Anal. Chim. Acta, 516*, 35–41.
115. Cai, H., Cao, X., Jiang, Y., He, P., & Fang, Y., (2003). Carbon nanotube-enhanced electrochemical DNA biosensor for DNA hybridization detection. *Anal. Bioanal. Chem., 375*, 287–293.
116. Wang, S. G., Wang, R., Sellin, P. J., & Zhang, Q., (2004). DNA biosensors based on self assembled carbon nanotubes. *Biochem. Biophys. Res. Comm., 325*, 1433–1437.
117. Kerman, K., Morita, Y., Takamura, Y., Ozsoz, M., & Tamiya, E., (2004). DNA-directed attachment of carbon nanotubes for enhanced label-free electrochemical detection of DNA hybridization. *Electroanal., 16*, 1667–1672.
118. Wang, J., Liu, G., & Jan, M. R., (2004). Ultrasensitive electrical biosensing of proteins and DNA: Carbon-nanotube derived amplification of the recognition and transduction events. *J. Am. Chem. Soc., 126*, 3010–3011.
119. Britto, P. J., Santhanam, K. S. V., & Ajayan, P. M., (1996). Carbon nanotube electrode for oxidation of dopamine. *Bioelectroch. Bioener., 41*, 121–125.
120. Guiseppi-Elie, A., Lei, C. H., & Baughman, R. H., (2002). Direct electron transfer of glucose oxidase on carbon nanotubes. *Nanotechnol., 13*, 559–564.

121. Zhao, Y. D., Zhang, W. D., Chen, H., & Luo, Q. M., (2002). Direct electron transfer of glucose oxidase molecules adsorbed onto carbon nanotube powder microelectrode. *Anal. Sci., 18*, 939–941.
122. Wang, J., Li, M., Shi, Z., Li, N., & Gu, Z., (2002). Direct electrochemistry of cytochrome c at a glassy carbon electrode modified with single-wall carbon nanotubes. *Anal. Chem., 74*, 1993–1997.
123. Musameh, M., Wang, J., Merkoci, A., & Lin, Y., (2002). Low-potential stable NADH detection at carbon-nanotube-modified glassy carbon electrodes. *Electrochem. Commun., 4*, 743–746.
124. Li, G., Liao, J. M., Hu, G. Q., Ma, N. Z., & Wu, P. J., (2005). Study of carbon nanotube modified biosensor for monitoring total cholesterol in blood. *Biosens. Bioelectron., 20*, 2140–2144.
125. Wu, K., Sun, Y., & Hu, S., (2003). Development of an amperometric indole-3-acetic acid sensor based on carbon nanotubes film coated glassy carbon electrode. *Sens. Actuators B-Chem., 96*, 658–662.
126. Besteman, K., Lee, J., Wiertz, F. G. M., Heering, H. A., & Dekker, C., (2003). Enzyme coated carbon nanotubes as single-molecule biosensors. *Nano Lett., 3*, 727–730.
127. Chen, R. J., Bangsaruntip, S., Drouvalakis, K. A., Kam, N. W. S., Shim, M., Li, Y., Kim, W., et al., (2003). Noncovalent functionalization of carbon nanotubes for highly specific electronic biosensors. *Proc. Nat. Acad. Sci. USA, 100*, 4984–4989.
128. Chen, R. J., et al., (2004). An investigation of the mechanisms of electronic sensing of protein adsorption on carbon nanotube devices. *J. Am. Chem. Soc., 126*, 1563–1568.
129. Star, A., et al., (2006). Label-free detection of DNA hybridization using carbon nanotube network field-effect transistors. *Proc. Nat. Acad. Sci. USA, 103*, 921–926.
130. Tang, X. W., et al., (2006). Carbon nanotube DNA sensor and sensing mechanism. *Nano Lett., 6*, 1632–1636.
131. Modi, A., Koratkar, N., Lass, E., Wei, B. Q., & Ajayan, P. M., (2003). Miniaturized gas ionization sensors using carbon nanotubes. *Nature, 424*, 171.
132. Victor, P., et al., (2007). Electromechanical properties of a macroscale carbon nanotube block. *Appl. Phys. Lett.* In review.
133. Baughman, R. H., Cui, C., Zhakhidov, A. A., Iqbal, Z., Barisci, J. N., Spinks, G. M., Wallace, G. G., et al., (1999). Carbon nanotube actuators. *Science, 284*, 1340.
134. Hamberg, M. W., et al., (1995). An electrochemical micro actuator. In: *Micro–Electro Mechanical Systems, Proc. IEEE, 106.*
135. Rinzler, A. G., et al., (1995). Unraveling nanotubes: Field emission from an atomic wire. *Science, 269*, 1550–1553.
136. Heer, W. A. D., Chatelain, A., & Ugarte, D., (1995). A carbon nanotube field emission electron source. *Science, 270*, 1179, 1180.
137. Saito, Y., & Uemura, S., (2000). Field emission from carbon nanotubes and its applications to electron sources. *Carbon, 38*, 169–182.
138. Fan, S. S., Chaplind, M. G., Franklin, N. R., Tombler, T. W., Cassel, A. M., & Dai, H., (1999). Self oriented regular arrays of carbon nanotubes and their field emission properties. *Science, 283*, 512–514.
139. Saito, Y., Uemura, S., & Hamaguchi, K., (1998). Cathode ray tube lighting elements with carbon nanotube field emitters. *Jpn. J. Appl. Phys., 37*, L346–L348.
140. Choi, W. B., Chung, D. S., Kang, J. H., Kim, H. Y., Jin, Y. W., Han, I. T., Lee, Y. H., et al., (1999). Fully sealed high brightness carbon nanotube field emission display. *Appl. Phys. Lett., 75*, 3219–3231.

141. Bonard, J. M., Salvetat, J. P., Stockli, T., Deheer, W. A., Forro, L., & Chatelain, A., (1998). Field emission from single-wall carbon nanotube film. *Appl. Phys. Lett., 73*, 918–920.
142. Kurachi, H., Uemura, S., Yotani, J., Nagasako, T., Yamada, H., Ezaki, H., Maesoba, T., et al., (2001). In: *Proc. 21st Int. Display Res. Conf./8th Int. Display Workshops: Soc. Inf. Display* (pp. 1245–1248).
143. Seko, K., Kinoshita, J., & Saito, Y., (2005). In situ transmission electron microscopy of field-emitting bundles of double wall carbon nanotubes. *Jpn. J. Appl. Phys., 44*, L743–L745.
144. Son, Y. W., Oh, S., Ihm, J., & Han, S., (2005). Field emission properties of double-wall carbon nanotubes. *Nanotechnol., 16*, 125–128.
145. Hiraoka, T., Yamada, T., Hata, K., Futaba, D. N., Kurachi, H., Uemura, S., Yumura, M., & Iijima, S., (2006). Synthesis of single and double walled carbon nanotubes forests on conducting metal foils. *J. Am. Chem. Soc., 128*, 13338, 13339.
146. Charlier, J. C., Terrones, M., Baxendale, M., Meunier, V., Zacharia, T., Rupesinghe, N. L., Hsu, W. K., et al., (2002). Enhanced electron field emission in B-doped carbon nanotubes. *Nano Lett., 2*, 1191.
147. Golberg, D., Dorozhkin, P. S., Bando, Y., Dong, Z. C., Tang, C. C., Uemura, Y., Grobert, N., et al., (2003). Structure, transport and field-emission properties of compound nanotubes: CNx vs. BNCx (x <0.1). *Appl. Phys. A-Mater., 76*, 499.
148. Doytcheva, M., Kaiser, M., Reyes-Reyes, M., Terrones, M., & De Jonge, N., (2004). Electron emission from individual nitrogen-doped multi-walled carbon nanotubes. *Chem. Phys. Lett., 396*, 126.
149. Heller, D. A., et al., (2006). Optical detection of DNA conformational polymorphism on single-walled carbon nanotubes. *Science, 311*, 508–511.
150. Cherukuri, P., et al., (2004). Near-infrared fluorescence microscopy of single-walled carbon nanotubes in phagocytic cells. *J. Am. Chem. Soc., 126*, 15638, 15639.
151. Cherukuri, P., et al., (2006). Mammalian pharmacokinetics of carbon nanotubes using intrinsic near-infrared fluorescence. *Proc. Nat. Acad. Sci. USA, 103*, 18882–18886.
152. Heller, D. A., et al., (2005). Single-walled carbon nanotube spectroscopy in live cells: Towards long-term labels and optical sensors. *Adv. Mater., 17*, 2793–2799.
153. Klumpp, C., et al., (2006). Functionalized carbon nanotubes as emerging nanovectors for the delivery of therapeutics. *BBA-Biomembranes, 1758*, 404–412.
154. Pantarotto, D., et al., (2004). Translocation of bioactive peptides across cell membranes by carbon nanotubes. *Chem. Commun., 2004*, 16, 17.
155. Kam, N. W. S., et al., (2004). Nanotube molecular transporters: Internalization of carbon nanotube-protein conjugates into mammalian cells. *J. Am. Chem. Soc., 126*, 6850–6851.
156. Cai, D., et al., (2005). Highly efficient molecular delivery into mammalian cells using carbon nanotube spearing. *Nature Method., 2*, 449–454.
157. Lu, Q., et al., (2004). RNA polymer translocation with single-walled carbon nanotubes. *Nano Lett., 4*, 2473–2477.
158. Kam, N. W. S., & Dai, H. J., (2005). Carbon nanotubes as intracellular protein transporters: Generality and biological functionality. *J. Am. Chem. Soc., 127*, 6021–6026.
159. Wu, W., et al., (2005). Targeted delivery of amphotericin B to cells by using functionalized carbon nanotubes, *Angew. Chem. Int. Edit., 44*, 6358–6362.
160. Pantarotto, D., et al., (2004). Functionalized carbon nanotubes for plasmid DNA gene delivery, *Angew. Chem. Int. Edit., 43*, 5242–5246.

161. Kam, N. W. S., Liu, Z., & Dai, H. J., (2005). Functionalization of carbon nanotubes via cleavable disulfide bonds for efficient intracellular delivery of siRNA and potent gene silencing. *J. Am. Chem. Soc., 127*, 12492, 12493.
162. Zheng, et al., (2003). Structure-based carbon nanotube sorting by sequence dependent DNA assembly. *Science, 302*, 1545–1548.
163. Panhuis, M. I. H., et al., (2003). Characterization of an interaction between functionalized carbon nanotubes and an enzyme. *J. Nanosci. Nanotechnol., 3*, 209–213.
164. Chen, R. J., et al., (2001). Noncovalent sidewall functionalization of single-walled carbon nanotubes for protein immobilization. *J. Am. Chem. Soc., 123*, 3838, 3839.
165. Liu, J., et al., (1998). Fullerene pipes. *Science, 280*, 1253–1256.
166. Hu, H., et al., (2004). Chemically functionalized carbon nanotubes as substrates for neuronal growth. *Nano Lett., 4*, 507–511.
167. Liu, Y., et al., (2005). Polyethylenimine-grafted multiwalled carbon nanotubes for secure noncovalent immobilization and efficient delivery of DNA. *Angew. Chem. Int. Ed., 44*, 4782–4785.
168. Pastorin, G., et al., (2006). Double functionalisation of carbon nanotubes for multimodal drug delivery. *Chem. Commun., 11*, 1182–1184.
169. Singh, R., et al., (2005). Binding and condensation of plasmid DNA onto functionalized carbon nanotubes: Toward the construction of nanotube-based gene delivery vectors. *J. Am. Chem. Soc., 127*, 4388–4396.
170. Salvador-Morales, C., et al., (2006). Complement activation and protein adsorption by carbon nanotubes. *Mol. Immunol., 43*, 193–201.
171. Kam, N. W. S., et al., (2005). Carbon nanotubes as multifunctional biological transporters and near-infrared agents for selective cancer cell destruction. *Proc. Nat. Acad. Sci. USA, 102*, 11600–11605.
172. Lopez, C. F., et al., (2004). Understanding nature's design for a nanosyringe. *Proc. Nat. Acad. Sci. USA, 101*, 4431–4434.
173. Lynch, I., (2007). Are there generic mechanisms governing interactions between nanoparticles and cells? Epitope mapping the outer layer of the protein material interface. *Physica A, 373*, 511–520.
174. Chithrani, B. D., Ghazani, A. A., Chan, & W, W. C., (2006). Determining the size and shape dependence of gold nanoparticle uptake into mammalian cells. *Nano Lett., 6*, 662–668.
175. Kam, N. W. S., Liu, Z. A., & Dai, H. J., (2006). Carbon nanotubes as intracellular transporters for proteins and DNA: An investigation of the uptake mechanism and pathway. *Angew. Chem. Int. Ed., 45*, 577–581.
176. Osaki, F., et al., (2004). A quantum dot conjugated sugar ball and its cellular uptake on the size effects of endocytosis in the subviral region. *J. Am. Chem. Soc., 126*, 6520, 6521.
177. Gao, H. J., Shi, W. D., & Freund, L. B., (2005). Mechanics of receptor-mediated endocytosis. *Proc. Nat. Acad. Sci. USA, 102*, 9469–9474.
178. Woolley, A. T., Guillemette, C., Cheung, C. L., Housman, D. E., & Lieber, C. M., (2000). Direct haplotyping of kiobase size DNA using carbon nanotube probes. *Nature Biotechnol., 18*, 760–763.
179. Hafner, J. H., Cheung, C. L., Woolley, A. T., & Lieber, C. M., (2001). Structural and functional imaging with carbon nanotube AFM probes-review. *Prog. Biophys. Mol. Bio., 77*, 73–110.
180. URL: http://www.xidex.com (accessed on 19 September 2022).

181. Dai, H., Franklin, N., & Han, J., (1998). Exploiting the properties of carbon nanotubes for nanolithography. *Appl. Phys. Lett., 73*, 1508–1510.
182. Hinds, B. J., Chopra, N., Rantell, T., Andrews, R., Gavalas, V., & Bachas, L. G., (2004). Aligned multiwalled carbon nanotube membranes. *Science, 303*, 62–65.
183. Holt, J. K., Park, H. G., Wang, Y. M., Stadermann, M., Artyukhin, A. B., Grigoropoulos, C. P., Noy, A., & Bakajin, O., (2006). Fast mass transport through sub-2-nanometer carbon nanotubes. *Science, 312*, 1034–1037.
184. Srivastava, A., Srivastava, O. N., Talapatra, S., Vajtai, R., & Ajayan, P. M., (2004). Carbon nanotube filters. *Nature Mater., 3*, 610–614.
185. Kordas, K., et al., (2006). Inkjet printing of electrically conductive patterns of carbon nanotubes. *Small, 2*, 1021–1025.
186. Ericson, L. M., et al., (2004). Macroscopic, neat, single-walled carbon nanotube fibers. *Science, 305*, 1447–1450.
187. Y.-Li, L., Kinloch, I. A., & Windle, A. H., (2004). Direct spinning of carbon nanotube fibers from chemical vapor deposition synthesis. *Science, 304*, 276–278.
188. Wei, J., Zhu, H., Wu, D., & Wei, B., (2004). Carbon nanotube filaments in household light bulbs. *Appl. Phys. Lett., 84*, 4869–4871.
189. Cao, A., Veedu, V. P., Li, X., Yao, Z., Ghasemi-Nejhad, M. N., & Ajayan, P. M., (2005). Multifunctional brushes made from carbon nanotubes. *Nature Mater., 4*, 540–545.
190. Yurdumakan, B., Raravikar, N. R., Ajayan, P. M., & Dhinojwala, A., (2005). Synthetic gecko foot-hairs from multiwalled carbon nanotubes. *Chem. Commun., 30*, 3799–3801.
191. Dalton, A. B., et al., (2004). Continuous carbon nanotube composite fibers: Properties, potential applications and problems. *J. Mater. Chem., 14*, 1.
192. Sun, L., Banhart, F., Krasheninnikov, A. V., Rodriguez-Manzo, J. A., Terrones, M., & Ajayan, P. M., (2006). Carbon nanotubes as high-pressure cylinders and nanoextruders. *Science, 312*, 1199–1202.
193. Ajayan, P. M., Stephan, O., Redlich, P., & Colliex, C., (1995). Carbon nanotubes as removable templates for metal oxide nanocomposites and nanostructures. *Nature, 375*, 564.
194. Ou, F. S., Shaijumon, M. M., Ci, L., Benicewicz, D., Vajtai, R., & Ajayan, P. M., (2006). Multisegmented one-dimensional hybrid structures of carbon nanotubes and metal nanowires. *App. Phys. Lett., 89*, 243122.
195. Meng, G. W., Jung, Y. J., Cao, A., Vajtai, R., & Ajayan, P. M., (2005). Controlled fabrication of hierarchically branched nanopores, nanotubes and nanowires. *Proc. Nat. Acad. Sci., 102*, 707.
196. Sano, N., (2004). Low cost synthesis of single-walled carbon nanohorns using the arc in water method using gas injection. *J. Phys. D: Appl. Phys., 37*, L17–L20.
197. Shvedova, A. A., Castranova, V., Kisin, E. R., Schwegler-Berry, D., Murray, A. R., Gandelsman, V. Z., Maynard, A., & Baron, P., (2003). Exposure to carbon nanoutbe material: Assessment of nanotube cytoxicity using human keratinocyte cell. *J. Toxicol. Environ. Health, 66*, 1909–1926.
198. Lam, C. W., James, J. T., McCluskey, R., & Hunter, R. L., (2004). Pulmonary toxicity of single-wall carbon nanotubes in mice 7 and 90 days after intratracheal instillation. *Toxicol. Sci., 77*, 126–134.
199. Warheit, D. B., Laurence, B. R., Reed, K. L., Roach, D. H., Reynolds, G. A., & Webb, T. R., (2004). Comparative pulmonary toxicity assessment of single-wall carbon nanotubes in rats. *Toxicol. Sci., 77*, 117–125.

200. Maynard, A. D., Baron, O. P. A., Foley, M., Shvedova, A. A., Kisin, E. R., & Castranova, V., (2004). Exposure to carbon nanotube material: Aerosol release during the handling of unrefined single walled carbon nanotube material. *J. Toxicol. Environ. Heath, 67*, 87–107.
201. Koyama, S., Endo, M., Kim, Y. A., Hayashi, T., Yanagisawa, T., Osaka, K., Koyama, H., et al., (2006). Role of systemic T-cells and histopathological aspects after subcutaneous implantation of various carbon nanotubes in mice. *Carbon, 44*, 1079–1092.
202. Jia, G., Wang, H., Yan, L., Wang, X., Pei, R., Yan, T., Zhao, Y., & Guo, X., (2005). Cytoxicity of carbon nanomaterials: Single wall nanotube, multi-wall nanotube, and fullerene, Environ. *Sci. Technol., 39*, 1378–1383.
203. Cui, D., Tian, F., Ozkan, C., Wang, M., & Gao, H., (2005). Effect of single wall carbon nanotubes on human HEK293 cells. *Toxicol. Lett., 155*, 73–85.
204. Sato, Y., Shibata, K., Kataoka, H., Ogino, S., Bunshi, F., Yokoyama, A., Tamura, K., et al., (2005). Mol. BioSyst. 1, 142–145.
205. Service, R. F., (2006). Science policy: Priorities needed for nano-risk research and development. *Science, 314*, 45.
206. Cui, D. X., et al., (2005). Effect of single wall carbon nanotubes on human HEK293 cells. *Toxicol. Lett., 155*, 73–85.
207. Manna, S. K., et al., (2005). Single walled carbon nanotube induces oxidative stress and actiaves nuclear transcription factor – kappa B in human keratinocytes. *Nano Lett., 55*, 1676–1684.
208. Kagan, V. E., et al., (2006). Direct and indirect effects of single walled carbon nanotubes on RAW 264.7 macrophages: Role of iron. *Toxicol. Lett., 165*, 88–100.
209. Sayes, C. M., et al., (2006). Functionalization density dependence of single-walled carbon nanotubes cytotoxicity in vitro. *Toxicol. Lett., 161*, 135–142.
210. Dumortier, H., et al., (2006). Functionalized carbon nanotubes are non-cytotoxic and preserve the functionality of primary immune cells. *Nano Lett., 6*, 1522–1528.
211. Kam, N. W. S., et al., (2006). Carbon nanotubes as multifunctional biological transporters and near-infrared agents for selective cancer cell destruction. *Proc. Nat. Acad. Sci. USA, 102*, 11600–11605.

CHAPTER 10

CNT-BASED SOLID COMPOSITES FOR WATER TREATMENT

RONY RAJAN PAUL and MARIA MATHEW

Department of Chemistry, CMS College, Kottayam, Kerala, India

ABSTRACT

Lack of accessibility to safe and clean water is one of the greatest challenges that mankind is facing today. The increasing population and rapidly progressing urbanization and industrialization have led to a great demand for fresh water. Since the sustainability of the human race is largely dependent on the availability of fresh water, researchers are always in search for a better technology for wastewater treatment. Among the various methods developed, water treatment applications based on carbon nanotube (CNT) appears to be the most dependable technological advancement. The large surface area, high chemical reactivity, mesoporous structure, tunable surface properties, low cost, and less impact on the environment prompted researchers to work on it. This chapter discusses the various potential applications of CNT as adsorbents, catalysts, antimicrobial surfaces, and sensors. Also, the recent developments in the field of CNTs in this line are discussed.

10.1 INTRODUCTION

The shortage of availability of fresh and clean water is one of the major universal problems which the researchers are trying to address. With the increasing population, urbanization, and climate changes water crisis has

Nanotechnology-Enhanced Solid Materials: Design, Synthesis, Properties, Applications, and Perspectives. Lionello Pogliani, Ann Rose Abraham, A. K. Haghi, & Prabhat Ranjan (Eds.)

become a major issue. According to the reports of the World Bank, about 40% of world population is badly facing the issue of water crisis. It has been predicted that by 2025; about 3 billion people will be hardly struggling for drinking water [1]. Water scarcity can affect both economic growth and human health, as it threatens industrial production and at the same time, a shortage of hygienic food and drinks can cause various epidemic diseases.

The increasing pressure on water scarcity issue highly requires cost-effective water treatment technologies to produce high-quality clean water. Some processes, such as reverse osmosis (RO), forward osmosis (FO), membrane distillation, and capacitive deionization are favorable in the desalination of both sea and brackish water. The conventional water treatment technologies include physical methods (boiling, distillation, filtration, ultrafiltration, and coagulation); chemical degradation (ozone, chlorine, ultraviolet, solar water disinfection, etc.); and biological methods (microbial water sludge treatment) have failed because of their dependence on influent water qualities. However, the advent of nanotechnology introduces impressive nanomaterials such as CNT, fullerenes, nanofibers, NWs, zeolites, and various nanoparticles–Fe_3O_4, MnO_2, Co_3O_4, and TiO_2. Amongst these, CNT-based water treatment technologies are one of the leading and most effective ones. It is capable of purifying water from organic, inorganic, and biological pollutants [2]. The high surface area, low cost, greater reactivity, low chemical mass, and lower impact on the environment adds to its properties to be utilized in this manner. CNTs are made of graphite sheets rolled up in the form of a cylinder (Figure 10.1). Mainly there are two types of CNTs-multi-walled carbon nanotubes (MWCNTs) composed of multiple layers of graphene sheets and single-walled carbon nanotubes (SWCNTs) consisting of a single graphene sheet. These structures help in the easy transport of water molecules and thus purify water efficiently [3].

Figure 10.2 illustrates some of the most important functions of CNTs in water purification. Generally, they are used as an adsorbent for the adsorption of various organic (volatile organic compounds, drug molecules, dyes, insecticides, and fats), inorganic (heavy metal ions), and biological pollutants (bacteria, viruses, fungi). Secondly, CNT catalysts support enhancing the catalytic reactivity toward contaminant mitigation [4]. Also, the mechanistic action of CNTs toxicity towards bacterial cells in water disinfection is not worthy. Here, CNT act by interrupting bacterial cell membrane causing its death. The other major applications include the desalination of seawater and brackish water using CNT membrane and the sensing and monitoring of water pollutants using nanosensors modified with CNTs.

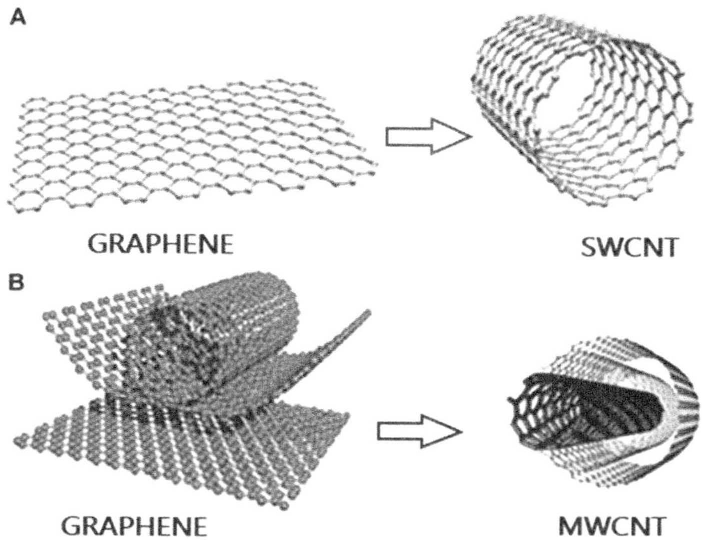

FIGURE 10.1 Structural representation of (A) MWCNT; and (B) SWCNT.

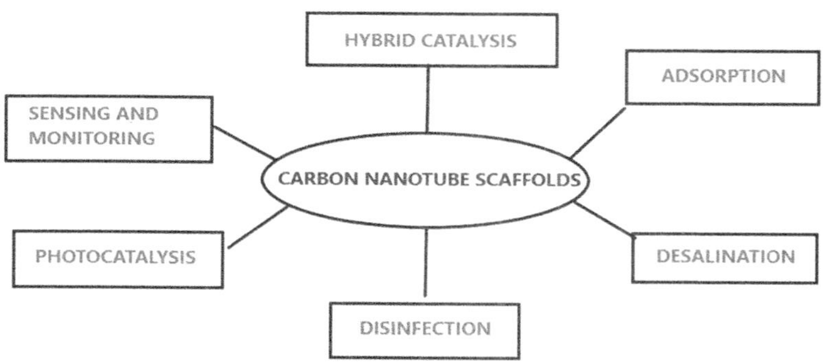

FIGURE 10.2 Water treatment applications of CNTs.

This chapter aims to discuss the mechanism of action of CNTs for adsorption, catalysis, disinfection, desalination, and sensing of various types of water pollutants.

10.2 ADSORPTION

Adsorption is a surface phenomenon in which adsorbate molecules are adsorbed and creates a film on adsorbent surface. The surface chemistry, morphology, physical, and chemical interaction with organic, inorganic, and biological pollutants have made CNT an effective and preferred material for water remediation [2, 3, 5–7]. CNT is now considered as a better material than usual adsorbents such as zeolite, clay, and activated charcoal.

Rate of pollutant adsorption depends upon several factors, such as available sites, pore density, outer surface area, functional groups present, purity, etc. Figure 10.3 illustrates the four major CNT adsorption sites for the adsorption of water pollutants (a) interstitial channels, (b) inner CNT holes, (c) grooves, and (d) CNT outer surfaces. Most of the organic pollutants are adsorbed at the outer surface and inner sites of open-ended SWCNTs. Functionalization of the external surface of CNTs helps in attaching various organic and inorganic pollutants [3]. Small pollutants are easily trapped by the interstitial channels. However, the rate of adsorption depends upon the type of nanotube. SWCNTs create more interstitial channels than MWCNTs. Because, SWCNTs have greater aggregation effects than MWCNTs. Rate of aggregation decreases with increase in the number of nanotube layers [8]. Aggregation of CNTs has both favorable and unfavorable effects on adsorption [9–18]. On one side, it reduces the total surface area of SWCNTs which helps in bulk pollutant adsorption, and on the other side; it creates interstitial channels between the tubes that would significantly increase the pore volume and groves on the external surface of CNT bundles.

10.2.1 ORGANIC POLLUTANTS

Organic compounds present in water are engulfed by microorganisms that will decrease the oxygen content of water during the decomposition process. Similarly, the organic effluents ejected from industries also cause the deterioration of the quality of freshwater sources. In order to overcome this problem, various organic pollutants present in water has to be removed.

10.2.1.1 FACTORS AFFECTING THE ADSORPTION OF ORGANIC MOLECULES

1. **Morphology and Active Sites of CNTs:** Surface topography has a huge impact on the adsorption capabilities of CNTs. Rough surfaces have large surface areas than smooth ones; hence, rough surfaces show greater adsorption than smooth-surfaced CNTs like pristine [10].

2. **Effect of pH:** Several comparative studies have been done in order to understand the effect of pH on the adsorption rate. Ghaedi et al. [11] found out that MWCNTs have higher adsorption capacity and they reach equilibrium faster than SWCNTs.
3. **Effect of Temperature:** Adsorption capacity increases with increase in temperature, because CNTs are endothermic in nature which increases adsorption with temperature [12]. The highest result obtained is at around 133 mg/g at 310 K [13].
4. **Magnetic MWCNT:** Gong et al. have studied the adsorption of cationic dyes such as methylene blue (MB), neutral red (NR), and gleaming cresyl blue (GCB) from aqueous solution by means of magnetic MWCNT nanocomposite. Similar studies were also done by Gaog et al. [14–21] for the removal of anionic azo dyes and found out that magnetic polymer MWCNTs are very effective for the separation process.

10.2.2 FUNCTIONALIZATION AND ADSORPTION

Functionalization of CNTs proved to be very effective for the specific pollutant adsorption. It increases CNT colloidal stability as well as the chemical reactivity [15, 16]. Three different methods for CNT functionalization are: (i) physical non-covalent wrapping [17]; (ii) covalent wet chemical agent treatment [18]; (iii) endohedral filling of CNT hollow cavity. Major interactions between CNT functionalities and water pollutants are covalent bonding, hydrogen bonding, electrostatic interactions, hydrophobic interactions, ion exchange, and π-π electron coupling. In Figure 10.4, we have depicted different functionalities added to the CNT surface through different routes.

CNTs are hydrophobic in nature because of the presence of non-polar graphene sheets. These hydrophobic natures make them capable of adsorbing aromatic pollutants such as benzene, naphthalene, pyrene, and so on. The π electrons available on the individual graphene sheets help in the formation of strong chemical complex between organic aromatic pollutants and CNT surface. Functionalities can enhance CNT solubility and bring CNTs repel each other. This allows the free movement of individual CNTs through homogeneous solution, thus avoiding aggregation. In addition, charged functional groups on the CNT surface can bind ionized water pollutants through electrostatic interactions.

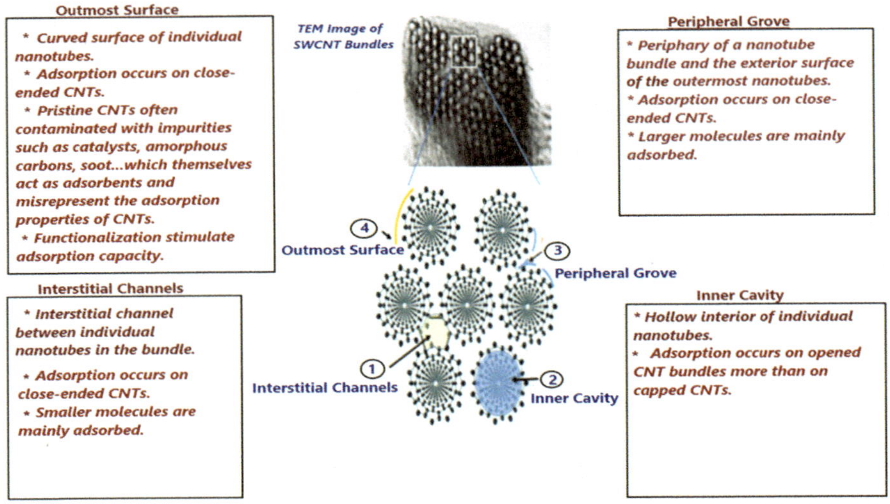

FIGURE 10.3 Four major adsorption sites on a homogeneous bundle of partially opened SWCNT.

10.2.3 INORGANIC POLLUTANTS

CNTs are found to be used for the adsorption of toxic heavy metal ions such as Cd(II), Cu(II), Pb(II), and Hg(II) from water as they are harmful to living beings [19]. Sorption capacities can be improved by oxidizing CNT surfaces with H_2O_2, $KMnO_4$, and HNO_3 [20]. Pristine CNTs have shown an adsorption capacity of only 1.1 mg/g, however, on oxidation, it reached nearly 5.0–11.0 mg/g. Cd (II) adsorption by CNT is highly dependent on pH and this pH dependence leads to the conclusion that, adsorption by CNT is not only dependent on its surface chemistry but also regulated by other parameters like pH, temperature, nature of adsorbate, concentration of pollutants, contact time, and so on.

10.3 CATALYSIS

The unique electrical properties of CNTs, which resulted from the presence of surface functional groups, curvature pattern, topological defects, and edge plane sites [21], make CNTs an ideal candidate for catalysis (photocatalysis, catalytic wet air oxidation (CWAO), enzyme biocatalysis, and so on). High electrical conductivity and excellent charge-transferring ability of one-dimensional CNTs make them highly effective for the photo-degradation of organic pollutants [22, 23]. In CWAO, dissolved and suspended water contaminants are oxidized by molecular oxygen from the air. In addition, CNTs are also used for immobilizing enzymes. The known atomic structure, larger surface area, mass transfer ability, and good enzyme loading capacities make CNTs an effective biocatalyst for water purification. The mechanisms of three CNT-based catalytic process – photocatalysis, CWAO, and enzyme-biocatalysis are discussed in subsections.

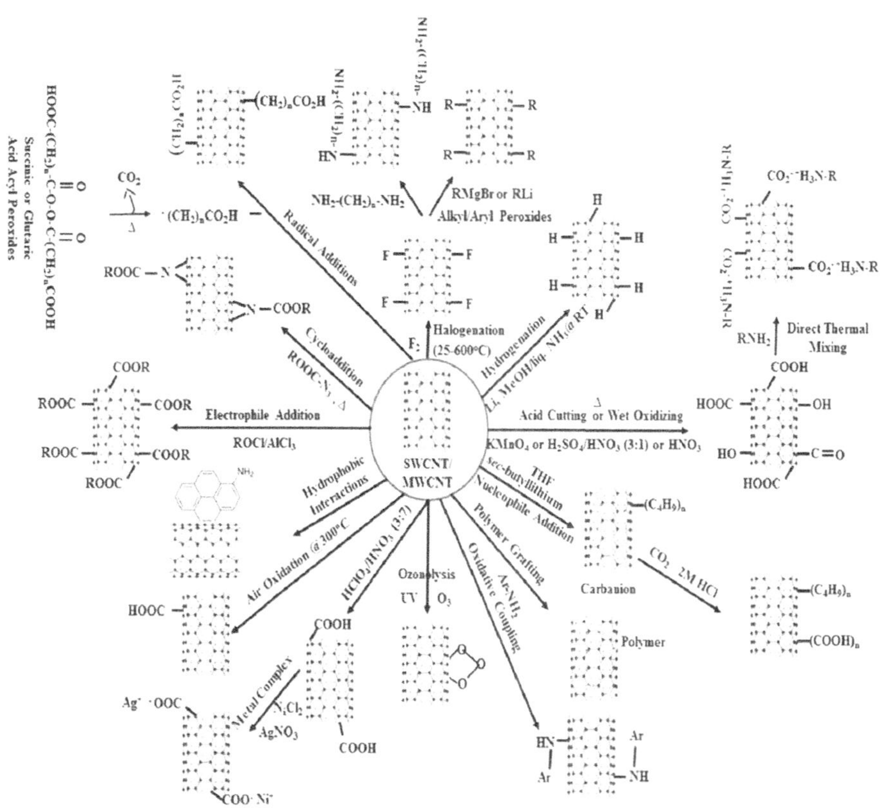

FIGURE 10.4 Surface functionalization of CNT.

10.3.1 PHOTOCATALYSIS

Initially, semiconductors such as TiO_2, ZnO, CdS, and ZnS have been used as photocatalyst to remove various organic compounds from water. However, these semiconductors have a major drawback that the photogenerated charge carriers will undergo rapid recombination resulting in low quantum efficiency [24]. In order to overcome these drawbacks, researchers developed a solid photocatalyst with improved photocatalytic activity [25, 26]. CNTs have been used as reinforced photocatalytic composite material along with these semiconductors [27]. The major advantages of CNTs, achieved by functionalization are their tunable surface properties, large surface area, and thermodynamic stability, which make them suitable for photodegradation.

The two methods used for the synthesis of CNT-based photocatalytic composites are: (i) attachment of desired nanoparticles on the CNT surface *via* covalent and noncovalent bonding [40]; (ii) *in-situ* synthesis of desired nanoparticle followed by electrochemical deposition [41].

Figure 10.5 illustrates the possible mechanism for CNT-supported photocatalysis. When a catalyst nanoparticle is irradiated with photons from different light sources, electrons from its VB are excited to its conduction band (CB) creating holes in the VB. These holes will then react with H_2O and OH^- ions followed by the formation of hydroxyl radicals (•OH) which will directly oxidize the adsorbed pollutants on the CNT surface. At the same time, the photoexcited electrons from the CB will be trapped by MWCNTs which then react with molecular oxygen to form reactive species such as O_2^-, H_2O_2 and OH. These highly reactive species can attack and oxidize the target pollutant immediately. Thus, the recombination of charge carriers can be reduced, thereby improving the photocatalytic activity [42].

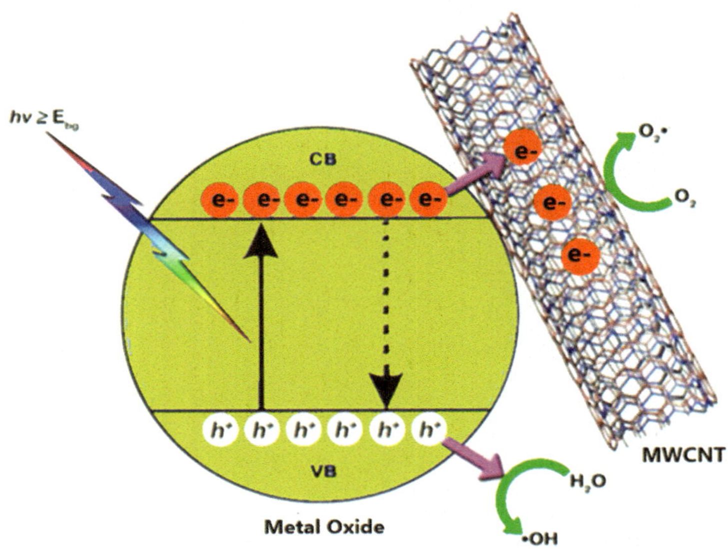

FIGURE 10.5 Charge separation efficiency in MWCNTs nanohybrid photocatalytic system.

Also, CNT as catalyst support makes photocorrosion unfavorable enhancing the photodegradation rate (Table 10.1) [43, 44].

TABLE 10.1 Photocatalytic Decomposition of Organic Pollutants by CNT-Supported Photocatalyst

Hybrid Photocatalyst	Organic Pollutant	Optimum Loading (MWCNT)	Degradation Efficiency	References
Ag_3PO_4/ MWCNTs	Rhodamine B (RhB)	0.1 wt.%	Ag_3PO_4: 68% (12 min) Ag_3PO_4/MWCNTs: 92% (12 min)	[28]
BiOBr/ MWCNTs	Rhodamine B (RhB)	0.05 wt.%	BiOBr: 70% (75 min) BiOBr/MWCNTs: 97% (75 min)	[29]
BiOCOOH/ MWCNTs	Rhodamine B (RhB)	3 wt.%	BiOCOOH: 53% (60 min) BiOCOOH/MWCNTs: 92% (60 min)	[30]
BiOI/ MWCNTs	Acid Orange II (AOII)	1%	BiOI: 60% (180 min) BiOI/ MWCNTs: >80% (180 min)	[31]
Bi_2MoO_6/ MWCNTs	Methylene blue (MB)	Not provided	Bi_2MoO_6: 50% (30 min) Bi_2MoO_6/MWCNTs: 95% (30 min)	[32]
$Bi_2O_2CO_3$/ MWCNTs	Rhodamine B (RhB)	3 wt.%	The decomposition rate of RhB by $Bi_2O_2CO_3$/MWCNTs is 4.5 times higher than bare $Bi_2O_2CO_3$	[33]
Bi_2WO_6/ MWCNTs	Tetracycline (TC)	3%	The degradation rate of TC by Bi_2WO_6/MWCNTs is 35% higher than bare Bi_2WO_6	[34]
$Bi_4O_5Br_2$/ MWCNTs	Tetracycline hydrochloride (TCH)	0.1%	$Bi_4O_5Br_2$: 42% (30 min) $Bi_4O_5Br_2$/MWCNTs: 63% (30 min)	[35]
C_3N_4/ MWCNTs	Methylene blue (MB)	Not provided	C_3N_4: 35% (90 min) C_3N_4/ MWCNTs: 67% (90 min)	[36]
$CaIn_2S_4$/ MWCNTs	X-3B	1 wt.%	$CaIn_2S_4$: 89% (90 min) $CaIn_2S_4$/MWCNTs: 98% (90 min)	[37]
CdS/MWCNTs	Methyl orange (MO)	3%	CdS: 30% (60 min) CdS/ MWCNTs: 90% (60 min)	[38]
$CuFe_2O_4$/ MWCNTs	Diethyl phthalate (DEP)	33.3%	$CuFe_2O_4$: 30% (30 min) $CuFe_2O_4$/MWCNTs: 75% (30 min)	[39]

10.3.2 CATALYTIC WET AIR OXIDATION (CWAO)

Catalytic wet air oxidation (CWAO) is one of the efficient advanced oxidation processes (AOP) used to convert highly concentrated, hazardous, and toxic organic pollutants to CO_2 and H_2O [45]. WAO process takes place under high temperature (200–300°C) and high pressure (20–200 bar) using O_2 as the oxidant without producing any toxic emissions. It is also an economically feasible process without utilizing ozone and hydrogen peroxide (H_2O_2). Despite these advantages, its high operating temperature and pressure limit its practical applications. Here comes the necessity of developing a process under mild conditions using a suitable catalyst.

Thus, CWAO has emerged as one of the excellent water treatment methods. In this process, CNT act as a very good catalyst due to the presence of carboxyl and hydroxyl functional groups introduced *via* surface functionalization. It also offers several other advantages: (i) increased catalytic efficiency; (ii) simplified operation conditions; (iii) reduced cost; (iv) enhanced reaction rate, and so on. Yang et al. [46] reported functionalized MWCNTs for CWAO of phenol in a batch reactor (Figure 10.6). Carboxylic groups (–COOH) on CNTs exhibited excellent catalytic activity in phenol oxidation (1,000 mg/L) at 160°C and 2.0 MPa. The adsorbed liquid O_2 will get dissociated on MWCNTs and produce dissociated oxygen atoms (DOA) [47]. Carboxylic functional groups are introduced by treating MWCNTs with H_2SO_4/HNO_3 and DAO will produce OH_2 radical. Then the radicals decompose phenol to CO_2 and H_2O and in this way, organic compounds are removed in CWAO process.

10.3.3 NANO-BIOHYBRID CATALYSIS

Another important application of CNT is as a biocatalyst support matrix for the removal of recalcitrant pollutants such as dyes, polyaromatic hydrocarbons and other agricultural and pharmaceutical wastes. Enzymes are naturally occurring complex protein structures with specific biochemical properties useful for the biodegradation of various recalcitrant pollutants. CNT-enzyme biocatalysts or simply "Nanobiohybrid catalyst" formed by the immobilization of enzymes on CNT surfaces are found to be highly selective and efficient [48] for sensing, monitoring, and degrading multiple pollutants from water. Formation of nanobiohybrid by immobilizing enzymes on CNT surface takes place via three major routes: (i) binding to a support (physical adsorption and covalent bonding); (ii) cross-linking (career free); (iii) encapsulation or entrapment [49].

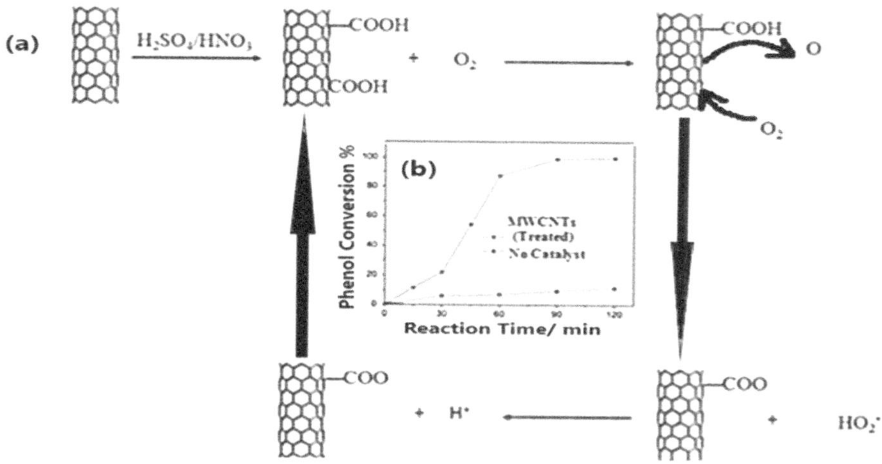

FIGURE 10.6 (a) Mechanism of producing the radical ($HO_2\bullet$) in CWAO for phenol breakdown by $H_2SO_4^-/HNO_3$ treated MWCNTs; (b) efficiency of phenol removal.

In physical adsorption, enzymes are spontaneously adsorbed on to CNT surface *via* hydrophobic and electrostatic interaction (Figure 10.7(I)) [50]. Enzymes are covalently bonded to the CNT surface by inducing the reaction between the free amine groups on the surface of protein with the –COOH group on the CNT surface (Figure 10.7(II)) [51, 52]. Cross-linking polymers such as chitosan, poly(diallyldimethylammonium chloride) can also be used for immobilizing enzymes on CNT surface. Finally, multilayer enzyme films are produced on the CNT surface by the encapsulation of enzymes (Figure 10.7(III)). Proper considerations should be taken while selecting the immobilization technique; else, it will lead to the denaturation of the enzyme [53].

The immobilization of enzymes onto CNT surfaces offers a lot of practical benefits, such as cost-effective post-treatment removal, increased reusability, broad pH and temperature ranges, enhanced storage, and operational stability [54]. Nanohybrid catalysis for water purification is particularly effective in those cases in which the reaction products are less toxic, the enzymatic reaction is straightforward, i.e., no co-factors are required, and the composition of the contaminated water is not too complex.

The properties of CNT-based nanobiohybrids can be enhanced by the chemical modifications of enzymes and surface functionalization of CNT surfaces.

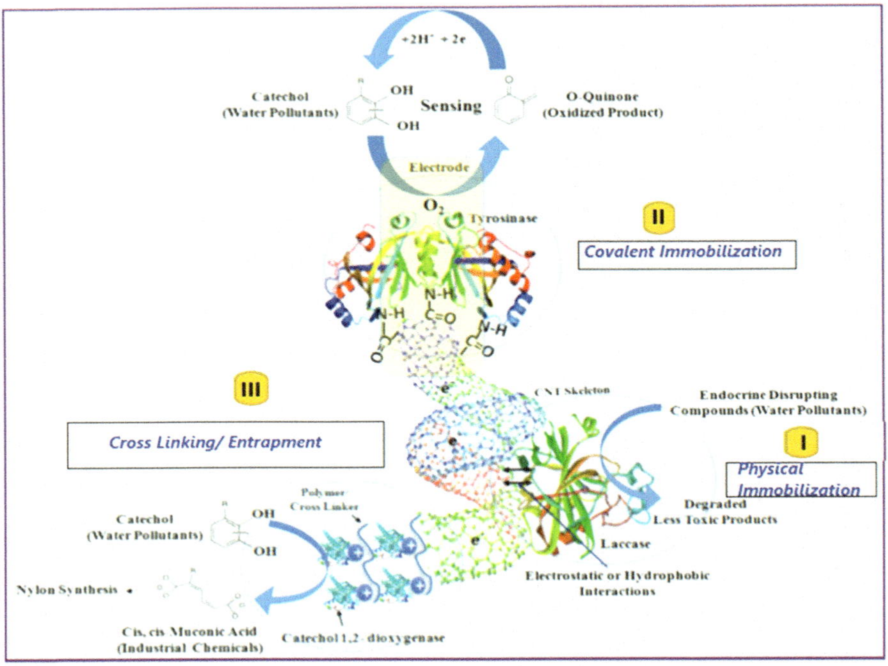

FIGURE 10.7 Mechanism of a potential nano-biohybrid catalyst for water purification.

10.4 DISINFECTION

Disinfection is one of the most powerful and heavily practiced techniques to remove various chemical and biological contaminants from freshwater sources. The requirement of high-quality water for drinking purposes demanded the removal of pathogens from freshwaters, which is a very difficult process due to the varying concentrations and types of pathogens present in water [55].

Studies have revealed that CNTs, specifically SWCNTs are highly efficient for the selective adsorption of bacterial species compared to other available adsorbents. This is due to their fibrous structure and external surface accessibility [56]. Liu et al. [57] have studied the toxic effects of pristine SWCNTs on both gram-positive bacteria like *B. subtilis* and *Staphylococcus aureus* (*S. aureus*), and gram-negative bacteria like *Escherichia coli* (*E. coli*) and *Pseudomonas aeruginosa* [58]. They have reported that the dispersed individual CNTs puncture the cell membrane rapidly in comparison with the aggregated CNTs. Soft and smooth cells such as gram-positive bacteria were more vulnerable and are easily attacked by CNTs than gram-negative bacteria. It has been shown that the membrane-penetrating effect can be enhanced by increasing CNTs concentration, controlling shaker speed augmentation during incubation, and using dispersed CNTs solution.

10.4.1 CHARACTERISTICS OF CNT FOR WATER DISINFECTION

Extremely high surface area-to-volume ratio and optimally controlled pore size distribution have provided CNTs exceptional and efficient sorption capabilities compared to other conventional sorbents such as activated charcoal. Powerful antimicrobial activity and high-water flux are some other important features. CNTs have been compared with porous materials of comparable sizes. Several methods have been developed in this line to improve the adsorption characteristics of CNTs, like functionalization of CNTs which in turn enhanced their dispersibility and specific surface area.

The application of CNT membrane in disinfection and bio-fouling reduction for conventional water treatment emerged as one of the focused research areas in the 21st century. By modifying the structural characteristics of CNT membranes, such as pore size, granule shape, and tortuosity, CNT membranes can provide better performance in water treatment applications. The anti-biofouling capability of the CNT membrane can inactivate bacteria based on its effect on bacterial adhesion. The resistance offered by CNT to the well-mature biofilm formed by a large number of bacteria adhered to the membrane is an add-on benefit in water disinfection.

The role of CNT in the removal of pathogenic bacteria has been well studied. The development of PGLa antimicrobial peptide and glutathione-conjugated CNTs with porous graphene oxide (GO) membrane has been manifested to be highly efficient in disinfecting the *E. coli* O157: H7 bacteria and for the elimination of As(III), As(V) and Pb(II) from the water (Figure 10.8) [59] involving synergetic mechanism. Silver nanoparticles/MWCNTs coated on polyacrylonitrile (PAN) hollow fiber membrane is an alternative and a safer way to eliminate pathogens. The disinfection properties of these composite membranes are achieved by the proper dispersion of Ag nanoparticles on the outer surface of CNTs, which can react with the bacteria cells. Also, Vecitis et al. [55] have studied the inactivation of bacteria (*E. coli*) and viruses (MS2) *via* anodic MWCNTs. Filtration followed by electrolysis significantly enhanced the inactivation. Fan et al. [60] recently developed an electropolished CNT/ceramic membrane that possesses the high capability for the reduction of bio-fouling in water. The performance of the electropolished membrane is dependent on the synergistic effects of electrostatic repulsion, electrokinetic behavior, and electrochemical oxidation.

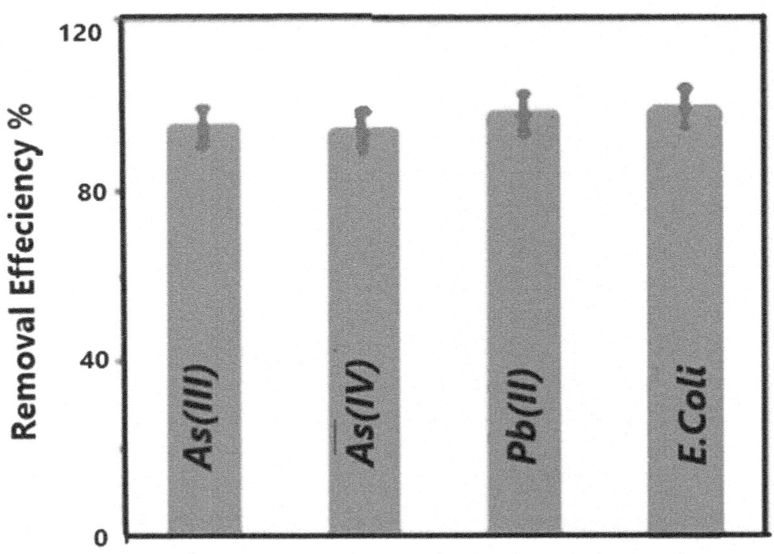

FIGURE 10.8 Removal efficiency of As(III), As(IV), Pb(III), and *E. coli* from water by CNT.

10.4.2 ANTIBACTERIAL MECHANISM

An understanding of the antimicrobial action of CNT is very crucial to reduce the occurrence of antibiotic-resistant bacteria spread through contaminated water sources. The most important mechanistic action of CNT's toxicity towards bacterial cells is the disruption of bacterial membrane integrity through interaction between bacterial cells and CNTs [61]. This loss of membrane integrity will cause the death of cells through the leakage of bacterial DNA and RNA content [62]. However, the interaction with gram-positive and gram-negative bacteria differs due to the difference in the thickness of the peptidoglycan layer. The peptidoglycan layer is thicker in gram-positive bacteria and in gram-negative bacteria there is an additional outer membrane layer that consists of lipid bilayer, which is responsible for its resistance to CNTs.

Oxidative stress is another mechanism by which bacterial cells are destroyed by the production of reactive oxygen species (ROS) through CNT exposure. ROS is generated through the photoinduced chemical reactions of CNTs in aqueous media in which large amounts of singlet oxygen ($^1O_2$), hydroxyl radicals (•OH), and superoxide anions ($O_2•^-$) are continuously produced. •OH radicals initiate lipid peroxidation process with the poly-unsaturated fatty acids, sugars, and proteins in biological components [63] which cause disintegration of membrane integrity facilitated through oxidative stress mediated by hydroxyl groups, ultimately resulting in the mortality of the bacterial cells.

10.5 DESALINATION

In the context of the global water crisis, CNT's role in the desalination of both sea and brackish waters cannot be overruled [64]. Currently, polymer reverse-osmosis membranes are used to filter out dissolved salts and fine solids. At present, there are mainly two types of nanotube membranes: (i) vertically aligned (VA); and (ii) mixed matrix (MM) CNT membranes [65] (Figure 10.9). These membranes can operate at lower energy because of CNT's frictionless water transport ability through the hydrophobic hollow cavity. It has been proved both theoretically and experimentally that water transport through CNTs is several orders of magnitude faster than other porous materials of comparable size. Ordered hydrogen bonds are created in the water-CNT interface, which in turn introduces a vapor phase barrier between the chain of water molecules and the non-polar CNT wall that allows the frictionless flow of water. Dumee et al. [66] observed the highest water permeability in (5,5) and (8,8) SWCNTs and the highest sodium ion retention in (5,5) and (6,6) SWCNTs at flux rates of ca. 12 kg m^{-2} h^{-1}.

CNTs acquire ion exclusion ability *via* tip functionalization. Functionalities such as positive ($-NH_3^+$), negative ($-COO^-$), and hydrophobic (aromatic) groups introduced on the CNT interface [67, 68] using different wet oxidizing agents has the ability to reject a particular pollutant, and at the same time it can attract water molecules through nanotube hole. Fornasiero et al. reported that negatively charged groups introduced by plasma treatment resulted in aligned CNT membranes which reject as high as 98% ions, and the ion exclusion was dominated by the charge exclusion mechanism. The defect-free CNT membranes with sub-nanometer pores will be, for sure, a future desalination technology.

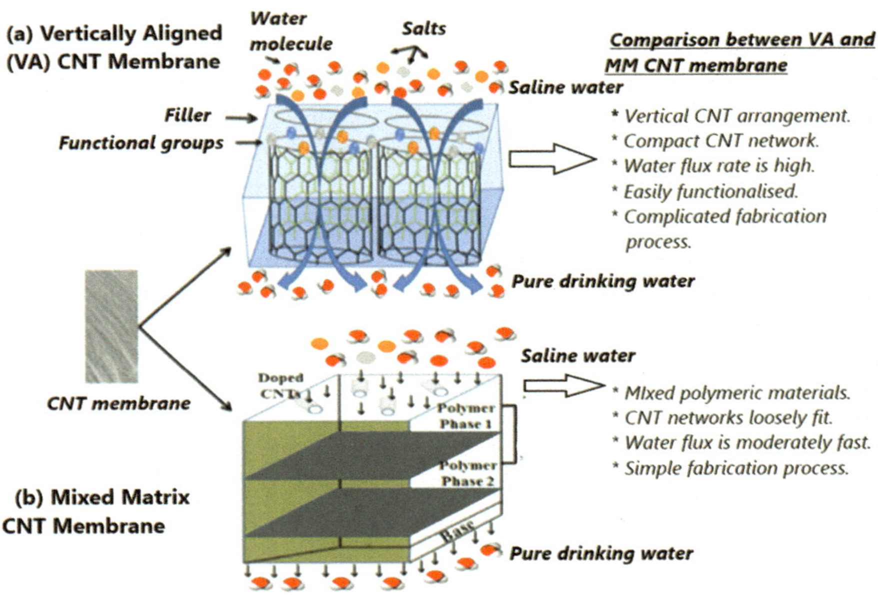

FIGURE 10.9 Schematic illustrations of the two types of CNT membranes: (a) vertically aligned; (b) mixed matrix.

10.6 SENSING AND MONITORING

Analysis of the concentration of pollutants is essential for investigating the quality of drinking water [69–71]. For many years, high-performance liquid chromatography (HPLC), gas chromatography/mass spectrometry (GC-MS), and different detectors such as the electron capture detector (ECD), flame ionization detector (FID) were used for the analysis of water pollutants [72–75]. Even though these techniques possess a lot of advantages, the need for derivatization for analyzes, large purification setups, and highly expensive equipment remained as drawbacks. To an extent, the development of highly sensitive and fast responsive CNT-based sensors addressed these backdrops. The chemically modified electrochemical sensors are found to be able to analyze compounds of weak oxidation/reduction signals. The high electrical conductivity of CNTs makes them the best conductive materials for the modification of electrochemical sensors. The electrical conductivity of CNT is a function of its diameter and carbon atom bonds in a hexagonal lattice, which is useful for fabricating electrochemical sensors of controlled electrical conductivity [76–79]. Unlike in graphite, carbon atoms in CNTs bind to the other three carbon atoms with sp^2 molecular orbitals. The remaining fourth valence electron can be hybridized to form π-bands, thus increasing the electrical conductivity of CNTs.

Surface functionalization is another important method for customizing CNTs for use in electrochemical sensors. There are several reports on the surface modification of CNTs with –COOH, $-NH_2$, –SH, and other active groups [80, 81]. And this will improve the coupling of CNTs with electroactive mediators like metal oxide nanoparticles or biomolecules [82, 83]. The sensitivity and selectivity of electrochemical sensors for the trace analysis of water pollutants can be enhanced in this manner. Karimi-Maleh et al. reported that CNTs modified with ZnO nanoparticles could improve the sensitivity of the resulting electrodes for the voltammetric determination of hydrazine in the presence of phenol, which in turn, are two important water pollutants [84].

10.6.1 PROCEDURE FOR THE MODIFICATION OF ELECTROCHEMICAL SENSORS WITH CNTS

Two approaches are mainly utilized for the modification of working electrodes with CNT:

i. In the case of solid electrodes such as glassy carbon and gold electrode, a thin film of CNT is coated on the electrode surface by the dropping method or by electro-polymerization of CNTs in polymer solutions (Figure 10.10(a)). Neflon, sulfuric acid, acetone, and DMF are the solvents used for preparing suspensions of CNTs [85–87].
ii. Second approach involves the addition of CNTs to the carbon paste matrix, which is then used as a conductive binder in the composite electrode material (Figure 10.10(b)) [88].

10.6.2 TYPES OF ELECTROCHEMICAL SENSORS

There are mainly three types of electrochemical sensors modified with CNTs that are used for the analysis of water pollutants viz: (i) potentiometric; (ii) voltammetric; and (iii) amperometric sensors. Potentiometric sensors are useful for the determination of toxic ions, while voltammetric and amperometric sensors are used for the determination of organic and inorganic pollutants.

Arguably, CNT-modified electrochemical sensors have developed as an effective tool for the determination of water pollutants.

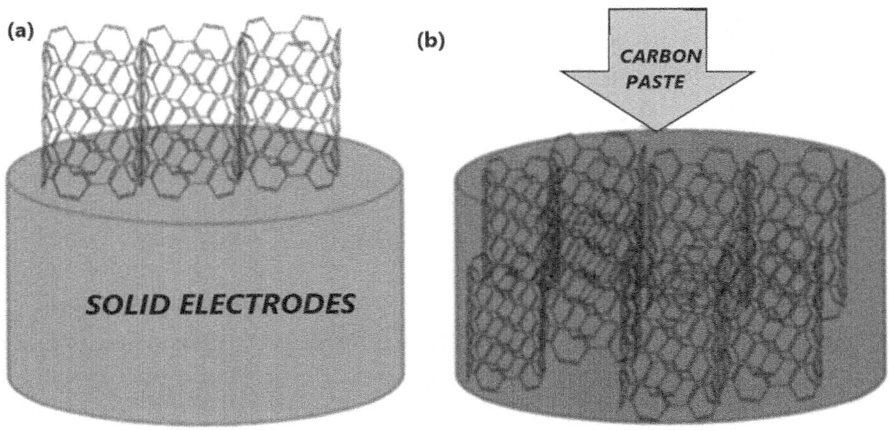

FIGURE 10.10 Diagram showing the process of modification of (a) solid electrodes; (b) carbon paste electrodes with CNTs.

10.7 CONCLUSION

The changes in the global scenario, as part of industrialization and urbanization have increased the scarcity of fresh and safe water. The need of fresh water for human sustainability has become a major concern worldwide. CNT-based water purification has emerged as the technology of the future, particularly in wastewater purification. CNT-based adsorption, catalysis, disinfection, desalination, as well as sensing of organic, inorganic, and biological pollutants has opened up a new era for wastewater treatment. Interesting physiochemical properties, low cost, and less impact on the environment are the driving forces that prompted researchers to exploit CNT-based water treatment. Further interdisciplinary research in this line, as well as timely decisions, will definitely help mankind to drink safe water and sustain life in this world.

KEYWORDS

- **adsorption**
- **carbon nanotubes**
- **catalysis**
- **desalination**
- **disinfection**
- **sensing and monitoring**

REFERENCES

1. Urban Urgency, (2007). *Water Caucus Summary*. World Water Council (WWC), Marseille, France.
2. Upadhyayula, V. K. K., Deng, S. G., Mitchell, M. C., & Smith, G. B., (2009). *Science of the Total Environment, 408*, 1–13.
3. Ren, X., Chen, C., Nagatsu, M., & Wang, X., (2011). *Chemical Engineering Journal*. 170 395–410.
4. Qu, X. L., Alvarez, P. J. J., & Li, Q. L., (2013). Applications of nanotechnology in water and wastewater treatment. *Water Res., 47*, 3931–3946.
5. Hosaain, F., Perales-Perez, O. J., Hwang, S., & Roman, F., (2014). Antimicrobial nanomaterials as water disinfectant: Applications, limitations and future perspectives. *Science of the Total Environment, 466*, 1047–1059.
6. Yu, J. G., Zhao, X. H., Yu, L. Y., Jiao, F. P., Jiang, J. G., & Chen, X. Q., (2014). Removal, recovery and enrichment of metals from aqueous solutions using carbon nanotubes, *J. Radioanal. Nucl. Chem., 299*, 1155–1163.
7. Mubarak, N., Sahu, J., Abdullah, E., & Jayakumar, N., (2014). Removal of heavy metals from wastewater using carbon nanotubes, *Sep. Purif. Rev., 43*, 311–338.
8. Yang, S., Hu, J., Chen, C., Shao, D., & Wang, X., (2011). Mutual effects of Pb (II) and humic acid adsorption on multiwalled carbon nanotubes/polyacrylamide composites from aqueous solutions. *Environ. Sci. Technol., 45*, 3621–3627.
9. Chi, W., Shi, H., Shi, W., Guo, Y., & Guo, T., (2012). 4-Nitrophenol surface molecularly imprinted polymers based on multiwalled carbon nanotubes for the elimination of paraoxon pollution. *J. Hazard. Mater., 227*, 243–249.
10. Peng, X., Li, Y., Luan, Z., Di, Z., Wang, H., Tian, B., & Jia, Z., (2003). *J. Chem. Phys. Lett., 736*, 154–158.
11. Ghaedi, M., Shokrollahi, A., Hossainian, H., & Kokhdan, S. N., (2011). *J. Chem. Eng. Data., 56*, 3227–3235.
12. Huaiyu, Z., Chumxiang, L., Jiajun, Z., Minghua, L., & Shiyu, F., (2010). Trends, *Carbohy. Res., 2*, 1–9.
13. Shahryari, Z., Goharrizi, A. S., & Azadi, M., (2010). *Int. J. Water Resour. Environ. Eng., 2*, 16–28.

14. Gong, J., Wang, B., Zeng, G., Yang, C., Niu, C., Niu, Q., Zhou, W., & Liang, Y., (2009). *J. Hazard. Mater., 164*, 1517–1722.
15. Das, R., Hamid, S. B. A., Ali, M., Annuar, M., Samsudin, E. M. B., & Bagheri, S., (2015). Covalent functionalization schemes for tailoring solubility of multi-walled carbon nanotubes in water and acetone solvents. *Sci. Adv. Mater., 7*, 2726–2737.
16. Ali, M., Das, R., Maamor, A., & Hamid, S. B. A., (2014). Multifunctional carbon nanotubes (CNTs): A new dimension in environmental remediation. *Adv. Mater. Res., 832*, 328–332.
17. Das, R., (2017). Nanobiohybrid preparation. in: *Nanohybrid Catalyst Based on Carbon Nanotube* (pp. 105–128). Springer.
18. Das, R., Hamid, S. B. A., & Annuar, M. S. M., (2016). Highly efficient and stable novel nanobiohybrid catalyst to avert 3, 4-dihydroxybenzoic acid pollutant in water. *Sci. Rep., 6*, 33572.
19. Srivastava, S., (2013). *Mater. Lett., 4*, 2–8.
20. Li, Y. H., Hung, T. H., & Chen, C. W., (2009). A first-principles study of nitrogen-and boron-assisted platinum adsorption on carbon nanotubes, *Carbon, 47*, 850–855.
21. Rivas, F., Kolaczkowski, S., Beltran, F., & McLurgh, D., (1998). Development of a model for the wet air oxidation of phenol-based on a free radical mechanism, *Chem. Eng. Sci., 53*, 2575–2586.
22. Czech, B., & Buda, W., (2015). Photocatalytic treatment of pharmaceutical wastewater using new multiwall-carbon nanotubes/TiO2/SiO2 nanocomposites. *Environ. Res., 137*(Supplement C), 176–184.
23. Da Dalt, S., Alves, A. K., & Bergmann, C. P., (2013). Photocatalytic degradation of methyl orange dye in water solutions in the presence of MWCNT/TiO_2 composites. *Mater. Res. Bull., 48*(5), 1845–1850.
24. Ananthakumar, S., Ramkumar, J., & Babu, S. M., (2016). Semiconductor nanoparticles sensitized TiO_2 nanotubes for high efficiency solar cell devices. *Renewable and Sustainable Energy Reviews, 57*(Supplement C), 1307–1321.
25. Kim, W., Tachikawa, T., Kim, H., Lakshminarasimhan, N., Murugan, P., Park, H., et al., (2014). Visible light photocatalytic activities of nitrogen and platinum-doped TiO_2: Synergistic effects of co-dopants. *Appl. Catal. B: Environ., 147*(Supplement C), 642–650.
26. Tanabe, I., & Ozaki, Y., (2014). Consistent changes in electronic states and photocatalytic activities of metal (Au, Pd, Pt)-modified TiO_2 studied by far-ultraviolet spectroscopy. *Chem. Commun., 50*(17), 2117–2119.
27. Di Paola, A., Garcia-Lopez, E., Marci, G., & Palmisano, L., (2012). A survey of photocatalytic materials for environmental remediation, *J. Hazard. Mater., 211*, 3–29.
28. Xu, H., Wang, C., Song, Y., Zhu, J., Xu, Y., Yan, J., et al., (2014). CNT/Ag_3PO_4 composites with highly enhanced visible light photocatalytic activity and stability. *Chem. Eng. J., 241*, 35–42.
29. Xia, J., Di, J., Yin, S., Li, H., Xu, L., Xu, Y., et al., (2014). Improved visible light photocatalytic activity of MWCNT/BiOBr composite synthesized via a reactable ionic liquid. *Ceram. Int., 40*(3), 4607–4616.
30. Hu, S., Li, S., Xu, K., Jiang, W., Zhang, J., & Liu, J., (2017). MWCNTs/BiOCOOH composites with improved sunlight photocatalytic activity. *Mater. Lett., 191*(Supplement C), 157–160.
31. Su, M., He, C., Zhu, L., Sun, Z., Shan, C., Zhang, Q., et al., (2012). Enhanced adsorption and photocatalytic activity of BiOI–MWCNT composites towards organic pollutants in aqueous solution. *J. Hazard. Mater., 229*, 72–82.

32. Dai, K., Li, D., Geng, L., Liang, C., & Liu, Q., (2015). Facile preparation of Bi_2MoO_6/ multi-walled carbon nanotube nanocomposite for enhancing photocatalytic performance. *Mater. Lett., 160 (Supplement C)*, 124–127.
33. Zhang, X., Li, S., Hu, S., Chen, J., Jiang, W., & Zhang, J., (2016). Flower-like MWCNTs/ $Bi_2O_2CO_3$ composites with enhanced photocatalytic activity under simulated solar light irradiation. *Mater. Lett., 185*, 50–53.
34. Yue, L., Wang, S., Shan, G., Wu, W., Qiang, L., & Zhu, L., (2015). Novel MWNTs–Bi_2WO_6composites with enhanced simulated solar photoactivity toward adsorbed and free tetracycline in water. *Appl. Catal. B, 176*, 11–19.
35. Di, J., Ji, M., Xia, J., Li, X., Fan, W., Zhang, Q., et al., (2016). $Bi_4O_5Br_2$ ultrasmall nanosheets in situ strong coupling to MWCNT and improved photocatalytic activity for tetracycline hydrochloride degradation. *J. Mol. Catal. A: Chem., 424*(Supplement C), 331–341.
36. Xu, Y., Xu, H., Wang, L., Yan, J., Li, H., Song, Y., et al., (2013). The CNT modified white C_3N_4 composite photocatalyst with enhanced visible-light response photoactivity. *Dalton Trans., 42*(21), 7604–7613.
37. Xia, Y., Li, Q., Wu, X., Lv, K., Tang, D., & Li, M., (2017). Facile synthesis of CNTs/ $CaIn_2S_4$ composites with enhanced visible-light photocatalytic performance. *Appl. Surf. Sci., 391*, 565–571.
38. Ye, A., Fan, W., Zhang, Q., Deng, W., & Wang, Y., (2012). CdS–graphene and CdS–CNT nanocomposites as visible-light photocatalysts for hydrogen evolution and organic dye degradation. *Catal. Sci. Technol., 2*(5), 969–978.
39. Zhang, X., Feng, M., Qu, R., Liu, H., Wang, L., & Wang, Z., (2016). Catalytic degradation of diethyl phthalate in aqueous solution by persulfate activated with nano-scaled magnetic $CuFe_2O_4$/ MWCNTs. *Chem. Eng. J., 301*, 1–11.
40. Peng, X., Sfeir, M. Y., Zhang, F., Misewich, J. A., & Wong, S. S., (2010). Covalent synthesis and optical characterization of double-walled carbon nanotube-nanocrystal heterostructures, *J. Phys. Chem. C, 114*, 8766–8773.
41. Yang, J., Jiang, L. C., Zhang, W. D., & Gunasekaran, S., (2010). A highly sensitive non-enzymatic glucose sensor based on a simple two-step electrodeposition of cupric oxide (CuO) nanoparticles onto a multi-walled carbon nanotube arrays. *Talanta, 82*, 25–33.
42. Vamvasakis, I., Georgaki, I., Vernardou, D., Kenanakis, G., & Katsarakis, N., (2015). Synthesis of WO3 catalytic powders: Evaluation of photocatalytic activity under NUV/ visible light irradiation and alkaline reaction pH. *J. Sol-Gel. Sci. Technol., 76*(1), 120–128.
43. Cui, X., Wang, Y., Jiang, G., Zhao, Z., Xu, C., Duan, A., et al., (2014). The encapsulation of CdS in carbon nanotubes for stable and efficient photocatalysis. *J. Mater. Chem. A., 2*(48), 20939–20946.
44. Tang, Y., Tian, J., Malkoske, T., Le, W., & Chen, B., (2017). Facile ultrasonic synthesis of novel zinc sulfide/carbon nanotube coaxial nanocables for enhanced photodegradation of methyl orange. *J. Mater. Sci., 52*(3), 1581–1589.
45. Zimmermann, F., (1958). New waste disposal process. *Chem. Eng., 65*(8), 117–121.
46. Yang, S., Zhu, W., Li, X., Wang, J., & Zhou, Y., (2007). Multi-walled carbon nanotubes (MWNTs) as an efficient catalyst for catalytic wet air oxidation of phenol. *Catal. Commun., 8*(12), 2059–2063.
47. Mestl, G., Maksimova, N. I., Keller, N., Roddatis, V. V., & Schlögl, R., (2001). Carbon nanofilaments in heterogeneous catalysis: An industrial application for new carbon materials? *Angew. Chem. Int. Ed., 40*(11), 2066–2068.

48. Nelson, D. L., Lehninger, A. L., & Cox, M. M., (2008). *Lehninger Principles of Biochemistry* (pp. 369–420). Macmillan.
49. Hwang, E. T., & Gu, M. B., (2013). Enzyme stabilization by nano/microsized hybrid materials. *Eng. Life. Sci., 13*, 49–61.
50. Gao, Y., & Kyratzis, I., (2008). Covalent immobilization of proteins on carbon nanotubes using the cross-linker 1-ethyl-3-(3-dimethylaminopropyl) carbodiimide-a critical assessment. *Bioconjug. Chem. 19*, 1945–1950.
51. Asuri, P., Karajanagi, S. S., Sellitto, E., Kim, D. Y., Kane, R. S., & Dordick, J. S., (2006). Water-soluble carbon nanotube-enzyme conjugates as functional biocatalytic formulations *Biotechnol. Bioeng., 95*, 804–811.
52. Jiang, K., Schadler, L. S., Siegel, R. W., Zhang, X., Zhang, H., & Terrones, M., (2004). Protein immobilization on carbon nanotubes via a two-step process of diimide-activated amidation. *J. Mater. Chem., 14*, 37–39.
53. Sheldon, R., & Van, P. S., (2013). Enzyme immobilization in biocatalysis: Why, what and how. *Cheml. Soc. Rev.*, 6224–6225.
54. Dai, Y., Yao, J., Song, Y., Liu, X., Wang, S., & Yuan, Y., (2016). Enhanced performance of immobilized laccase in electrospun fibrous membranes by carbon nanotubes modification and its application for bisphenol A removal from water. *J. Hazard. Mater., 317*(Supplement C), 485–493.
55. Vecitis, C. D., Schnoor, M. H., Rahaman, M. S., Schiffman, J. D., & Elimelech, M., (2011). Electrochemical multiwalled carbon nanotube filter for viral and bacterial removal and inactivation. *Environ. Sci. Technol., 45*, 3672–3679.
56. Upadhyayula, V. K., Deng, S., Mitchell, M. C., & Smith, G. B., (2009). Application of carbon nanotube technology for removal of contaminants in drinking water: A review. *Sci. Total Environ., 408*, 1–13.
57. Liu, H., Ru, J., Qu, J., Dai, R., Wang, Z., & Hu, C., (2009). Removal of persistent organic pollutants from micro-polluted drinking water by triolein embedded absorbent. *Bioresour. Technol., 100,* 2995–3002.
58. Karthikairaj, K., Isaiarasu, L., & Sakthivel, N., (2014). Efficacy of some herbal extracts on microbes causing flacherie disease in mulberry silkworm, *Bombyx mori* L. *J. Biopesticides, 7*, 89.
59. Viraka, N. B. P., Kanchanapally, R., Pedraza, F., Sinha, S. S., Pramanik, A., Hamme, A. T., et al., (2015). Bio-conjugated CNT-bridged 3D porous graphene oxide membrane for highly efficient disinfection of pathogenic bacteria and removal of toxic metals from water. *ACS Appl. Mater. Interfaces, 7*, 19210–19218.
60. Fan, X., Zhao, H., Quan, X., Liu, Y., & Chen, S., (2016). Nanocarbon-based membrane filtration integrated with electric field driving for effective membrane fouling mitigation. *Water Res., 88*, 285–292.
61. Maas, M., (2016). Carbon nanomaterials as antibacterial colloids. *Materials, 9*, 617.
62. Kang, S., Mauter, M. S., & Elimelech, M., (2008). Physicochemical determinants of multiwalled carbon nanotube bacterial cytotoxicity. *Environ. Sci. Technol., 42*, 7528–7534.
63. Rajavel, K., Gomathi, R., Manian, S., & Rajendra, K. R. T., (2014). In vitro bacterial cytotoxicity of CNTs: Reactive oxygen species mediate cell damage edges over direct physical puncturing. *Langmuir, 30*, 592–601.
64. Das, R., Ali, M. E., Abd, H. S. B., Ramakrishna, S., & Chowdhury, Z. Z., (2014). Carbon nanotube membranes for water purification: A bright future in water desalination, *Desalination, 336*, 97–109.

65. Ahm, C. H., Baek, Y., Lee, C., Kim, S. O., Lee, S., Kim, S. H., Bae, S. S., et al., (2012). Carbon nanotube-based membranes: Fabrication and application to desalination, *J. Ind. Eng. Chem., 18*, 1551–1559.
66. Dumee, L. F., Sears, K., Suhutz, J., Finn, N., Huynh, C., Hawkins, S., Duke, M., & Gray, S., (2010). Characterization and evaluation of carbon nanotube Bucky-paper membranes for direct contact membrane distillation, *J. Membr. Sci., 351*, 36–43.
67. Goh, P., Ismail, A., & Ng, B., (2013). Carbon nanotubes for desalination: Performance evaluation and current hurdles. *Desalination, 308*, 2–14.
68. Kar, S., Bindal, R., & Tewari, P., (2012). Carbon nanotubes membrane for desalination and water purification: Challenges and opportunities. *Nano Today, 7*, 385–389.
69. Mijangos, L., Ziarrusta, H., Olivares, M., Zuloaga, O., Möder, M., Etxebarria, N., & Prieto, A., (2018). Simultaneous determination of 41 multiclass organic pollutants in environmental waters by means of polyethersulfone microextraction followed by liquid chromatography-tandem mass spectrometry. *Anal. Bioanal. Chem., 410*(2), 615–632.
70. Mohamed, M. E., Frag, E. Y., & Mohamed, M. A., (2018). A newly validated and characterized spectrophotometric method for determination of a three water pollutants metal ions. *Spectrochim. Acta Part A Mol. Biomol. Spectrosc., 189*, 239–249.
71. Do Nascimento, M. T. L., De Oliveira, S. A. D., Felix, L. C., Gomes, G., e Sá, M. D. O., Da Cunha, D. L., Vieira, N., Hauser-Davis, R. A., Neto, J. A. B., & Bila, D. M., (2018). Determination of water quality, toxicity and estrogenic activity in a nearshore marine environment in Rio de Janeiro, Southeastern Brazil. *Ecotoxicol. Environ. Saf., 149*, 197–202.
72. Magi, E., Di Carro, M., Mirasole, C., & Benedetti, B., (2018). Combining passive sampling and tandem mass spectrometry for the determination of pharmaceuticals and other emerging pollutants in drinking water. *Microchem. J., 136*, 56–60.
73. Tankiewicz, M., & Biziuk, M., (2018). Fast, sensitive and reliable multi-residue method for routine determination of 34 pesticides from various chemical groups in water samples by using dispersive liquid-liquid microextraction coupled with gas chromatography-mass spectrometry. *Anal. Bioanal. Chem., 410*(5), 1533–1550.
74. Miyawaki, T., Tobiishi, K., Takenaka, S., & Kadokami, K., (2018). A rapid method combining microwave-assisted extraction and gas chromatography-mass spectrometry with a database, for determining organochlorine pesticides and polycyclic aromatic hydrocarbons in soils and sediments. *Soil Sedim. Contam. Int. J., 27*(1), 31–45.
75. Háková, M., Havlíková, L. C., Chvojka, J., Solich, P., & Šatínský, D., (2018). An on-line coupling of nanofibrous extraction with column-switching high-performance liquid chromatography–a case study on the determination of bisphenol A in environmental water samples. *Talanta, 178*, 141–146.
76. Ebbesen, T., Lezec, H., Hiura, H., Bennett, J., Ghaemi, H., & Thio, T., (1996). Electrical conductivity of individual carbon nanotubes. *Nature, 382*(6586), 54.
77. Lekawa-Raus, A., Patmore, J., Kurzepa, L., Bulmer, J., & Koziol, K., (2014). Electrical properties of carbon nanotube based fibers and their future use in electrical wiring. *Adv. Func. Mater., 24*(24), 3661–3682.
78. Dresselhaus, M. S., Dresselhaus, G., Eklund, P., & Rao, A., (2000). *Carbon Nanotubes, the Physics of Fullerene-Based and Fullerene-Related Materials* (pp. 331–379). Springer, Berlin.
79. Ramanathan, T., Fisher, F., Ruoff, R., & Brinson, L., (2005). Amino-functionalized carbon nanotubes for binding to polymers and biological systems. *Chem. Mater., 17*(6), 1290–1295.

80. Balasubramanian, K., & Burghard, M., (2005). Chemically functionalized carbon nanotubes. *Small, 1*(2), 180–192.
81. Zhao, Q., Gan, Z., & Zhuang, Q., (2002). Electrochemical sensors based on carbon nanotubes. *Electroanalysis, 14*(23), 1609–1613.
82. Ramnani, P., Saucedo, N. M., & Mulchandani, A., (2016). Carbon nanomaterial-based electrochemical biosensors for label-free sensing of environmental pollutants. *Chemosphere, 143*, 85–98.
83. Liu, G., Wang, S., Liu, J., & Song, D., (2012). An electrochemical immunosensor based on chemical assembly of vertically aligned carbon nanotubes on carbon substrates for direct detection of the pesticide endosulfan in environmental water. *Anal. Chem., 84*(9), 3921–3928.
84. Karimi-Maleh, H., Moazampour, M., Ensafi, A. A., Mallakpour, S., & Hatami, M., (2014). An electrochemical nanocomposite modified carbon paste electrode as a sensor for simultaneous determination of hydrazine and phenol in water and wastewater samples. *Environ. Sci. Pollut. Res., 21*(9), 5879–5888.
85. Musameh, M., Wang, J., Merkoci, A., & Lin, Y., (2002). Low-potential stable NADH detection at carbon-nanotube-modified glassy carbon electrodes. *Electrochem. Commun., 4*(10), 743–746.
86. Tsai, Y. C., Chen, J. M., Li, S. C., & Marken, F., (2004). Electroanalytical thin film electrodes based on a Na fion™–multi-walled carbon nanotube composite. *Electrochem. Commun., 6*(9), 917–922.
87. Wu, K., Hu, S., Fei, J., & Bai, W., (2003). Mercury-free simultaneous determination of cadmium and lead at a glassy carbon electrode modified with multi-wall carbon nanotubes. *Anal. Chim. Acta, 489*(2), 215–221.
88. Ensafi, A. A., & Karimi-Maleh, H., (2010). Modified multiwall carbon nanotubes paste electrode as a sensor for simultaneous determination of 6-thioguanine and folic acid using ferrocene dicarboxylic acid as a mediator. *J. Electroanal. Chem., 640*(1), 75–83.

CHAPTER 11

INVESTIGATION OF SELECTED ANTI-BACTERIALLY MODIFIED MICRO/NANO-POLYMERS USING COLD PLASMA

IGOR NOVÁK[1], ANTON POPELKA[2], AHMED ASANIDEZHAD[3], MARIÁN LEHOCKÝ[3], IVAN CHODÁK[1], JÁN SEDLIAČIK[4], JÁN MATYAŠOVSKÝ[5], PETER JURKOVIČ[5], KATARÍNA VALACHOVÁ[6], and LADISLAV ŠOLTÉS[6]

[1]*Polymer Institute, Slovak Academy of Sciences, Bratislava, Slovakia*

[2]*Center for Advanced Materials, Qatar University, Doha, Qatar*

[3]*Tomas Bata University in Zlín, T.G.M., Zlín, Czech Republic*

[4]*Technical University in Zvolen, T.G. Masaryka, Zvolen, Slovakia*

[5]*VIPO, A.S., Gen. Svobodu, Partizánske, Slovakia*

[6]*Centre of Experimental Medicine SAS, Institute of Experimental Pharmacology and Toxicology, Bratislava, Slovakia*

ABSTRACT

Medical-grade polymers, i.e., polyvinylchloride and low-density polyethylene (LDPE), were modified by antibacterial compounds through a multistep physicochemical approach. The polymer initial surface activation was performed foremost via diffuse coplanar surface barrier discharge (DCSBD)

Nanotechnology-Enhanced Solid Materials: Design, Synthesis, Properties, Applications, and Perspectives. Lionello Pogliani, Ann Rose Abraham, A. K. Haghi, & Prabhat Ranjan (Eds.)

plasma in air at ambient temperature and atmospheric pressure. Then, radical graft copolymerization of acrylic acid (AA) through grafting-from pathway was directed to render a well-defined brush of high density, and finally the selected antibacterial compounds were immobilized onto the functionalized polymer surfaces. This modification led to inhibition of *Escherichia coli* and *Staphylococcus aureus* bacterial strains adhesion to the polymer surface.

11.1 INTRODUCTION

Medical-grade polymers are widely used in many biomedical applications including, e.g., the production of catheters for coronary angioplasty [1–9]. However, infections may cause implant failure or removal, complex revision processes, which can lead to patient suffering, prolonged hospitalization, and in some cases, even death. Biocompatibility depends on the many surface characteristics such as wettability, roughness, chemistry, surface charge, and density of functional groups. The presence of hydrophobic and hydrophilic domains, charge, the functional group densities, and their conformation play ascendant roles in affecting cell behavior. Many problems can be eliminated by surface modification of polymers using an appropriate method as well as cold plasma when surface free energy is increasing due to the introduction of polar functional groups on treated surface thereby the surface of polymers becomes more hydrophilic. The development of discharge plasma techniques belong to important surface modifications of polymers when a very thin layer of polymer surface is treated without various changes. Processing of discharge plasma can significantly enhance adhesion by removing surface contamination and have impact on surface morphology changes through increased roughness due to etching. Anti-infective properties of polymers can be achieved as follows: (a) anti-infection agent added into the polymer; (b) copolymerization of an anti-infection agent with monomer; (c) surface treatment of medical polymer. Antibacterial surface modification is controlled by physicochemical interactions between bacteria and polymer surface. This treatment has several advantages due to no impact on the volume of the polymer, no release of antibacterial agents from the polymer, and relative simplicity and effectivity of this technique.

It is eligible that surface coatings such as hospital furniture, kitchen worktops, wood-based floorings, and many other surfaces used in public areas possess antibacterial properties. Cold plasma can be used on various surfaces where their self-cleaning and antibacterial ability is very important. A multistep physicochemical process allowing the use of plasma technology combined with wet chemistry, which is considerable interesting due to delivery of surface-active anti-adherence materials [10]. In the first step of the approach, concerning an inherent lack of functional groups on a pristine substrate, plasma treatment at low temperature and atmospheric pressure has been substantiated to be productive to yield reactive entities on the surface [11–15]. Low temperature atmospheric pressure plasma is put into the spotlight by virtue of rapid, efficient in-line processing possibility without costly vacuum apparatus [6]. However, the necessity to shorten the duration of the treatment to a few seconds remains a pressing obstacle to extensive applications of this type of plasma [16–19]. To eliminate the underscored drawbacks, a novel technology coined as diffuse coplanar surface barrier discharge (DCSBD) has been devised [20], which enables the generation of a uniform plasma layer under atmospheric pressure with a high surface power density in the very close contact of treated specimen. DCSBD is established on surface barrier discharge with a high density of thin discharge channels generated on a dielectric surface in parallel with the sample surface. The plasma is in a tight contact with the sample thereof [20, 21], which appreciably shortens the treatment time and lessens energy consumption [21–26]. In the second phase of the approach, an end-functionalized polymer brush is synthesized on the surface via surface-initiated polymerization. Surface-initiated polymerization encourages the formation of surface-confined thick brush layers of high grafting density on the surface using generation of appropriate initiators anchored to substrate in which monomers are able to easily make their way through already-grafted layer and contribute to the chain growth [27–29].

11.2 EXPERIMENTAL PART

11.2.1 MATERIALS

PVC pellets, extrusion medical-grade RB1/T3M of 1.25 $g \cdot cm^{-3}$ density, were obtained from ModernPlast (Italy) and used as received. Low-density polyethylene (LDPE) Bralen FB 2–17: Slovnaft MOL (Slovakia), containing no additives, the thickness of LDPE film was 20 μm, density = 0.918 $g \cdot cm^{-3}$, mass flow rate, MFR (190°C, 2.16 kg) = 2 g per 10 min, Vicat softening temperature = 96°C. This type of LDPE is suitable for food contact. The product complies with Food Contact Regulations and the grade is suitable for manufacturing of pharmaceutical packing-product. Pectin from apple (BioChemika, with esterification of 70–75%), acrylic acid (99.0%, anhydrous), and *N*-(3-dimethyl aminopropyl)-*N'*-ethyl carbodiimide hydrochloride (EDAC, 98.0%) were supplied by Fluka (USA). Chitosan from crab shells with medium molecular weight and deacetylation degree of 75–85%. Triclosan (5-chloro-2-(2,4-dichlorophenoxy)phenol): Irgasan, $C_{12}H_7Cl_3O_2$, Fluka Analytical (Italy), white powder, assay ≥ 97.0% (HPLC), M_r = 289.54 g·mol^{-1}, ash ≤ 0.1%, melting temperature = 56°C. Chlorhexidine (N,'N"'-hexane-1,6-diylbis[N-(4-chloropheny:): imidodicarbonimidic diamide, $C_{22}H_{30}Cl_2N_{10}$, Aldrich Chemistry (Spain), white powder, assay = 98%, M_r = 505.46 g·mol^{-1}, melting temperature = 134°C. Acrylic acid (propenoic acid): $C_3H_4O_2$, colorless liquid, Acros Organics (Belgium), assay = 99.5%, extra pure, stabilized by 180 to 220 pm monomethyl ether of hydroquinone (MEHQ), M_r = 72.06 g·mol^{-1}, flash temperature = 48°C, density = 1.050 g·cm^{-3}, boiling temperature = 139°C.

11.2.2 PLASMA TREATMENT

Plasma treatment was implemented in static conditions by DCSBD technology (Scheme 11.1) of laboratory scale with air as the gaseous medium at atmospheric pressure and room temperature. A schematic profile of the plasma system is illustrated in Scheme 11.1. It basically comprises a series of parallel metallic electrodes inset inside a ceramic dielectric located in a glass chamber which allows the carrier gases to flow. All samples were treated on both sides with plasma power of 200 W for 15 sec.

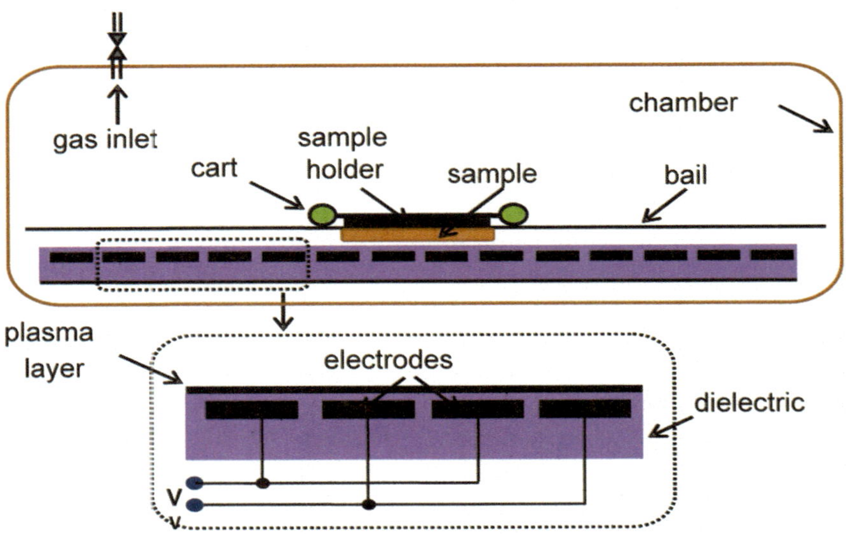

SCHEME 11.1 Scheme of device for DSBD plasma treatment.

11.2.3 MULTISTEP MODIFICATION TECHNIQUE

PVC and/or LDPE substrates, upon treatment with plasma, were immersed into spacer solutions containing 10 vol.% acrylic acid (AA) in aqueous solution. The reaction was allowed to proceed for 24 h at 30°C. PAA grafted polymer samples were immersed into EDAC aqueous solution at 4°C for 6 h in order to activate the carboxyl groups on the surface. The highly active key intermediate, *O*-acylisourea, is produced having the potential to react with reducing agents. Subsequently, they were transferred to chitosan and stored for 24 h at 30°C. This way pre-prepared LDPE sample was then immersed into the solution of triclosan and chlorhexidine (Scheme 11.2).

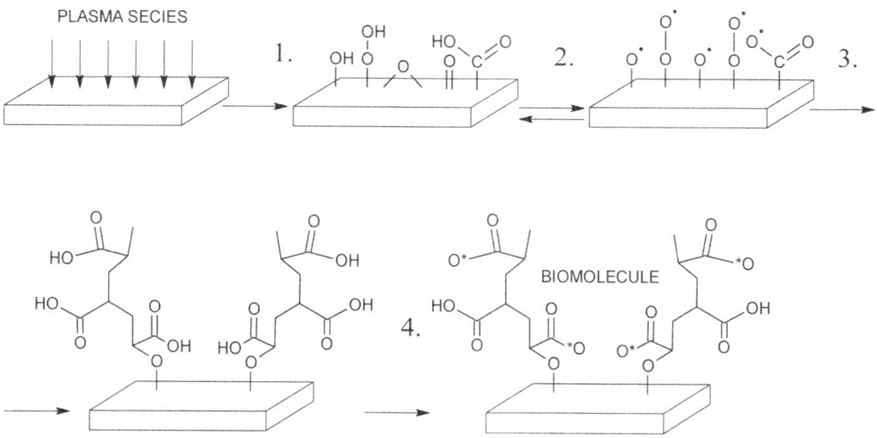

SCHEME 11.2 Multistep approach of biomolecules binding: (1) Plasma treatment, (2) Generation radicals, (3) AA radical graft polymerization, (4) Antibacterial deposition.

The wettability of LDPE treated by multistep process via PAA plasma grafted and antibacterial immobilization were carried out by the measurement of contact angle using sessile drop technique via surface energy evaluation system (SEE system with CCD camera, Advex Instruments, Czech Republic). As testing liquids deionized water, ethylene glycol (EG), glycerol, formamide, diiodomethane were used at volume 5 μl (elimination of influence of gravity) and a static contact angle was captured shortly after its creation when thermodynamic equilibrium is reached among the three phases: solid, liquid, and gas. Surface energy (γ^{tot}), its polar acid-base (γ^{AB}), dispersive (γ^{LW}), electron-acceptor (γ^{-}) and electron-donor (γ^{+}) components were calculated by Acid-Base regression model using the method of least squares.

11.2.4 SEM

Scanning electron microscopy (SEM) was carried out on VEGA II LMU (TESCAN) operating in the high vacuum/secondary electron imaging mode at an accelerating voltage of 5–20 kV.

11.2.5 ADHESIVE PROPERTIES

The adhesive properties, namely peel strength (force per unit width) of adhesive joint of antibacterially-treated LDPE and immobilized by antibacterial compounds via DCSBD to poly-(2-ethylhexyl acrylate) deposited onto polypropylene foil (15 mm width), were carried out by measurements of 90° peel test at a rate of peel 10 mm per minute using 100 N universal Instron 4,301 dynamometer (England). Ends of the polymer film were firmly fixed into the jaws of dynamometer so that the tension was evenly distributed across the entire width of the surface.

11.2.6 FTIR

Attenuated total reflectance FTIR measurements were performed by FTIR NICOLET 8,700 spectrometer (thermo-scientific) through the single bounce ATR accessory equipped with Ge crystal at an angle of incidence 45°. For each measurement, the spectral resolution and the number of scans were 2 cm^{-1} and 64, respectively.

11.2.7 BACTERIAL ADHESION AND BIOFILM TESTS

Bacterial adhesion and biofilm experiments were performed using gram-positive (*S. aureus* 3953) and gram-negative (*E. coli* 3954) bacteria. The circular shape specimens ($d \approx 8$ mm) were cut from the polymer before further investigation. After 24 h incubation at 37°C under continuous shaking at 100 rpm, the bacteria adhered on the surface of the specimens were removed by vigorous shaking of the test tube at 2,000 rpm for 30 s and quantified by serial dilutions and spread plate technique.

11.3 RESULT AND DISCUSSION

11.3.1 MEDICAL-GRADED POLY(VINYL CHLORIDE) (PVC)

11.3.1.1 SURFACE WETTABILITY

A highly surface sensitive technique is contact angle analysis which enables a convenient assessment of the surface wettability. Table 11.1 includes the contact angle values of deionized water (θ_w) recorded on different PVC samples (numbered from 1 to 5). Based on the given data, Sample 1 exhibits hydrophobic character with an evident change in θ_w and hydrophilicity ascends is anticipated after being treated by plasma. This trend continues up to Sample 3 on which polyacrylic acid (PAA) chains are grafted where more hydrophilic propensity is inferred from θ_w value. The elevated hydrophilicity upon multistep modifications is assumed to originate from the inclusion of superficial hydrophilic entities. Hydrophilicity then decreases as polysaccharides are coated onto the surface, though it is higher than that of Sample 1, as the inherent hydrophilicity of chitosan is beyond doubt. Furthermore, Sample 5 exhibits higher wettability than Sample 4 implying a more effective binding of chitosan onto the surface, as remarked in other efforts as well.

TABLE 11.1 Surface Properties of Poly(Vinyl Chloride) Treated by Multistep Process*

Sample	θ_w	θ_e	θ_d	θ_f	γ^-	γ^+	γ^{AB}	γ^{LW}	γ^{tot}
1	85.9 (±2.5)	60.5 (±3.0)	43.5 (±3.5)	64.2 (±6.0)	5.1	0.0	1.0	37.8	38.8
2	64.9 (±3.0)	49.4 (±4.0)	36.2 (±5.5)	51.0 (±6.0)	6.6	0.1	1.1	41.4	42.6
3	46.5 (±4.0)	32.1 (±2.4)	38.0 (±5.0)	47.7 (±4.5)	10.4	0.5	4.5	43.7	48.1
4	63.7 (±5.5)	43.4 (±3.0)	28.2 (±2.5)	44.9 (±5.0)	2.4	0.7	2.5	44.0	46.5
5	50.5 (±3.5)	40.0 (±2.5)	31.5 (±4.5)	31.0 (±3.5)	5.2	0.2	2.0	44.4	46.4

Surface free energies: γ^-, γ^+, γ^{AB}, γ^{LW}, γ^{tot} in mJ/m^2; contact angle θ in °.

w = deionized water, e = ethylene glycol, d = diiodomethane, f = formamide.

**Sample 1*: pristine LDPE/control; Sample 2: plasma-treated; Sample 3: PAA grafted; Sample 4: chitosan immobilized; Sample 5: chitosan/pectin immobilized (mean±standard deviation).

To further explore the physicochemical parameters of the examined surfaces, an extensively used theory, Lifshitz-van der Waals/acid-base (LW/

AB), has been exploited for free surface energy evaluation whose outputs with reference to diiodomethane, EG, and deionized water as wetting liquids are supplied in Table 11.1. Sample 1 exhibits a basic character ($\gamma^{-}>\gamma^{+}$) as proposed by the data, even though the acidity or basicity of neat PVC is yet controversial. Subsequent to the plasma treatment, the total surface free energy (γ^{tot}) is raised in terms of the contact angle values. This increase is principally assisted by the polar (acid-base) component (γ^{AB}), rather than the apolar one (γ^{LW}), implying an incorporation of superficial polar oxygen-containing entities thanks to the air plasma treatment. A significant rise in γ^{tot} and γ^{AB} values is noticed for Sample 3, in comparison with samples 1 and 2, indicative of the presence of carboxyl-containing units on the surface. As for samples 4 and 5, a reduction in γ^{AB} and γ^{tot} values is observed compared to Sample 3, however, their γ^{tot} values rise above that of Sample 1. The minimum values of θ_E and θ_F are found for Sample 5 which reflect that the surface is seemingly coated by alcoholic and amine containing moieties which in fact points to the more efficient binding of chitosan when compared to Sample 4.

11.3.1.2 SURFACE MORPHOLOGY

The surface topography of samples 1–5, investigated by SEM as a common surface qualitative technique, is presented in Figure 11.1. Sample 1 shows uniform morphology which goes through a significant alteration ensuing the plasma treatment taking on an etched pattern with an unevenly shaped texture. The generated morphology is advantageous for the next coupling processes due to an enhanced surface area and roughness. The developed pattern on Sample 2 is indeed an outcome of competing functionalization and ablation phenomena which brings on reorganization of the surface microstructure.

The incident of the ablation is validated by gravimetric analysis where a weight loss of 4 $\mu g \cdot cm^{-2}$ was observed due to the plasma treatment for 15 s implying an approximate etching rate of 2 nm/s in terms of the used PVC grade density. Based on micrograph from Sample 3, PAA chains develop superficial domains of submicron dimension and brush-like features are then recognizable on the surface. As the grafting moves forward, clustering takes place because of the domains size growth. An additional compelling factor in controlling the surface microstructure is the grafting mechanism which is actually initiated by generated surface radicals.

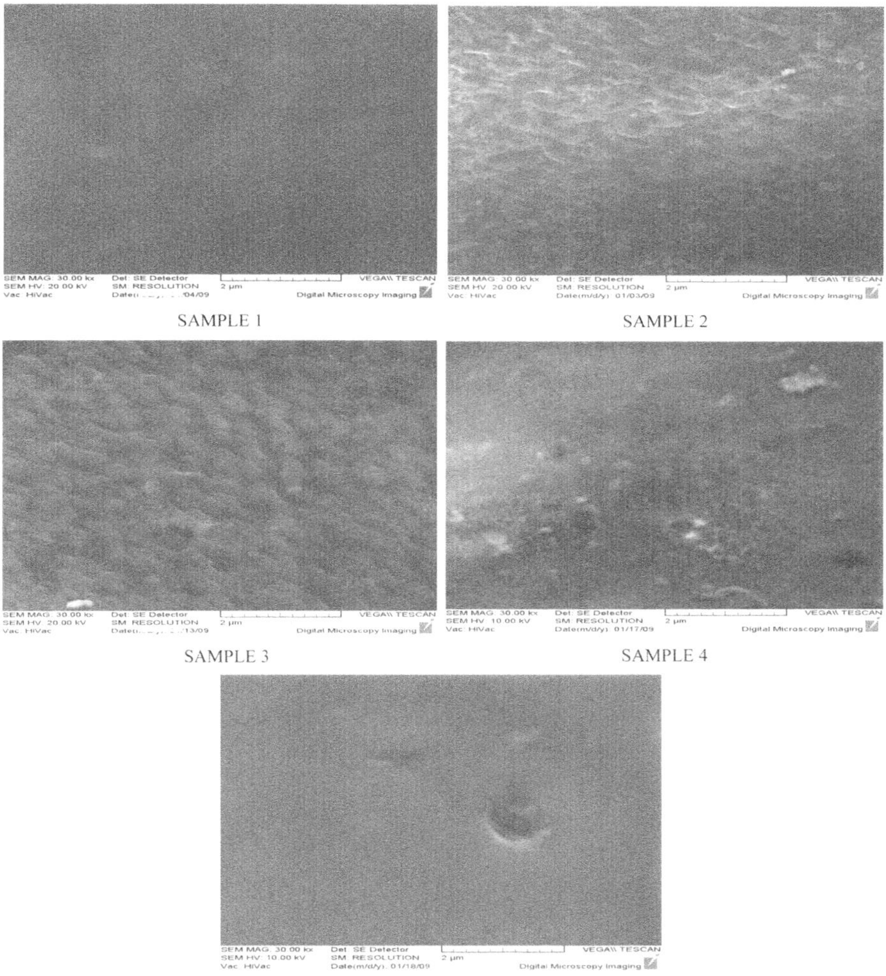

FIGURE 11.1 SEM micrographs of PVC samples 1–5 taken at 30,000 times magnification.

11.3.1.3 FTIR

FTIR is extensively utilized as a surface analysis technique to provide semi-quantitative information on the chemistry of the near-surface region. The mean probe depth of FTIR equipped with ZnSe crystal for PVC of refractive index measures 1.5 lies on the order of 4 μm which goes beyond the regular thickness of modified layers on a substrate. Figure 11.2 displays the infrared spectra of samples 1–5 separated over three wavenumber ranges for clarity purpose. In the spectrum of Sample 1, one can readily identify the characteristic bands of ester, and carbonyl-containing groups, besides neat PVC typical signals. This shows the existence of some additives in the current medical-grade PVC. After exposing the samples to plasma, no pronounced change is detected regarding the spectrum of Sample 2, compared to Sample 1. This happens due to overlapping of signals, which leads to disguising of the plasma effects, and also due to the limited depth of plasma modification to outermost layers, which thus cannot be well explored by FTIR. Nevertheless, a new absorption of weak strength comes into view around 1,630 cm^{-1} which relates to C=C stretching vibration as a result of the dehydrochlorination process.

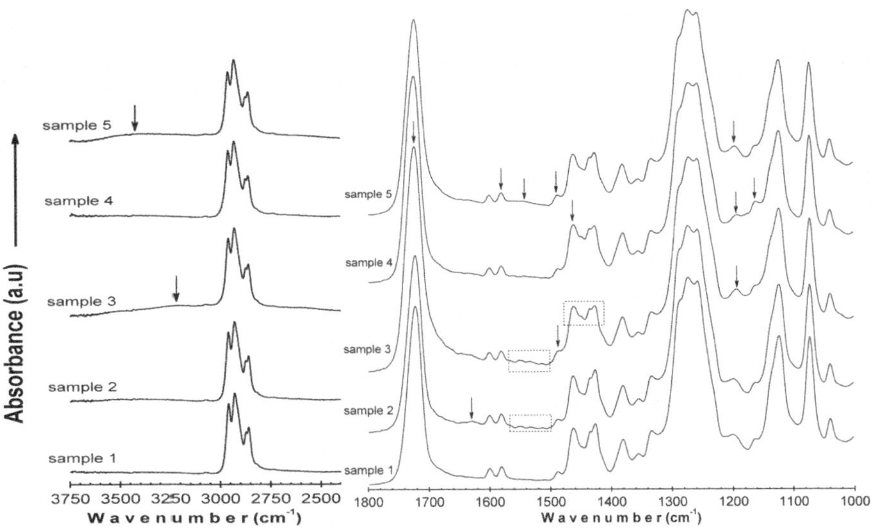

FIGURE 11.2 FTIR spectra of PVC samples 1–5 with different range of wavenumbers.

11.3.1.4 X-RAY PHOTOELECTRON SPECTROSCOPY (XPS) ANALYSIS

X-ray photoelectron spectroscopy (XPS), with a probe depth measuring around 5 nm, was selected to monitor thoroughly the bearings of the surface modifications by picking up a quantitative perception into the surface elemental composition. The survey spectra along with the corresponding surface atomic compositions and ratios of samples 1–5 are reported in Figure 11.3. Carbon (C), oxygen (O), chlorine (Cl), and silicon (Si) elements are found on the Sample 1 surface whose composition and elemental ratios are presented in the legend of the respective graph. The Cl2p atomic content is substantially lower than the amount found for neat PVC containing no additives which refers to the existence of several additives and also X-ray degradation. The same rationale accounts for the considerable amount of O1s detected in Sample 1 which is not a typical element in standard PVC. XPS also recognizes some trivial quantity of silicon which may be associated with the production/molding process as a contaminant. An ostensible change in C1s, O1s, and Cl2p core-level signals intensity arises for Sample 2 quantitatively expressed in the legend of the graph. It is obvious that the Cl2p intensity abates while that of O1s noticeably increases which alludes to the introduction of superficial oxygenous entities together with dehydrochlorination as a result of the ablation. Nitrogen (N1s) is also found on the surface because the plasma treatment was implemented in an air medium. The O/C ratio increases while Cl/C declines when compared to those of Sample 1. The amount of Si2p on the Sample 2 surface increases supposedly coming from plasma parts as a pollutant.

Upon binding chitosan on the surface (Sample 4), pronounced changes appear in the surface chemistry, as O1s content and O/C fraction increase and also N1s signal emerges, while Cl2p and Si2p bands abate due to the surface coverage by polysaccharide species. This trend yet continues for Sample 5 as higher O1s and N1s as well as O/C and N/C atomic rations are detectable compared to Sample 4 giving support to the notion that chitosan can be more stable, i.e., in higher quantity, attached onto the surface when layered along with pectin. In other words, the use of pectin can promote the ability of chitosan binding.

11.3.1.5 BACTERIAL ADHESION AND BIOFILM ASSAY

The most crucial step of the biofilm formation is bacterial adhesion, considered as a sophisticated topic in biointerface science whose plenty of aspects have not yet been well conceived. As a matter of fact, adhesion phenomenon is an interplay of myriad factors. Figure 11.4 shows the histograms of bacterial adhesion extent for samples 1–5 after 24 h incubation. As regards the adherence degree of *S. aureus* onto the samples 2–4, no reduction is evident in the number of viable adhered colonies, compared to Sample 1, signifying an inability of the modifications in hampering the *S. aureus* adhesion to the surface. From samples 1–3, both hydrophilicity and roughness rise, as remarked earlier, and then decrease in samples 4 and 5.

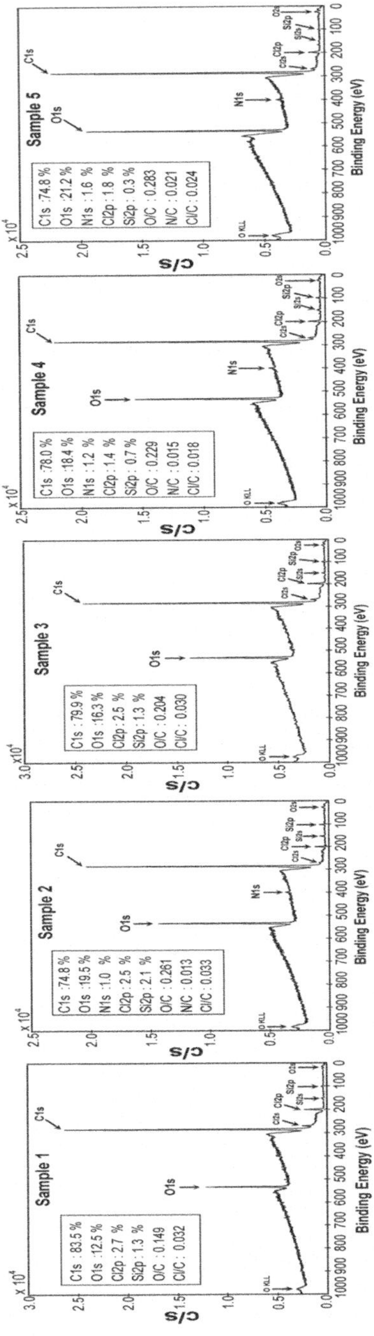

FIGURE 11.3 XPS survey-scan spectra of PVC samples 15 along with atomic compositions.

The adhesion degrees vary with a similar trend as well. Considering Sample 5, it is inferred that chitosan/pectin assembly imparts biocidal effects against *S. aureus*. Chitosan single layer and chitosan/pectin multilayer restrain the adherence degree by 50% and 20%, respectively. Chitosan/pectin multilayer is found to be effective against both gram-positive and gram-negative strains which can be translated as a higher quality of chitosan coating when it is applied along with pectin.

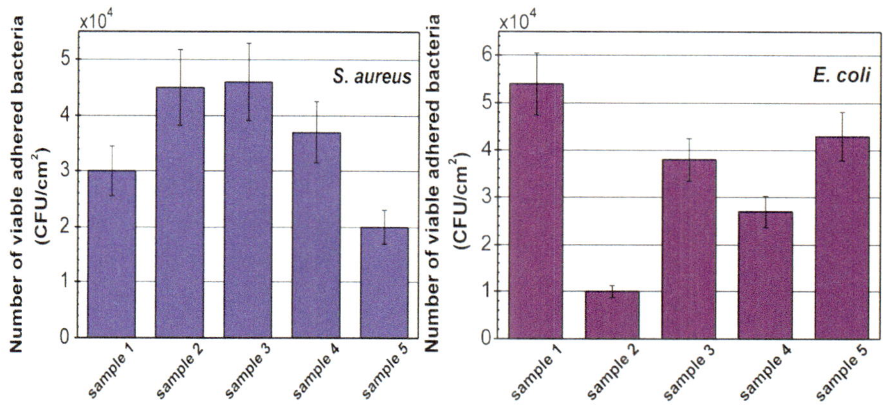

FIGURE 11.4 Histograms of bacterial adhesion degree for PVC samples 1–5 after 24 h incubation against two strains of microorganisms.

11.3.1.6 CONCLUSION (PVC)

These research results highlight the functionality of the adopted multistep physicochemical approach to bind polysaccharide species onto the medical-grade PVC surface. DCSBD plasma is capable of raising roughness, surface energy, and introducing oxygen-containing functionalities anchored onto the surface. A structured PAA brush of high graft density is synthesized using surface-initiated approach to further improve hydrophilicity and develop a stable brush-like assembly to yield a platform for biomolecular binding. *In vitro* bacterial adhesion and biofilm formation assays indicate incapability of single chitosan layer in hindering the adhesion of *S. aureus* bacterial strain, while up to 30% reduction is achieved by chitosan/pectin layered assembly. On the other hand, chitosan and chitosan/pectin multilayer could retard *E. coli* adhesion by 50%, and 20%, respectively. Furthermore, plasma treated and graft copolymerized samples are also found effective to diminish the adherence degree of *E. coli*.

11.4 LOW-DENSITY POLYETHYLENE (LDPE)

11.4.1 SURFACE WETTABILITY

Wetting can be defined by the degree of a wetted solid. When a drop is totally spread on a solid surface, and the contact angle approaches 0 deg, then the surface is completely wet. However, in many cases, it is only a partial wettability (or non-wettability). The extent of wettability of a solid surface can be evaluated by contact angle measurements. Wetting can be expressed as relative strength of cohesion (liquid/liquid) and adhesion (solid/liquid) forces. Weak cohesion with strong adhesion due to the very low contact angle approaches the full wettability. When solid/liquid interactions decrease and liquid/liquid interactions increase, the wettability decreases. A drop with a large contact angle about 90 deg is hydrophobic, which is associated with poor wettability, poor adhesion, and low surface free energy of solid. On the other side, a drop with a small contact angle is more hydrophilic, which reflected the higher wettability, higher adhesion, and higher surface energy of an investigated material. Changes in contact angles of the testing liquid set, graft yield (GY), surface free energy (γ^{tot}), and its components of anti-bacterially treated LDPE are shown in Table 11.2. The GY was calculated by the following equation:

$$GY[\%] = \left(\frac{W_2 - W_1}{W_1} \right) \times 100\%$$

where; W_1 and W_2 represent the weight of the samples before and after surface treatment. The water contact angle (θ_w) of untreated LDPE (Sample 1) achieves the highest values from all samples because it is polymer with hydrophobic and chemical inert surface. The parameter θ_w significantly decreased after exposition of plasma to Sample 1 when different functional groups in plasma species were introduced and therefore this treatment surface acquired more polar or hydrophilic character. The most decreased contact angle was observed in PAA (Sample 3) which relates to its hydrophilic character. To investigate other physicochemical parameters of the treated surface LW/AB theory was used whereby γ^{tot} and non-polar LW (γ^{LW}) and polar AB (γ^{AB}) components can be obtained. LW indicates the total dispersive Lifhitz-Van der Walls interaction and AB refers to the acid-base or electron-acceptor/electron donor interaction according to Lewis. LDPE belongs to group of low-energy polymeric materials and therefore γ^{tot} of Sample 1 achieves very low values which related to its worse further processing, such as dyeing, printing, and bonding (low adhesion). This lack can be removed by plasma treatment of LDPE when γ^{tot} can significantly increase as in Sample 2. Compared with other samples, the highest increase of γ^{tot} and γ^{AB} was shown in Sample 3 which had high polarity due to the presence of polar oxygen groups. Samples 4 and 5 showed a similar increase in surface free energy values thereby confirming an increase in wettability.

TABLE 11.2 Surface Properties of Low-Density Polyethylene Treated by Multistep Modification Process*

Sample	θ_w	θ_e	θ_d	θ_f	γ^-	γ^+	γ^{AB}	γ^{LW}	γ^{tot}	GY(%)
1	99.2 (±0.6)	70.9 (±1.2)	48.4 (±1.2)	80.7 (±0.9)	1.0	0.1	0.7	34.5	35.2	-
2	77.53 (±1.1)	51.0 (±2.8)	36.0 (±1.2)	52.8 (±1.5)	6.6	0.1	1.1	41.4	42.6	0.0
3	66.9 (±0.7)	32.1 (±2.4)	32.5 (±1.6)	37.0 (±2.0)	10.4	0.5	4.5	43.7	48.1	0.5
4	79.9 (±1.5)	36.1 (±0.7)	30.5 (±1.5)	48.3 (±1.2)	2.4	0.7	2.5	44.0	46.5	1.8
5	76.7 (±0.5)	38.1 (±2.5)	30.0 (±1.6)	50.4 (±1.5)	5.2	0.2	2.0	44.4	46.4	2.0

Surface free energies: γ^-, γ^+, γ^{AB}, γ^{LW}, γ^{tot} in mJ/m^2; contact angle θ in °.

w=deionized water, e=ethylene glycol, d=diiodomethane, f=formamide.

**Sample 1*: untreated LDPE; Sample 2: plasma-treated; Sample 3: Poly(acrylic acid) grafted; Sample 4: triclosan immobilized; Sample 5: chlorhexidine immobilized (mean ± standard deviation).

11.4.2 ADHESIVE PROPERTIES

The results of measurements peel strength of adhesive joint LDPE–poly(acrylate) is shown in Figure 11.5. Surface energy changes are closely related to adhesion between two materials in contact. Therefore, the increased wettability has resulted in an increase in adhesion strength of adhesive joint to more polar poly(acrylate). The adhesive properties depend on chemical composition as well as on surface morphology (roughness), which allows rougher surface, and higher adhesion. Adhesion is thus a complex of several related chemical and physicochemical properties. Therefore, in Sample 3 even though the surface energy reaches its highest value, the peel strength is lower than in samples 4 and 5. Crosslinking occurring in Sample 5 (via glutaraldehyde) was another factor that contributed to the increase in the adhesion strength.

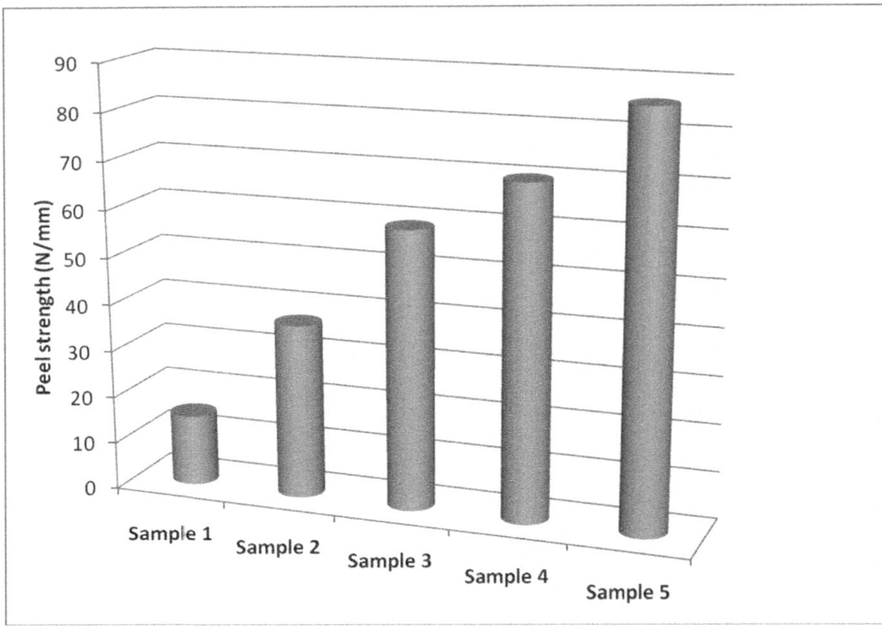

FIGURE 11.5 LDPE peel strength vs. surface treatment.

11.4.3 SURFACE CHEMISTRY

The FTIR measurements provide mostly semi-quantitative information on the chemical changes of the near-surface region because of the average sampling depth (depending on refractive indices of polymers, approximately 4 μm. For better visualization, the infrared spectra of pure LDPE and the modified material together with the pure triclosan were splitted into three different wavenumber regions. The significant changes in spectra are observed also in (both) cases of modification of LDPE-PAA grafting and after the subsequent triclosan coating. On the spectrum of grafted material one can observe some characteristic peaks of PAA, i.e., the most intense peak at 1,712 cm^{-1} (carbonyl band, C=O stretching) and also some unresolved peaks in the fingerprint region (1,300–1,100 cm^{-1}, C-O stretching and CH_2 bending). After triclosan treatment, the shape of the spectrum changes, as can be seen in Figure 11.6. These changes are significant almost in a whole mid-infrared region, especially in the region below 1,700 cm^{-1}. Because of the simple spectrum of LDPE (small numbers of peaks) in comparison with the spectrum of triclosan, we can assume that almost all changes in the spectrum of the triclosan immobilized LDPE originate from triclosan. The presence of triclosan in the treated sample is confirmed by an appearance of a number of peaks which are also present in the spectrum of the pure triclosan (e.g., at 1,491 cm^{-1} (benzene ring vibration), and also undoubtedly by a peak at 752 cm^{-1} (stretching mode of C-Cl in triclosan molecule). The shifts in the maxima of individual peaks and the changes in their shapes can probably be assigned to the bonding of triclosan molecules to the LDPE surface and inhibition of their unrestricted motion.

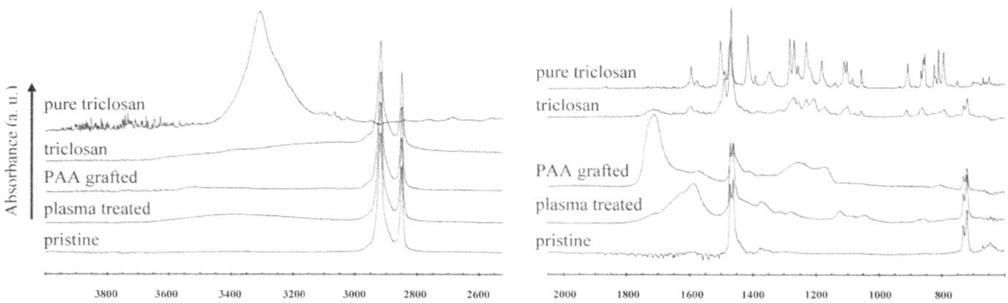

FIGURE 11.6 FTIR spectra LDPE treated by plasma, grafted by AA, and immobilized by triclosan.

The more significant changes in spectra are observed in (both) cases of modification of LDPE-PAA grafting and subsequently immobilized by chlorhexidine. After chlorhexidine treatment, the shape of the spectrum changes, as can be seen in Figure 11.7. These changes are significant almost in a whole mid-infrared region, especially in the region below 1,700 cm^{-1}. The presence of chlorhexidine is confirmed by an appearance of a peak at 1,640 cm^{-1} (C=N vibration) and also undoubtedly by a peak at 1,530 cm^{-1} (stretching mode of aromatic ring in chlorhexidine molecule).

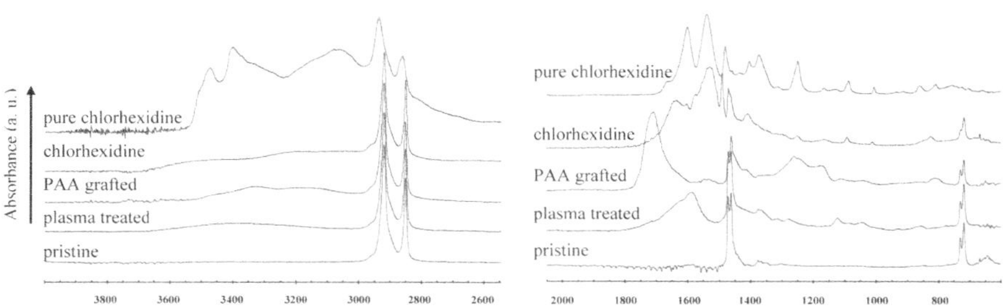

FIGURE 11.7 FTIR spectra LDPE treated by plasma, grafted by AA, and immobilized by chlorhexidine.

11.4.4 ANTIBACTERIAL ACTIVITY

Inhibition zone area was calculated from the average diameter of inhibition zone, whereas the area of the sample was not taken into account. Each experiment was performed in triplicate (Figure 11.8). The untreated sample, plasma treated sample and sample grafted by PAA with chitosan together with glutaraldehyde did not show any antibacterial activity against *E. coli* and *S. aureus* strains. Chitosan/pectin coated sample showed minor activity only against *S. aureus,* their inhibition zone moves around 70 mm^2. Similar results were gained for chitosan/pectin coated with glutaraldehyde sample. This sample also showed activity against *E. coli*. However, the antibacterial activity of these samples is not significant. The highest and most clear inhibition zones were given by samples grafted by PAA and coated by chitosan. Their levels were on average 35 mm^2 for *E. coli* and 275 mm^2 for *S. aureus*.

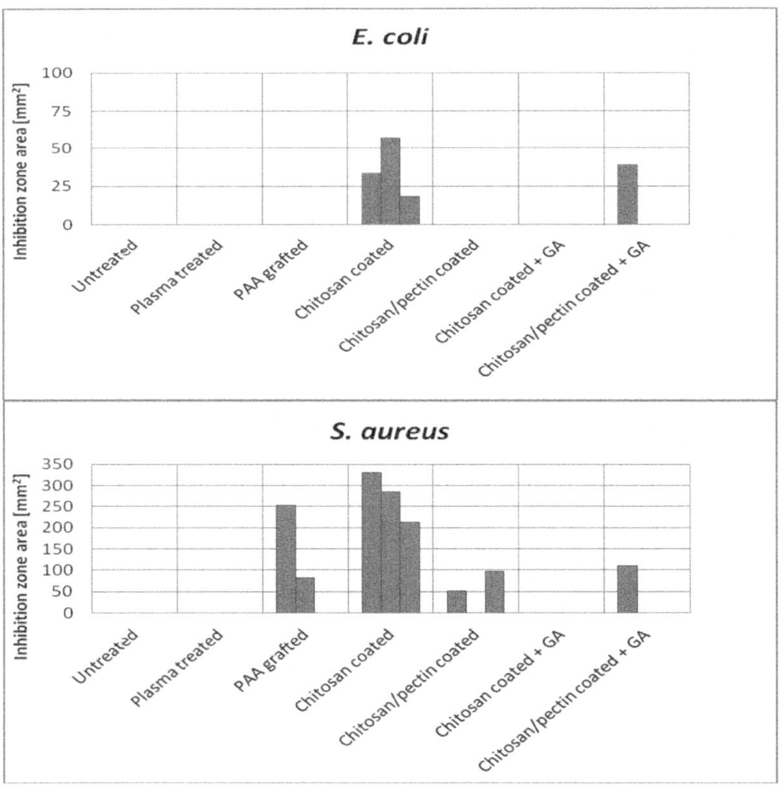

FIGURE 11.8 Inhibition zone area on LDPE surface for *S. aureus* and *E. coli* strains.

PAA grafted sample showed no inhibition zone for *E. coli*, nevertheless the same sample indicated antibacterial activity for *S. aureus*. This could be explained by the high sensitivity of PAA brush and its ability to easily absorb impurities during manipulation. As it can be seen from the results, only sample grafted by PAA and coated by chitosan demonstrated active antibacterial properties against both bacterial strains.

11.4.5 CONCLUSION (LDPE)

The aim of this work was to examine the impact of selected antibacterial agents, namely triclosan and chlorhexidine bound to the surface of LDPE with acrylic acid via DCSBD plasma. As an activator of LDPE surface for efficient binding of acrylic acid DCSBD plasma generator was used. The plasma was used for transformation of acrylic acid to form the polymer by radical polymerization. The grafted acrylic acid creates polymer brushes on the surface, that was capable of binding antibacterial agent in an effective manner. Antibacterial effect of such treated LDPE film was confirmed against *E. coli* as well as for *S. aureus*.

ACKNOWLEDGMENTS

The authors are grateful for financial support to the Slovak Research and Development Agency projects No. APVV-18-0378, APVV-19-0269 and APVV-20-0159 APVV 20-0593, and VEGA grant 2/0019/19.

KEYWORDS

- **antibacterial properties**
- **cold plasma**
- **FTIR**
- **low-density polyethylene**
- **medical-grade polyvinylchloride**
- **multi-step modification**
- **scanning electron microscopy**
- **surface energy**
- **X-ray photoelectron spectroscopy**

REFERENCES

1. Desmet, T., Morent, R., Geyter, N. D., Leys, C., Schacht, E., & Dubruel, P., (2009). Nonthermal plasma technology as a versatile strategy for polymeric biomaterials surface modification: A review. *Biomacromolecules, 10*, 2351–2378.
2. Novák, I., Števiar, M., Popelka, A., Chodák, I., Mosnáček, J., Špírková, M., Janigová, I., et al., (2013). Surface modification of polyethylene by diffuse barrier discharge plasma. *Polym. Eng. Sci., 53*, 516–523.
3. Popelka, A., Novák, I., Lehocký, M., Junkar, I., Mozetic, M., Kleinová, A., Janigová, I., et al., (2012). A new route for chitosan immobilization onto polyethylene surface. *Carbohydr. Polym., 90*, 1501–1508.
4. Popelka, A., Novák, I., Lehocký, M., Chodák, I., Sedliačik, J., Gajtanská, M., Sedliačiková, M., et al., (2012). Antibacterial treatment of polyethylene by cold plasma for medical purposes. *Molecules, 17*, 762–785.
5. Kenawy, E. R., Worley, S. D., & Broughton, R., (*2007*). The chemistry and applications of antimicrobial polymers: A state-of-the-art review. *Biomacromolecules, 8*, 1359–1384.
6. Denes, F. S., & Manolache, S., (2004). Macromolecular plasma-chemistry: An emerging field of polymer science. *Prog. Polym. Sci., 29*, 815.
7. Černák, M., Černáková, L., Hudec, I., Kováčik, D., & Zahoranová, A., (2009). Diffuse coplanar surface barrier discharge and its applications for in-line processing of low-added-value materials. *Eur. Phys. J. Appl. Phys., 47*, 22806.
8. Černák, M., Ráhel, J., Kováčik, D., Šimor, M., Brablec, A., & Slavíček, P., (2004). Generation of thin surface plasma layers for atmospheric-pressure surface treatments. *Contrib. Plasma Phys., 44*, 492.
9. Asanidezhad, A., Novák, I., Lehocký, M., Bílek, F., Vesel, A., & Junkar, I., (2010). Polysaccharides coatings on medical-grade PVC: A probe into surface characteristics and the extent of bacterial adhesion. *Molecules, 15*, 1007–1027.
10. Asanidezhad, A., Novák, I., Lehocký, M., Sedlarik, V., Vesel, A., & Junkar, I., (2010). A. Physicochemical approach to render antibacterial surfaces on plasma-treated medical-grade PVC: Irgasan coating. *Plasma Processes Polym., 7*, 504–514.
11. Bhattacharya, A., & Misra, B. N., (2004). Grafting: A versatile means to modify polymers techniques, factors and applications. *Prog. Polym. Sci., 29*, 767–814.
12. Zhao, B., & Brittain, W. J., (2000). Polymer brushes: Surface-immobilized macromolecules. *Prog. Polym. Sci., 25*, 677.
13. Costa, F., Carvalho, I. F., Montelaro, R. C., Gomes, P., & Martins, M. C., (2011). Covalent immobilization of antimicrobial peptides (AMPs) onto biomaterial surfaces. *Acta Biomater., 7*, 1431–1440.
14. Goddard, J. M., & Hotchkiss, J. H., (2008). Tailored functionalization of low-density polyethylene surfaces. *J. Appl. Polym. Sci., 108*, 2940–2949.
15. Faucheux, N., Schweiss, R., Lutzow, K., Wemer, C., & Groth, T., (2004). Self-assembled monolayers with different terminating groups as model substrates for cell adhesion studies. *Biomaterials, 25*, 2721–2730.
16. Michael, K. E., Vernekar, V. N., Keselowsky, B. G., Meredith, J. C., Latour, R. A., & Garcia, A. J., (2003). Adsorption-induced conformational changes in fibronectin due to interactions with well-defined surface chemistries. *Langmuir, 19*, 8033–8040.

17. Keselowsky, B. G., Collard, D. M., & Garcia, A. J. J., (2003). Surface chemistry modulates fibronectin conformation and directs integrin binding and specificity to control cell adhesion. *J. Biomed. Mater. Res. A, 66*, 247–259.
18. Luk, Y. Y., Kato, M., & Mrksich, M., (2000). Self-assembled monolayers of alkanethiolates presenting mannitol groups are inert to protein adsorption and cell attachment. *Langmuir, 16*, 9604–9608.
19. Vesel, A., Junkar, I., Cvelbar, U., Kovac, J., & Mozetic, M., (2008). Surface modification of polyester by oxygen- and nitrogen-plasma treatment. *Surf. Interface Anal., 40*, 1444–1453.
20. Drnovská, H., Lapčík, L. Jr., Buršíková, V., Zemek, J., & Barros-Timmons, A. M., (2003). Surface properties of polyethylene after low-temperature plasma treatment. *Colloid Polym. Sci., 281*, 1025.
21. Novák, I., Števiar, M., Chodák, I., Krupa, I., Nedelčev, T., Špírková, M., Chehimi, M. M., et al., (2007). Study of adhesion and surface properties of low-density polyethylene pre-treated by cold discharge plasma. *Polym. Adv. Technol., 18*, 97.
22. Olifirenko, A. S., Novak, I., Rozova, E. Y., Saprykina, N. N., Mitilineos, A. G., & Elyashevich, G. K., (2009). Hydrophilization of porous polyethylene films by cold plasma of different types. *Polym. Sci., 51*, 247–255.
23. Lloyd, G., Friedman, G., Jafri, S., Schultz, G., Fridman, A., & Harding, K., (2010). Gas plasma: Medical uses and developments in wound care. *Plasma Process. Polym., 7*, 194–211.
24. Sanchis, R., Fenollar, O., García, D., Sánchez, L., & Balart, R., (2008). Improved adhesion of LDPE films to polyolefin foams for automotive industry using low-pressure plasma. *Int. J. Adh. Adhes., 28*, 445.
25. Pappas, D., (2011). Status and potential of atmospheric plasma processing of materials. *J. Vac. Sci. Technol. A, 29*, 020801:1–020801:17.
26. Yang, L., Chen, J., Guo, Y., & Zhan, Z., (2009). *Surface* modification of a biomedical polyethylene. *Appl. Surf. Sci., 255*, 4446–4451.
27. Černák, M., Černáková, Ľ., Hudec, I., Kováčik, D., & Zahoranová, A., (2009). Diffuse coplanar surface barrier discharge and its applications for in-line processing of low-added-value materials. *Eur. Phys. J. Appl. Phys., 47*, 22806.
28. Šimor, M., Ráheľ, J., Vojtek, P., Černák, M., & Brablec, A., (2002). Atmospheric-pressure diffuse coplanar surface discharge for surface treatments. *Appl. Phys. Lett., 81*, 2716–2718.
29. Černák, M., Ráheľ, J., Kováčik, D., Šimor, M., Brablec, A., & Slavíček, P., *(2004)*. Generation of thin surface plasma layers for atmospheric-pressure surface treatments. *P. Contrib. Plasma Phys., 44*, 492–495.

CHAPTER 12

STEADY-STATE CREEP DEFORMATION IN ROTATING DISCS AND CYLINDERS: FOR MACRO TO MICRO/NANO-SCALE SYSTEMS

SAVITA BANSAL[1] and SATYA BIR SINGH[2]

[1]*Research Scholar, Punjabi University, Patiala, Punjab, India*

[2]*Professor, Punjabi University, Patiala, Punjab, India*

ABSTRACT

The creep behavior of composites and functionally graded materials (FGMs) has become highly significant in engineering designs of rotating discs and cylinders. The chapter gives information about the steady state creep in discs and cylinders formed from monolithic material to FGM done by various researchers in the past eight decades. An overview of yield criteria, constitutive equations, creep laws, associated flow rules, and other related terms using classical theory is given. The empirical results have proved to be preferable for designing a model, and some authors have justified their results through experimental data. Setting up apparatuses, making samples of different materials, performing experiments by applying appropriate pressure/heat, etc., to obtain the values of yield strength, applied in longitudinal and transversal directions, is a time-consuming task. The aim of this chapter is to compare the impact of stresses, strains, and strain rates under internal and external pressure on the structural component cylinder done by many

Nanotechnology-Enhanced Solid Materials: Design, Synthesis, Properties, Applications, and Perspectives.
Lionello Pogliani, Ann Rose Abraham, A. K. Haghi, & Prabhat Ranjan (Eds.)

researchers over a period of time. Rotating discs having variable thickness has also been studied by many authors, but experimental validation of the results is very rare in these chapters.

12.1 INTRODUCTION

In the engineering of solids, the strength of the material is very important while making an object. The stronger the material is, the better the engineered object will be. However, in engineering designs of rotating objects, creep behavior of the materials also becomes very important. Creep refers to permanent deformation over a period of time on objects under constant stress or under stress for longer duration during the time period. Any engineer designing an object for her/his purpose would certainly look for greater longevity of the object thus created. If the object so designed is to withstand constant stress over a period of time, as in the case of fast-rotating objects like turbines, turbochargers, rotors in machines like pumps and motors, etc., the material so chosen will be checked for having a slower creep rate. The slower the creep rate, the longer the design object will last, and thus the greater the longevity of the design.

Over a period of time, people have researched the behavior of various types of materials to find their utility for an umpteen number of usages. The major utility of the researched objects has been found in the construction sector, manufacturing sector, in engineering of better and bigger machines, in aviation and in space, among others. The most researched materials for various purposes are wood, plastics, ceramics, metals, steel, alloys, organic materials, polymers, and so on and so forth.

Towards the end of the 20th century, it was realized that composite materials were a cost-effective method of finding the material for designing objects as compared to naturally occurring materials. Furthermore, it was researched that composite objects are better capable of exhibiting design-specific characteristics as compared to naturally occurring materials.

Such materials are generally formed by homogeneous mixing, graded mixing (fused) or binding of layers, such that different materials mixed, fused, and bound have different physical and chemical properties. The desired physical properties of a design object can be obtained if such materials are used for making the design objects.

Physically, such materials can be categorized into isotropic and anisotropic materials. Isotropic materials have the same physical, mechanical, thermal, and electrical properties in all directions, whereas anisotropic materials exhibit different behavior in different directions. Orthotropic materials are a subset of anisotropic materials which exhibit symmetric characteristics in three orthogonal planes. In the remaining planes, the object exhibits anisotropic behavior. The intersecting lines of these planes are known as axes of anisotropy.

The subset of composite materials, functionally graded materials (FGMs), is designed for acquiring graded variation in the material properties. The manufacturing technique of FGM leads to the manufacturing of unique materials which are not available naturally. For example, FGM is the most important non-homogeneous material formed for high-temperature application [33].

The overall objective of various research in this field is to create or discover such objects which exhibit the requisite physical, chemical, electromagnetic, and thermal characteristics without much deviation, and at the same time, withstand the rigorous test of time and force, such that the designed object has greater reliability and longevity (Figure 12.1).

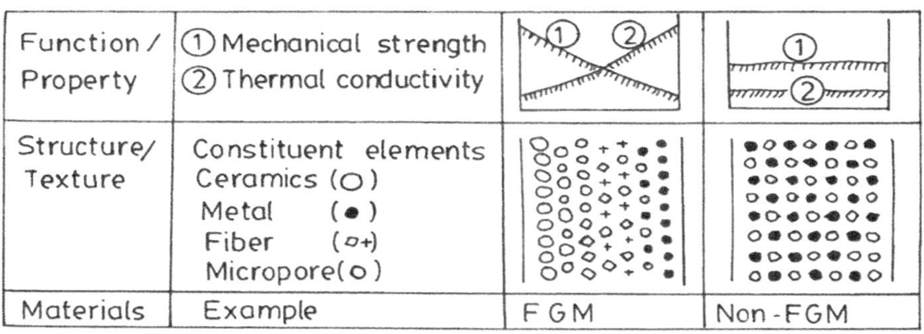

FIGURE 12.1 Characteristic representation of functionally gradient materials.
Source: Singh & Ray [33].

12.1.1 YIELDING CRITERIA

The yielding criterion is the limit of elasticity. It is the condition under which the object leaves its elastic behavior and starts behaving in a plastic manner. These criteria define the boundary conditions for the object to behave in an elastic manner. These criteria, coupled with constitutive equations, lead to assessing the creep rate of objects and materials. Some of the most popular yielding criteria preferred by researchers are von Mises and Tresca's Yield Criteria. Other important yield criteria are Hoffman's Yielding Criterion,

Hill's Yielding Criterion, and Singh and Ray's Yielding Criterion. Through these, we get to know when the material operating under the same load conditions start getting deformed in a plastic manner. These yield criteria are the functions of principal stresses σ_1, σ_2, σ_3, in such a way that $\sigma_1 > \sigma_2 > \sigma >$. The above-mentioned yielding criteria are as follows:

12.1.1.1 TRESCA YIELD CRITERION (OR MAXIMUM SHEAR-STRESS CRITERION)

Tresca yield criterion states that when τ_{max} (maximum shear stress) reaches a critical value (k), the material starts deforming. Here, τ_{max} is calculated as follows:

$$\tau_{max} = \frac{\sigma_1 - \sigma_3}{2}$$

Conditions for uniaxial tension are:

$$\sigma_1 = \sigma_e, \sigma_2 = \sigma_3 = 0, \tau_e = \frac{\sigma_e}{2}$$

Solving above two expressions, we get:

$$\tau_{max} = \frac{\sigma_1 - \sigma_3}{2} = \frac{\sigma_e - 0}{2} = \frac{\sigma_e}{2} = \tau_e$$

$$\Rightarrow \sigma_e = \sigma_1 - \sigma_3$$

Conditions of pure shear state are:

$$\sigma_1 = -\sigma_3 = k, \sigma_2 = 0,$$

Using these conditions in the above τ_{max} expression, we get:

$$\tau_{max} = \frac{\sigma_1 - \sigma_3}{2} = \frac{k + k}{2} = \frac{2k}{2} = k$$

$$\Rightarrow k = \frac{\sigma_1 - \sigma_3}{2} \Rightarrow \sigma_1 - \sigma_3 = 2k = \sigma_e$$

$$\Rightarrow k = \frac{\sigma_e}{2}$$

In Tresca yield criterion, intermediate principal stress is neglected, therefore, the following von-Mises yield criterion is preferred by researchers in most of their research papers.

12.1.1.2 VON-MISES YIELD CRITERION (OR DISTORTION-ENERGY PRINCIPLE)

In 1913, this criterion was proposed by von-Mises for isotropic material that yield exists when stress deviator exceeded some critical value, i.e.,

$$\frac{1}{6}\left[(\sigma_1-\sigma_2)^2+(\sigma_2-\sigma_3)^2+(\sigma_3-\sigma_1)^2\right]=k^2$$

Now, condition of yield in uniaxial tension is $\sigma_1=\sigma_e, \sigma_2=\sigma_3=0, \sigma_e^2+\sigma_e^2=6k^2$. After solving both the equations, we get von-Mises yield criterion as:

$$\sigma_e=\frac{1}{\sqrt{2}}\left[(\sigma_2-\sigma_3)^2+(\sigma_3-\sigma_1)^2+(\sigma_1-\sigma_2)^2\right]^{\frac{1}{2}}$$

12.1.1.3 HOFFMAN'S ISOTROPIC YIELD CRITERION

For isotropic material, Hoffman's yield criterion utilizes uniaxial compression (f_c) and tensile yield stresses (f_t) and is formulated as follows [31]:

$$\left(\sigma_1^2+\sigma_2^2+\sigma_3^2\right)-(\sigma_1\sigma_2+\sigma_2\sigma_3+\sigma_3\sigma_1)$$
$$+(f_c-f_t)(\sigma_1+\sigma_2+\sigma_3)-f_cf_t=0$$

12.1.1.4 HILL'S YIELD CRITERION

In 1948, Hill proposed this criterion for anisotropic material by modifying the von-Mises yield criterion (isotropic material) [31]:

$$2f=F(\sigma_2-\sigma_3)^2+G(\sigma_3-\sigma_1)^2+H(\sigma_1-\sigma_2)^2=1$$

Here, the potential function 2f can be found from effective stress expression given as:

$$\sigma_e=\frac{\sqrt{2f}}{\sqrt{G+H}}$$

In Hill's yield criterion, it is assumed that the material is orthotropic. This means there are 3 planes of symmetry which are mutually intersecting each other at 90° at every point inside material and these planes are known as the principal axes of anisotropy. When principal axes of anisotropy are taken in x, y, and z directions, Hill's yield criterion converts into the following form:

$$F\left(\sigma_2-\sigma_3\right)^2+G\left(\sigma_3-\sigma_1\right)^2+H\left(\sigma_1-\sigma_2\right)^2=1$$

where; F, G, and H are anisotropic constants. If F = G = H = 1, then this reduces to von-Mises yield criterion.

12.1.1.5 SINGH AND RAY'S YIELD CRITERION

In 2003, Singh & Ray proposed this criterion for anisotropic material using uniaxial compression (f_c) and tensile stress (f_t). This criterion is developed using Hill's criterion and Hoffman's criterion.

$$2f=F\left(\sigma_2-\sigma_3\right)^2+G\left(\sigma_3-\sigma_1\right)^2+H\left(\sigma_1-\sigma_2\right)^2+(k_1\sigma_1+k_2\sigma_2+k_3\sigma_3)=1;$$

$$k_i=\left[\frac{f_c-f_t}{f_c\,f_t}\right]_{i^{th}\text{ principal direction}} \qquad i=1,2,3$$

Since the difference between tensile stress and compressive stress has been calculated by thermal residual stresses, so:

$$k_1=k_2=k_3=\left[\frac{f_c-f_t}{f_c\,f_t}\right]$$

12.1.2 CONSTITUTIVE EQUATIONS

Constitutive equations of design objects define the behavior of a design object of a particular material when presented with external stimuli. The constitutive equations are the mathematical model of the object's behavior under various forces/stresses. Here the behavior implies strain on the object caused by external stresses.

The fundamental constitutive equations for the orthotropic theory of creep as given by Bhatnagar & Gupta [5] are as follows:

$$\dot{\varepsilon}_r=\frac{\dot{\varepsilon}}{2\sigma}\left[(G+H)\sigma_r-H\sigma_\theta-G\sigma_z\right]$$

$$\dot{\varepsilon}_\theta=\frac{\dot{\varepsilon}}{2\sigma}\left[(H+F)\sigma_\theta-F\sigma_z-H\sigma_r\right]$$

$$\dot{\varepsilon}_z=\frac{\dot{\varepsilon}}{2\sigma}\left[(F+G)\sigma_z-G\sigma_r-F\sigma_\theta\right]$$

12.1.3 CREEP LAWS ASSUMED BY DIFFERENT AUTHORS

12.1.3.1 NORTON'S LAW

$$\dot{\varepsilon} = A\sigma^{n}$$

12.1.3.2 TIME-HARDENING LAW

$$\dot{\varepsilon} = B\sigma^{n} t^{m-1}$$

12.1.3.3 STRAIN-HARDENING LAW

$$\dot{\varepsilon} = C\sigma^{\frac{n}{m}-1} \varepsilon^{1-\left(\frac{1}{m}\right)}$$

12.1.3.4 MARIN'S LAW

$$\dot{\varepsilon} = De^{-q}\sigma^{m} + E\sigma^{n}$$

where; A, B, C, D, E, n, m, q are the experimental constants in uniaxial cases by Finnie & Heller [15].

12.1.3.5 SHERBY'S LAW

$$\dot{\varepsilon} = \left[M(\sigma - \sigma_0)\right]^{n} \quad \text{where } M = \frac{1}{E}\left(A\exp\left(\frac{-Q}{RT}\right)\right)^{\frac{1}{n}}$$

where; A, Q, R, T, E, σ_0, n, M are constant, activation energy, gas constant, absolute temperature, Young's modulus, threshold stress, creep exponent, and estimated creep parameter, respectively.

12.2 LITERATURE REVIEW

Rotating discs are found in many fields, like engineering, biological sciences, mechanical engineering, etc. We can find the applications of rotating discs in the turbine section of an aircraft gas turbine engine, ship propellers, turbo generators, flywheels, gears, hub-mounted disc wheels, disc brakes on motorcycles, cars, aerospace structures, etc. Rotating discs can set up radial, circumferential, and axial stresses. To reduce these stresses, it is very important to understand the loads acting on the structural components. Rotating discs have been discussed in the form of solid discs and hollow discs (hollow shafts). This structural component has been actively studied by various researchers over the past years so that they can develop materials that are stronger, stiffer, and are less prone to creep. Researchers made the assumption that the axial stress is zero for a thin disc. Anisotropic, isotropic, and orthotropic rotating disc made of composites and functionally gradient materials have been taken into account by different researchers. This section of the chapter analyzes the stress and strains acting on rotating discs by various authors.

12.2.1 CREEP IN ROTATING DISCS

Wahl et al. [44] have presented steady state creep behavior and stress distribution in a rotating disk made of anisotropic material. The validity of theoretical results was observed experimentally. The authors compared the results formed by the Tresca yield criterion and the von Mises criterion and revealed that in the Tresca yield criterion, stresses in the rotating disc are in better agreement than in the von Mises criterion.

Wahl [43] calculated stress distributions with the help of Tresca and associated flow rules on inner rotating discs having different thicknesses. The stress distributions are calculated on discs undergoing steady-state creep at a very high temperature. The results obtained were compared with available data and were found to be reasonable. The methodology of Wahl when applied to transient states, where stress distribution changes from the starting period up to steady-state conditions, is also found satisfactory.

Wahl [42] simulated the real conditions of blades by subjecting the discs to peripheral load. He discovered that the power function of stress and time is directly proportional to the steady-state creep rate. Wahl formulated various design charts of stress distribution for different values of stress and creep exponent at various diameters of discs.

Saroja [26] analyzed the creep deformation in a thin rotating disc with radii of 1.25 inches and 6 inches of hole and disc, respectively, with constant thickness. Assumptions made are that a biaxial state of stress exists, axial stress is zero. The author compared her results of stress distribution with the results found by Wahl and presented them graphically. The author represented the creep at the inner and outer diameter of a rotating disk at a temperature of 1,000°F (degree Fahrenheit) and an initial speed of 15,000 r.p.m. It is concluded that we can eliminate the assumption that radial strain rate is equal to zero made by Wahl.

Ma [22] presented the analysis of creep in rotating disks in which the thickness and temperature are both variables, and finally compared the results with the distributions having uniform thickness and temperature varying. In this chapter, the author concluded that we can solve a complicated problem in a simpler way by using the closed form. With the help of Tresca's criteria, associated flow rule, exponential function creep law, equilibrium equations, the incompressibility equations, and compatibility equations, the stress equations are easily derived.

Venkatraman & Patel [41] analyzed creep in annular plates with uniform pressure by using the shearing stress criterion. The authors assumed that the creep rate is a power function of moment and time. They discussed four mathematical problems and concluded that in three problems, the size of the hole affects the nature of the moment distribution. Moreover, in the fourth mathematical problem, nature does not depend either on size or creep exponent.

Bhatnagar et al. [8] analyzed the torsion problem using a rod comprised of a substance exhibiting orthotropic characteristics. The rod (cylindrical in shape) is long, solid, and uniform in density, having a circular cross-sectional radius of 'a.' The axis of anisotropy is considered along the z-axis. The rod is applied with equal and opposite torque at both ends. The authors have derived the resultant stress as a function of for small displacements. The constitutive equations for creep were assumed to be similar to those defined by Bhatnagar & Gupta [45].

Chang [11] analyzed the anisotropic behavior of the material and stresses in rotating disks or orthotropic cylinders. The stresses and displacements established in a rotating disk are axially symmetrical and the deformation is not circular. To obtain a closed-form solution for the stresses in spinning circular and elliptical discs made of orthotropic material, the semi-inverse approach is used. The approach can be used to solve some types of boundary value problems.

Arya & Bhatnagar [2] presented the creep deformation of orthotropic and rotating disks by assuming that the material is incompressible, axial stress is negligible, and the creep rate is a function of stress into time, which is the time-hardening law. The strain rates and stresses were obtained with the help of the time-hardening law, constitutive equations created by Bhatnagar & Gupta for orthotropic creep theory and the equilibrium equation for a disk with constant thickness. The authors compared steady and non-steady state solutions. The author discusses the following cases to check the anisotropy of the material:

- **Case 1:** Greatest Anisotropy: Ratio of G and F = 0.5, Ratio of H and F = 1.5.
- **Case 2:** Smaller Anisotropy: Ratio of G and F = 0.75, Ratio of H and F = 1.25.
- **Case 3:** Anisotropy vanishes (Isotropic case): Ratio of G and F = 1, Ratio of H and F = 1.

Tangential stress versus radius (ranging from 1 inch to 6 inches) for all the three cases discussed above is represented graphically. The plotted graphs illustrate that tangential stress is decreased at all radii.

Guowei et al. [16] have developed a unified yield criterion (UYC) for analysis of a rotating annular disc considering variable thickness. The boundary conditions of the problem are attained by choosing a weight coefficient in the above-mentioned criterion and are as follows.

$$\sigma_1 - \frac{1}{1+b}(b\sigma_2 + \sigma_3) = \sigma_0 \quad \text{when}\ \sigma_2 \le \frac{1}{2}(\sigma_1 + \sigma_3),$$

$$\frac{1}{1+b}(\sigma_1 + b\sigma_2) - \sigma_3 = \sigma_0 \quad \text{when}\ \sigma_2 \ge \frac{1}{2}(\sigma_1 + \sigma_3)$$

Because of piecewise linearity, this criterion is very useful in the engineering branch. The formula derived here based on this criterion gives the exact solutions of rotating discs for separate yield criteria.

Singh & Ray [33] investigated the response of steady-state creep in a particle-reinforced FGM isotropic disc with linear variation of particle distribution along the radial distance and matched it with that of a disc containing particles which were distributed uniformly. They checked the results on a disc comprising of composite materials containing silicon carbide particles in a matrix of 6,061 aluminum alloy. Available experimental data was used to estimate the variation in the material parameters of creep using regression fit. The shearing stress rises in the area near the inner radius of the FGM disc due to increased density caused by higher particle content, and the creep rate is reduced by decreasing the value of creep parameters and increasing particle content.

Singh & Ray [32] have analyzed the anisotropy due to the formation of whiskers and found that short fiber formation actually helps in significantly lowering the tangential creep rate apart from a reduction in compressive radial strain. The results thus obtained in the chapter were significant from the point of view of the manufacturing of composite materials in the context of real-life engineering.

Singh & Ray [31] published two papers in the year. The first one was related to creep analysis in an isotropic aluminum composite rotating disc. They have assumed that there is a linear particle distribution in the composite for it to be isotropic. The authors deduced that the creep rates in the radial and tangential directions are smaller than those of an isotropic disc with uniform particle distribution. They have found that the steady-state creep response of this disc with linearly decreasing particle content with increasing radius is considerably higher than that of an isotropic rotating FGM disc with uniform particle content distribution.

Singh & Ray [31] in their second paper stated that many times, residual stress remains in the object due to the process of manufacturing, which may be because of the process involved, like extrusion, or because of sudden cooling of the object from high to low temperature. The authors modified the Hill Criterion to account for residual stresses. They found that the results are comparable when the residual stress is assumed to be negligible. The authors were able to thus establish that the use of aluminum composites under rotation leads to the formation of residual stresses.

Deepak, Gupta, & Dham [13] have analyzed the impact of stress exponents at 3, 5, and 8 of the creep law on creep performance of the aluminum-silicon matrix composites in rotating discs. They concluded that steady state creep in Al-SiC_p is better represented by assuming the stress exponent of the creep law at 5, rather than at 3 or 8. They also found that the distribution of stress varies with varying values of stress exponent and, thus, the strain rates are significantly affected by varying stress exponent. They found that radial stress corresponding to stress exponent 3 is the highest and stress exponent 8 is the lowest. Tangential stress is highest at the inner radius of stress exponent 3 and gradually becomes lowest at the outer radius. Similarly, for stress exponent 8, tangential stress is lowest at the inner radius, which gradually becomes highest at the outer radius. Radial and tangential strains are highest for stress component 8 and lowest for stress component 3. Values for stress exponent 5 lie in between the two stress exponent values for both stress and strain.

Chamoli, Rattan, & Singh [10] studied secondary creep in a rotating disc which is anisotropic in nature and is composed of Aluminum Silicon Carbide Particulate. The creepy behavior is expressed using Sherby's law. The study revealed that the anisotropy of a material significantly impacts the creep behavior of the rotating discs.

Vandana & Singh [39] analyzed the creep in a rotating disc composed of Al-SiC_p composite having different thicknesses at different radii using

Sherby's law. The chapter analyzed creep response for varying thicknesses as constant, linear variation and hyperbolic variation. The change in radial stress was not found to be very significant in the three cases. The authors have concluded that material anisotropy can help to keep the disc creep resistant.

Vandana & Singh [38] analyzed the effect of residual stress on creep in a rotating disc and its thickness varied. It was concluded that stress and strain distribution are affected by thermal residual stress in both types of discs. The authors observed that the thermal residual stress is less for anisotropic materials as compared to isotropic materials. The designer could keep in mind the residual stress while fabricating a design for a material.

Sharma & Sahni [29] analyzed the creep response of rotating discs which were thin and whose thickness and density varied with edge loading. The authors established that edge loading is safer in cases where density and thickness decrease with increasing radius of a rotating disc. They also found that deformation is much more significant in rotating discs which have edge loading as compared to those without edge loading.

Vandana & Singh [40] have created an analytical framework for creep analysis in an isotropic functionally graded rotating disc with varying thickness and uniform thickness. The authors also explained the analysis in non-FGM with uniform and varying thickness. Sherby's law has been used to determine the creep behavior in rotating discs and the value of creep exponent is taken as 8. Finally, they compared the results obtained for isotropic functionally graded rotating disc with constant thickness and with varying thickness.

Thakur, Gupta, & Singh [37] studied the impact of linear thermal gradient (non-classical approach) on creep behavior in discs comprised of aluminum silicon carbide and are isotropic in nature. The study is significant considering the fact that many engineering components resembling rotating discs are exposed to varying temperatures, and thus the creep response of such components can be best studied if the rotating discs are studied on a temperature gradient. The radial strain rate is lowest somewhere in the middle of the thickness of the cylinder. The results are quite significant, which show that the strain rates are reduced over the entire radius of the disc operating under a temperature gradient and is considered to be for safe designing of a disc under elevated temperatures.

Shahi, Singh, & Thakur [28] have presented the analysis of creep parameters in rotating discs made of Beryl, Magnesium, and Brass subjected to rigid shaft using Seth's theory, which is the non-classical approach. Authors have observed that the stress in the radial direction has more value than the stress in the circumferential direction at the internal surface of isotropic discs and creep rates for transversely isotropic material have the highest value at the inner surface.

12.2.2 CREEP IN ROTATING CYLINDERS

Rotating cylinders have many engineering applications in real life. The use of rotating cylinders has generated great interest in many research problems in the field of solid mechanics. Most researchers have used anisotropic, isotropic, and orthotropic rotating cylinders under constant internal pressure, varying internal pressure, and both internal and external pressure to predict the deformation. This section focuses on the analysis of creep stresses and creep strain rates in thick and thin rotating cylinders subjected to different loads. A good amount of phenomenological studies have been conducted in the field of creep analysis of rotating cylinders. The behavior of creep in cylinders has been discussed by the following authors:

Rimrott et al. [25] observed that in any hollow cylindrical vessel under constant inner pressure, the vessel will continue to expand at an increased rate until the wall thickness has decreased to the point that the material's strength is no longer adequate to support the weight of the constant internal pressure felt at the inner radius, and the vessel will crack due to the application of two simultaneous effects, i.e., constant internal pressure at the internal radius of the cylindrical vessel and reduced strength of the material due to expansion of the vessel. In this chapter, creep-failure time (the time where the true strain approaches infinity) has been examined for thin and thick-walled cylindrical vessels using Norton's law. The authors have represented their results through graphs.

Rimrott & Luke [24] analyzed the behavior of creep for a rotating cylinder using Norton's law and found the values of strain rates, strains, and stresses for creep exponent $(n) = 1$. The duo showed that we could find the values of strains, strain rates, and stresses for different values of the creep exponent. The authors said that the cylinders could be made from any material, and the findings are independent of inner and outer radius. The author derived the exact solution for isotropic material by assuming that strains are large. For this, the authors used the finite-strain theory to predict the time after which the cylinder would be deformed.

Schweiker & Sidebottom [27] developed an incompressible creep theory of a thick-walled cylinder to predict deformations under internal pressure. A hyperbolic sine function is used for the strain-stress-time expression. The general assumptions used are: the cylinder is closed-ended, the material is incompressible, the deformations are infinitesimal, and the validation of the total strain theory. The experiment was carried out on a 1.5-inch diameter high-density polyethylene bar. The authors have deduced that deformation in the cylinder cannot be changed by redistribution of stress with time.

Bhatnagar & Gupta [5] formulated constitutive equations based on the orthotropic theory of creep. Two applications are discussed. *One is the* tension of a bar (principal axes of stress and anisotropy do not coincide), and the other one is the variation in stresses taking load as constant with time. These constitutive equations are based on the anisotropic plastic model of Hill.

Pai [23] used a piecewise linear model and the von Mises criterion to get the solutions for the creep of an orthotropic cylinder under internal pressure. Under the hypothesis of plane strain, closed-end, and open-end, he found the expressions for stress and creep rate. In all three axial conditions, creep anisotropy of the material affected the strain rates. The chapter compares the creep behavior of tensile and compressive creep experiments, as well as allowing for the material's directional tendency in tensile creep tests. He observed that plane strain conditions are equivalent to closed-end conditions. The solution helped to predict the values of strain rate accurately. During steady-state creep conditions, isotropic materials in the elastic range or plastic range may become highly anisotropic.

King & Mackie [19] presented a detailed study of creep distortion under internal pressure and load on the isotropic material of a thick-walled cylinder. The authors observed that tensile stress would increase continuously during extension. They concluded that when the values of strains are greater than the elastic strains, then the creep distortion for cylinders and the experimental value from diametric measurements are correlated.

Bhatnagar & Gupta [6] found the different values of stresses and creep rates of a thick-walled cylinder made of an orthotropic material and pressure applied internally under the assumptions of:

i. Plane strain, i.e., $\dot{\varepsilon}_z = 0$
ii. Generalized plane strain, i.e., $\dot{\varepsilon}_z$ is not a function of r.
iii. Plane stress, i.e., $\sigma_z = 0$

They used Norton's law and the constitutive equations of anisotropy and compared their results with those obtained by Pai [23]. In-plane strain assumption, results show that axial stress is dependent on material anisotropy. The results give two independent anisotropic constants. In the plane stress case, the effect of anisotropy is small on radial stress. Lastly, they found that creep anisotropy has a notable effect on cylinder behavior and cylinders made from anisotropic material will be less deformed.

Bhatnagar & Arya [4] concluded that after applying external pressure on the thick-walled cylinder, the strain rate decreases and the durability of the cylinder increases, which is beneficial in designing a cylinder so that creep deformation will be less. They analyzed that the anisotropic behavior of the material affects the creep behavior of the cylinder. At the inner radius, axial stress is compressible and it is tensile at the outer radius. The chapter concludes that the strain rate increases with increasing deformation (strain). The authors also discussed the behavior of isotropic materials and compared their results with those of Rimrott for an isotropic cylinder.

Arya & Bhatnagar [3] investigated the analysis of creep in thick-walled cylinders composed of anisotropic material by applying internal and external pressure, both considering the elastic strains. The authors obtained fundamental equations of an orthotropic cylinder for the behavior of creep by assuming that the total axial strain is a function of time, not radius, displacement is radial (a function of radius), and the axial stress is uniform throughout. The solutions of these equations are discussed for planar anisotropy and these results are compared with the isotropic case.

Conditions for planar anisotropy are taken from Gupta [17]; and Lekhnitskii [21] as follows:

$$F = G, \quad S_{11} = \frac{1}{E} = S_{22}, \quad S_{12} = \frac{-v}{E},$$

$$S_{13} = \frac{-v^{1}}{E^{1}} = S_{23}, \quad S_{13} = \frac{1}{E^{1}}$$

where; E, E^1 are Young's moduli; v, v^1 are Poisson's ratio; F and G are anisotropic creep constants and S_{11}, S_{12}, …, S_{33} are anisotropic elastic constants.

The authors estimated the values of radial stress and tangential stress by taking different values of r (ranging from 1.0 inches to 2.0 inches) and time t (t = 100 hours, 1,000 hours, 10,000 hours). The values of strain rates are higher in the case of isotropic materials than in the case of anisotropic materials.

Bhatnagar, Arya, & Debnath [7] investigated the steady state creep analysis of a hollow rotating cylinder using power law by finding the values of different strain rates and corresponding stresses in an anisotropic material. Rotating cylinders are used in power generation and various industries. Assumptions made by the authors are that the material of the cylinder is homogeneous and orthotropic, no change in the volume after creep deformations (i.e., $\dot{\varepsilon}_r + \dot{\varepsilon}_\theta + \dot{\varepsilon}_z = 0$), condition of plane-strain occurs ($\dot{\varepsilon}_z = 0$). In numerical computations, the cylinder is made up of steel or steel alloys implemented in three cases of anisotropy. They concluded that the anisotropic behavior of the material and the creep exponent (n) both affected the strain rates and stresses. The authors suggested the anisotropic material is beneficial for the formulation of cylinders, giving reasons why the life of the cylinder will be longer (because the strain rate is less). Graphical representation of all the stresses and tangential strain rate for three cases of anisotropy is present and taking creep exponent (n) = 1.0, 3.0, 6.9, etc.

Bhatnagar, Kulkarni, & Arya [9] developed an application based on finite strain theory and demonstrated that if creep analysis is performed with respect to small strain theory, the structure of the cylinder would not be safe. The authors worked on the creep behavior of an orthotropic thick-walled revolving cylinder made up of anisotropic material. The Equilibrium equation can be obtained by using tangential strain and radial strains. By evaluating this equation, the expression for the radial stress is found. After that, the authors used boundary conditions by taking the axial force improvised on the cylinder as zero and testing for creep exponent n = 1.

Chen et al. [12] have compared the estimated solution of an equation of order five with the results of finite element analysis using the program ABAQUS as it allows to get an exact solution for the functionally graded cylinder's time-dependent behavior. The effects on the material are symmetric about an axis and depend on the radial coordinate. An approximated solution is derived using Taylor's expansion series for time-dependent creep response. Asymptotic solutions for all the three stresses under internal pressure and external pressure are plotted in this chapter. More attention is paid to the calculation when pressure is enforced on the functionally graded cylinder internally.

Abrinia et al. [1] applied an observational solution for calculating radial stress and tangential stress of a thick-walled functionally graded cylindrical vessel under internal pressure and applied temperature. The authors used Maple 9.5 software for plotting the graphs of stresses versus pressure and temperature. They have concluded that the radial stress and circumferential stress expressions of FGM and homogeneous material are the same whenever the value of the parameter is taken as zero. Lastly, the authors suggested that this chapter can be further extended by researchers in the yielding region to find an elastic-plastic solution.

Singh & Gupta [34] investigated the creep analysis of a cylinder having thick walls and it is formed of an Al-SiC_p composite under internal pressure. The stresses and strain rates are examined for different values of the size of the SiC_p, its content and temperature applied. It is concluded that there is very little impact of the above factors on stress. Moreover, the value of strain rate at the radius and tangent direction of the cylinder can be reduced by decreasing the size of SiC_p and applied temperature, and increasing its content.

Sherby's law (Creep exponent n = 5) is altered by the author as:

$$\dot{\varepsilon} = \left[M(\sigma - \sigma_0)\right]^n \quad \text{where } M = \frac{1}{E}\left(A^1 \exp\left(\frac{-Q}{RT}\right)\right)^{\frac{1}{n}}$$

The value of threshold stress will be found using a linear extrapolation technique. Assumptions made are: the material is incompressible, isotropic, and SiC_p is distributed uniformly in the aluminum matrix. Initially, less pressure is enforced and, after some time, it will be constant. At any point, stresses in this shape remain constant with time. Elastic deformation is neglected. Effective stress is taken by the von Mises yielding criterion given in Dieter [14]. We can see the variation in strain rates, stresses, and effective stress in the composite cylinder for various sizes of the SiC content and temperature. The strain distribution curves show that the strain is highest with the lowest reinforcement density and vice versa. It was also observed that the life of the cylinder increased significantly.

Habibi et al. [18] defined a numerical method for investigating the creep and elastic behavior of FGMs in rotating cylinders. Displacement equations are used to find the compatibility equation for the thermo-elasticity theory. Norton's law is used in this chapter as $\dot{\varepsilon} = K\sigma^{\varsigma}\tau^{q}$ where ς and q are experimental constants.

Equilibrium equations for a rotating functionally graded thick-walled cylinder based on thermos-elasticity theory are derived and nonlinear heat transfer equations in radial direction are used to get temperature distribution in a rotating cylinder. The solution was obtained by using three different cases on boundary conditions. Hollow cylinder with free edges (i.e.,σ_{rr} = 0, t = innerradius and outerradius), hollow cylinder with fixed-free edges (i.e., u = 0 when r = inner radius and σ_{rr} = 0, *t = inner radius and outer radius*), filled cylinder with free edges (i.e., u = 0 when r = 0; σ_{rr} = 0, *t* = inner radius). Here, u is the displacement, r is the radius of the cylinder. The expression of the radial stress is calculated by using the stress-strain relation for plain strain conditions. The authors validated their results with the help of examples and compared these results with the results generated by Singh & Ray graphically. The authors concluded that cylinder properties approach the properties of a metal base (it has a smaller modulus of elasticity) and radial strain changes from positive values to negative values and tangential strain increases in the positive direction as the value of n increases.

Singh & Singh [36] carried out the creep behavior analysis of an isotropic cylindrical vessel having thick walls and made of aluminum matrix reinforced with silicon carbide by using Sherby's law in which threshold stress is included.

i.e.,

$$\dot{\varepsilon} = \left[M(\sigma - \sigma_0)\right]^n$$

where; M is the creep parameter, $\dot{\varepsilon}, \sigma, \sigma_0$ are effective creep rate, effective stress, threshold stress respectively, n is the creep exponent and the values selected for creep exponent are 3, 5, and 8. The von Mises yielding criterion is used with constitutive equations and equilibrium equations to obtain the stress distribution, creep stress and strain rates in this vessel. Authors observed that all three stresses increase and the strain rates decrease along the radius. They concluded that for isotropic material, stress in the direction of radius is maximum at the inner radius and zero at the outer radius and the strain rates decrease at every point of the composite cylinder with a decrease in the value of creep exponent n from 8 to 3.

Kohli et al. [20] have described how the small strain theory leads to unsafe cylinder designs by using the finite strain theory. Small strain theory helps in designing a cylinder so that the cylinder can work under hard mechanical and thermal loadings. The chapter analyzed the behavior of creep of a thick-walled aluminum-based composite reinforced with silicon carbide under internal pressure. Sherby's law is used because of the large creep exponent and higher activation energy. The authors calculated the creep rate at inner radius, effective stress and then plotted all the stresses and strain rates for different values of the radii of a cylinder. They concluded that at the center of the cylinder, radial stress shows a slight change and other stresses and strain rates show a large change when the strain increases at the inner radius.

12.3 CONCLUDING REMARKS

The research done on FGM is either purely analytical or numerical-based methods. Many mathematical models adopted for the various analysis of FGM proved to be very efficient and convergent. Most work has been done on various aspects of FGM. However, there are still some gaps that need to be filled to take more in-depth advantage of FGM. Such gap areas are described below:

- In the existing studies, 3D analysis of FGM is mainly focused on linear buckling and free vibration analysis.
- Most of the research has focused on the properties of FGM without taking into account the environmental effects.
- Numerous applications of FGM exist in aerospace, defense, nuclear, automobiles, and other industries. For the betterment of human lives, more in-depth investigation needs to be done into FGM with respect to health care, such as bone implantation, dentistry, etc.

KEYWORDS

- **composite material**
- **constitutive equations**
- **creep**
- **functionally graded material**
- **organic materials**
- **stresses**
- **yielding**

REFERENCES

1. Abrinia, K., Naee, H., Sadeghi, F., & Djavanroodi, F., (2008). New analysis for the FGM thick cylinders under combined pressure and temperature loading. *American Journal of Applied Sciences, 5*(7), 852–859.
2. Arya, V. K., & Bhatnagar, N. S., (1979). Creep analysis of rotating orthotropic disks. *Nuclear Engineering and Design, 55*, 323–330.
3. Arya, V. K., & Bhatnagar, N. S., (1976). Creep of thick-walled orthotropic cylinders subject to combined internal and external pressures. *Journal Mechanical Engineering Science, 18*(1), 1–5.
4. Bhatnagar, N. S., & Arya, V. K., (1974). Large strain creep analysis of thick-walled cylinders. *Int. J. Non-Linear Mechanics, 9*, 127–140.
5. Bhatnagar, N. S., & Gupta, R. P., (1966). On the constitutive equations of the orthotropic theory of creep. *Journal of the Physical Society of Japan, 21*(5), 1003–1007.

6. Bhatnagar, N. S., & Gupta, S. K., (1969). Analysis of thick-walled orthotropic cylinder in the theory of creep. *Journal of the Physical Society of Japan, 27*(6), 1655–1661.
7. Bhatnagar, N. S., Arya, V. K., & Debnath, K. K., (1980). Creep analysis of orthotropic rotating cylinder. *Journal of Pressure Vessel Technology, 102*, 371–377.
8. Bhatnagar, N. S., Gupta, S. K., & Gupta, R. P., (1969). The torsion of an orthotropic rod in the theory of creep. *Wood Science and Technology, 3*, 167–174.
9. Bhatnagar, N S., Kulkarni, P. S., & Arya, V. K., (1986). Creep analysis of orthotropic rotating cylinders considering finite strains. *Int. J. Non-Linear Mechanics, 21*(1), 61–71.
10. Chamoli, N., Rattan, M., & Singh, S. B., (2010). Effect of anisotropy on the creep of a rotating disc of Al-SiC_p composite. *Int. J. Contemp. Math. Sciences, 5*(11), 509–516.
11. Chang, C. I., (1975). The anisotropic rotating disks. *Int. J. Mech. Sci., 17*, 397–402.
12. Chen, J. J., Tu Shan, T., Xuan, F. Z., & Wang, Z. D., (2007). Creep analysis for a functionally graded cylinder subjected to internal and external pressure. *J. Strain Analysis, 42*, 69–77.
13. Deepak, D., Gupta, V. K., & Dham, A. K., (2009). Impact of stress exponent on steady state creep in a rotating composite disc. *J. Strain Analysis, 44*, 127–135.
14. Dieter, G. E., (1988). *Mechanical Metallurgy*. McGraw-Hill, London.
15. Finnie, I., & Heller, W. R., (1959). *Creep of Engineering Materials*. McGraw-Hill, New York, NY.
16. Guowei, M., Hong, H., & Yukata, M., (2001). Limit angular velocity of rotating disc with unified yield criterion. *International Journal of Mechanical Sciences, 43*, 1137–1153.
17. Gupta, R. P., (1967). *Some Problems on the Theory of Creep*. Ph.D. Thesis, University of Roorkee.
18. Habibi, N., Samawati, S., & Ahmadi, O., (2016). Creep analysis of the FGM cylinder under steady state symmetric loading. *Journal of Stress Analysis, 1*(1), 9–21, Spring –Summer.
19. King, R. H., & Mackie, W. W., (1967). Creep of thick-walled cylinders. *Journal of Basic Engineering, 89*(4), 877–884.
20. Kohli, G. S., Singh, T., & Singh, H., (2020). Creep analysis in thick composite cylinder considering large strain. *Journal of the Brazilian Society of Mechanical Sciences and Engineering, 42*, Article No. 68, 1–8 of Chapter 68.
21. Lekhnitskii, S. G., (1963). *Theory of Elasticity of an Anisotropic Body*. Publisher: Holden & Day, OCLC Number: 803145982.
22. Ma, B. M., (1961). Creep analysis of rotating solid disks with variable thickness and temperature. *Journal of the Franklin Institute, 271*(1), 40–55.
23. Pai, D. H., (1967). Steady-state creep analysis of thick walled orthotropic cylinders. *Int. J. Mech. Sci., 9*, 335–348.
24. Rimrott, F. P. J., & Luke, J. R., (1961). Large strain creep of rotating cylinders. *ZAMM, 41*(12), 485–500.
25. Rimrott, F. P. J., Mills, E. J., & Marin, J., (1960). Prediction of creep failure time for pressure vessels. *Journal of Applied Mechanics, 27*, 303–308.
26. Saroja, B. V., (1960). The analysis of creep in a thin rotating disk with a central circular hole. *Aircraft Engineering and Aerospace Technology, 32*(2), 34–36.
27. Schweiker, J. W., & Sidebottom, O. M., (1965). Creep of thick walled cylinders subjected to internal pressure and axial load. *Experimental Mechanics, 5*, 186–192.
28. Shahi, S., Singh, S. B., & Thakur, P., (2019). Modeling creep parameter in rotating discs with rigid shaft exhibiting transversely isotropic and isotropic material behavior. *Journal of Emerging Technologies and Innovative Research, 6*(1), 387–395.

29. Sharma, S., & Sahni, M., (2013). Creep analysis of thin rotating disc having variable thickness and variable density with edge loading. *International Journal of Engineering, Tome XI- Fascicule-3*, 279–296.
30. Singh, S. B., & Ray, S., (2003). Newly proposed yield criterion for residual stress and steady-state creep in an anisotropic composite rotating disc. *Journal of Materials Processing Technology, 143, 144*, 623–628.
31. Singh, S. B., & Ray, S., (2003). Creep analysis in an isotropic FGM rotating disc of Al-SiC composite. *Journal of Materials Processing Technology, 143, 144*, 616–622.
32. Singh, S. B., & Ray, S., (2002). Modeling the anisotropy and creep in orthotropic aluminum-silicon carbide composite rotating disc. *Mechanics of Materials, 34*, 363–372.
33. Singh, S. B., & Ray, S., (2001). Steady-state creep behavior in an isotropic functionally graded material rotating disk of Al-SiC composite. *Metallurgical and Materials Transactions A, 32A*, 1679–1685.
34. Singh, T., & Gupta, V. K., (2009). Creep analysis of an internally pressurized thick cylinder made of a functionally graded composite. *J. Strain Analysis, 44*, 583–594.
35. Singh, T., & Gupta, V. K., (2009). Effect of material parameters on steady state creep in a thick composite cylinder subjected to internal pressure. *The Journal of Engineering Research, 6*(2), 20–32.
36. Singh, T., & Singh, I., (2016). Modeling to steady state creep in thick-walled cylinders under internal pressure. *International Journal of Mechanical and Mechatronics Engineering, 10*(5).
37. Thakur, P., Gupta, N., & Singh, S. B., (2017). Thermal effect on the creep in a rotating disc by using sherby's law. *Kragujevac Journal of Science, 39*, 17–27.
38. Vandana, & Singh, S. B., (2012). Creep modeling in a composite rotating disc with thickness variation in the presence of residual stress. *International Journal of Mathematics and Mathematical Sciences, 2012*, Article ID 924921, 1–14.
39. Vandana, & Singh, S. B., (2011). Modeling anisotropy and steady-state creep in a rotating disc of Al-SiC_p having varying thickness. *International Journal of Scientific & Engineering Research, 2*(10), 1–12.
40. Vandana, & Singh, S. B., (2016). Mathematical modeling of creep in a functionally graded rotating disc with varying thickness. *Regenerative Engineering and Translational Medicine, 2*(3, 4), 126–140. doi: 10.1007/s40883-016-0018-3.
41. Venkatraman, B., & Patel, S. A., (1963). Creep analysis of annular plates. *Journal of the Franklin Institute, 275*(1), 13–23.
42. Wahl, A. M., (1957). Stress distributions in rotating disks subjected to creep at elevated temperature. *Journal of Applied Mechanics, 24*, 299–306.
43. Wahl, A. M., (1956). Analysis of creep in rotating disks based on the Tresca criterion and associated flow rule. *Journal of Applied Mechanics, 23*, 231–238.
44. Wahl, A. M., Sankey, G. O., Manjoine, M. J., & Shoemaker, E., (1954). Creep tests of rotating disks at elevated temperatures and comparison with theory. *Journal of Applied Mechanics, 21*, 225–235.
45. Bhatnagar, N. S., & Gupta, R. P., (1967). On the constitutive equations of the orthotropic theory of creep. *Wood Science and Technology, 1*, 142–148.

INDEX

C

D

N

W

X

Y

Z